Macromolecule-Metal Complexes

Springer

Berlin
Heidelberg
New York
Barcelona
Budapest
Hong Kong
London
Milan
Paris
Santa Clara
Singapore
Tokyo

F. Ciardelli · E. Tsuchida · D. Wöhrle

Macromolecule-Metal Complexes

With 94 Figures and 19 Tables

 Springer

Editors:

Prof. F. Ciardelli
Universita di Pisa, Dipartimento di Chimica e
Chimica Industriale, Via Risorgimento 35,
56100 Pisa, Italy

Prof. Dr. E. Tsuchida
Waseda University, Department of Polymer Chemistry
Tokyo 169, Japan

Prof. Dr. D. Wöhrle
University of Bremen, Institute of Organic
and Macromolecular Chemistry, P.O. Box 330440,
28334 Bremen, Germany

ISBN 978-3-642-64642-3 Springer-Verlag Berlin Heidelberg New York

Library of Congress Cataloging-in-Publication Data
Macromolecule-metal complexes / F. Ciardelli, E. Tsuchida, D. Wöhrle [editors].
Includes bibliographical references and index.

ISBN 978-3-642-64642-3 (hardcover : alk. paper)   e-ISBN-13: 978-3-642-60986-2
DOI: 10.1007/978-3-642-60986-2

1. Macromolecules. 2. Metal complexes. I. Ciardelli, F. (Francesco) II. Tsuchida, E. (Eishun). 1990– .
III. Wöhrle, Dieter, 1989– .
QD381.8.M285 1995 574.19'283 – dc20

Typesetting: Fotosatz-Service Köhler OHG, Würzburg
SPIN: 10098364   2/3020 – 5 4 3 2 1 0 – Printed in acid-free paper

Preface

Macromolecular Metal Complexes (MMC) are essential for life on earth. Different MMCs regulate energy conversion in photosynthesis, multi-electron transfer and catalysis in cofactors and enzymes, gas transport in blood and other processes. Activity and selectivity is controlled strongly by specific structural arrangements of macromolecules and metals partly surrounded by a ligand. One field of research concentrates on the structure and function of biological macromolecular metal complexes. With better understanding of these systems increasing interest has developed in recent years for synthetic MMCs. This has been accompanied by more intensive research on low molecular weight metal complexes, organo-metallics and metal clusters.

The combination of macromolecules with metals could have enormous potential by improving the capacity of structural macromolecular materials. Not only organic polymers also inorganic high molecular weight compounds have to be considered. The metals can be metal ions, metal complexes, metal chelates or metal clusters. The unlimited structural possibilities provided by MMC can give rise to products with new static and dynamic properties on one side and electron, photon and small molecule interactions on the other. The area is highly interdisciplinary for fundamental studies and applications. Therefore it is not surprising that an increasing number of papers and international conferences concentrate on the subject of MMC.

This book attempts to provide for the first time an integrated coverage of various scientific aspects of MMC and their applications. The book includes the fundamental aspects on structure, synthesis and formation as well as different activities and properties. We are sure that MMC will more and more arouse the interest of chemists, physicists, biologists, physicians and engineers. Therefore it is necessary to provide knowledge of the state of the art regarding fundamental aspects, applications and trends. In other words, the scientific and technical challenge offered by MMC can be well-described with a sentence of Shakespeare's Hamlet:

"There are more things in heaven, Horatio, then are dreamt by your philosophy."

F. Ciardelli, E. Tsuchida, D. Wöhrle

List of Contributors

Ciardelli, Francesco
Universita di Pisa, Dipartimento di Chimica e
Chimica Industriale, Via Risorgimento 35,
56100 Pisa, Italy

Kaneko, Masao
Ibaraki University, Faculty of Science,
Department of Chemistry, 2-1-1 Bunkyo, Mito, Ibaraki 310, Japan

Komatsu, Teruyuki
Waseda University, Department of Polymer Chemistry,
Tokyo 169, Japan

Nishide, Hiroyuki
Waseda University, Department of Polymer Chemistry,
Tokyo 169, Japan

Pomogailo, Anatolii D.
Institute of Chemical Physics, Russian Academy of Sciences,
142432 Chernogolovka, Moscow Region, Russia

Reedijk, Jan
Leiden Institute of Chemistry, Gorlaeus Laboratories,
Leiden University, P.O. Box 9502, 2300 RA Leiden, The Netherlands

Takeoka, Shinji
Waseda University, Department of Polymer Chemistry,
Tokyo 169, Japan

Tsuchida, Eishun
Waseda University, Department of Polymer Chemistry,
Tokyo 169, Japan

Wöhrle, Dieter
University of Bremen, Institute of Organic
and Macromulecular Chemistry, P.O. Box 330 440,
28334 Bremen, Germany

Yamamoto, Kimihisa
Waseda University, Department of Polymer Chemistry,
Tokyo 169, Japan

Table of Contents

List of Abbreviations

AA	acrylic acid
AAm	acrylamide
AcAc	acetylacetonate
ADP	adenosine 5'-diphosphate
Asp	aspartic acid (D)[b]
ATP	adenosine 5'-triphosphate
Bchl-*a*	Bacteriochlorophyll *a*
bpy	dipyridyl, bipyridine
BQ	*p*-benzoquinone
15C5	15-crown-5
18C6	18-crown-6
21C7	21-crown-7
CMC	carboxylmethylcellulose
CMCS	chloromethylated copolymer of styrene and divinylbenzene
CN	metal ion coordination number
Cnd	coordinative node formation
CoP	cobalt-porphyrin
CpMn	Cyclopentadienyl-dicarbonylmanganese
CoP(py)4	5,10,15,20-tetrakis(4-pyridyl)porphyrinatocobalt(II)
CoPPh4	5,10,15,20-tetraphenylporphyrinatocobalt(II)
Cys	cysteine (C)[b]
Da	dalton; unit of molar mass
DA	dehydroxylated alumina
DDQ	2,3-dichloro-5,6-dicyano-*p*-benzoquinone

---

[a] Ligand abbreviations with the format H_n abc may lose one (or more) hydrons (H⁺) upon metal binding. Following IUPAC guidelines, lower-case abbreviations have been used for most ligands.

[b] For a complete list of one-letter abbreviations of the amino acids see: IUPAC-IUB, Recommendations 1983 "Nomenclature and Symbolism for Amino Acids and Peptides", such as published in e.g. Eur J Biochem 138:9–37 (1984) and J Biol Chem 260:14–42 (1985).

[c] Some of the above abbreviations are not generally accepted (although those recommended by IUPAC and IUB and relevant to the chapter are included) and are only used for the present purpose.

dib	*p*-diisocyanobenzene
DMF	*N,N*-dimethylformamide
dmgH	dimethylglyoximato
DMIm	1-dodecyl-2-methylimidazole
dmso	dimethylsulfoxide; Me_2SO
DMSO	dimethylsulfoxide
DPPC	1,2-bis(parmytoyl)-sn-glycero-3-phosphocholine
DS	dehydroxylated silica
EC	ethylene carbonate
EDTA	ethylenediaminotetraacetic acid
EPR	electron paramagnetic resonance
EXAFS	extended X-ray absorption fine structure
FAD	flavin-adenine dinucleotide
FeMoco	cofactor containing Fe and Mo in nitrogenases
gr	grafted
His	histidine (H)[b]
HSAB	Hard and Soft Acids and Bases
H_2TPP	metal free tetraphenylporphyrine
H_4edta	ethylenediaminetetraacetic acid
ITO	indium tin oxide glass
L	ligand
LADH	Liver alcohol dehydrogenase (EC 1.1.1.1)
M	metal
MAA	methacrylic acid
MAO	methylalumoxane
MCM	metal containing monomers
Met	methionine (M)[b]
MMA	methylmethacrylate
MMC	macromolecular metal complexes
MMCh	macromolecular metal chelate
MMO	methane monooxygenase (EC 1.14.13.25)
Moco	a cofactor containing molybdenum found in Mo enzymes other than nitrogenase
MV^{2+}	methylviologen, 1,1'-dimethyl-4,4'-bipyridinium dichloride
MX_n	metal compound, metal salt
NADH	reduced nicotinamide-adenine dinucleotide
NASICON	Na-super ionic conductor
OAc	acetate
OOE	oligo(oxyethylene)
OPS	oligo(*p*-phenylene sulfide)
PAN	polyacrylonitrile
P2VP	poly(2-vinylpyridinc)
P4VP	poly(4-vinylpyridine)
PAAc	polyacrylic acid
PAAl	polyallylalcohol
PB	Prussian Blue
PC	propylene carbonate

Pc	phthalocyanine
PChE	polymer chelate effect
PDA	partially dehydroxylated alumina
PDAA	polydiallylalcohol
PDS	partially dehydroxylated silica
PE	polyethylene
PEG	polyethylene glycol
PEI	polyethyleneimine
P_i	inorganic phosphate
PMA	polymethacryloylacetone
PMEOn	poly[α-methacryloyloxy-ω-methyloligo(oxyethylene)]
PMAAc	polymethacrylic acid
PMVK	polymethylvinylketone
POE	poly(oxyethylene)
POP	poly(oxypropylene)
PP	polypropylene
PPDPS	poly-p-diphenylphosphinostyrene
PPE	poly(2,6-dimethylphenylene ether)
PPG	polypropylene glycol
PPhS	poly(methyl-β-pheneethylsiloxane)
PPIX	protoporphyrin IX
PPP	poly(p-phenylene)
PPS	poly(p-phenylene sulfide)
PQQ	pyrroloquinoline quinone (2,7,9tricarboxy-1H-pyrrolo [2,3-f]-quinoline-4,5-dione
PS	polystyrene
PVAc	polyvinylacetate
PVAl	polyvinyl alcohol
PVC	polyvinylchloride
PVCz	poly(N-vinylcarbazole)
PVdf	polyvinylidene fluoride
PVPd	poly(N-vinylpyrrolidone)
pyz	pyrazine
RNR	ribonucleotide reductase (EC 1.17.4.1)
salen	N,N'-ethylenebis(salicylideneaminato)
smdpt	bis[(3-salicylideneamino)propyl]methylamine
SME	solid macromolecular electrolyte
SOD	superoxide dismutase
SRPEs	solid redox polymerization electrodes
Tap	alkylated tetrapyridino-tetraazaporphyrin
TCNQ	tetracyanoquinodimethane
terpy	tripyridyl
THF	tetrahydrofurane
TPP	5,10,15,20-tetraphenylporphyrin
TpyP	alkylated tetrapyridyloxyphthalocyanine
Tyr	tyrosine (Y)[b]
tz	tetrazine

VA	vinylacetate
VO(acac)2	bis(2,4-pentanedionato)oxovanadium(IV)
VO(Bzac)2	bis(1,3-phenylbutanedionato)oxovanadium(IV)
VO(salen)	(*N,N'*-ethylenebis(salicylideneaminato))oxovanadium(IV)
VO(TPP)	5,10,15,20-tetraphenylporphyrinatooxovanadium(IV)
VP	vinylpyridine
WLF	William-Landel-Ferry
XOH	2,6-dimethylphenol

1 Introduction and Fundamental Aspects

F. Ciardelli, E. Tsuchida and D. Wöhrle

1.1 Macromolecular Metal Complexes

Ten years ago the first International Conference on Macromolecular Metal Complexes (MMC I, 1985) was held in Beijng China as the result of efforts of several active research groups in the world who had discovered the enormous potentiality offered to material science by the systems obtained by combining organic and inorganic macromolecular compounds with metal salts, metal complexes, metal chelates or metal clusters.

Macromolecules are indeed very versatile for producing materials with largely varied structural properties and functionalities that allow applications in several areas.

Organic macromolecules are generally in some way limited by the relatively low thermal stability and limited capacity to be involved in important electron transfer and transport processes. On the other hand, inorganic polymers, even if they have better thermal stability, very often suffer from limited versatility and processability.

It resulted clearly that the combination of macromolecules with metal in a broad sense could provide enormous potentiality by largely improving the capacity of structural macromolecular material to participate in electron transport and transfer, thus extending their possible applications. On the other hand, metals and their derivatives could acquire unusual mechanical properties, due to their wrapping in a platic or elastic polymeric matrix. Also the more or less extended order of the macromolecular chain or network could provide a new sort of organization to metal atoms along the polymer backbone. In this connection macromolecular complexes are defined very generally as complexes composed of macromolecules. Typical examples are MMC and molecular assemblies are also included in the category of macromolecular complexes.

The expected and practically unlimited structural possibilities offered by this new approach were further expanded by the observation that the interaction between macromolecular compounds and metal derivatives in many cases giving rise to products with new structure and new static and dynamic properties. Also, the area appeared to be highly interdisciplinary for both applications and fundamental studies.

In this general situation this book attempts for the first time to provide an integrated coverage of various scientific aspects of MMC as well as of their applications. Indeed, a large number of papers have been dedicated to MMC.

Thus, in the past 6 years the proceedings of the second, third, fourth and fifth conference on macromolecular metal complexes [1–4] have provided a broad description of the main scientific results concerning MMCs. Moreover, monographs [5–23] and review papers on selected aspects [24–37] have been published particularly since the early 1980s. However, none of the above publications contained a full and consequential integrated description as attemped in the present book.

1.2 Types, Formation and Structural Features of MMC

In order to understand structure and function of MMCs it is necessary to consider in a first view natural systems consisting on the combination of biological polymers and metal ions/complexes/chelates/clusters (for more details see Chap. 3):

– hemoglobin, myoglobin → gas transport
– cofactors → electron interaction
– metalloenzymes → catalysis
– apparatus of photosynthesis → energy conversion
– metalloproteins and other → various functions

It is fundamental to mention that only the combinations of macromolecules with metals and their derivatives are responsible for the selective properties. To understand in detail the function of MMCs the complicated molecular arrangements on different levels must be considered:

– Primary structure (composition of an MMC)
– Secondary structure (steric orientation of an MMC unit)
– Tertiary structure (orientation of the whole MMC)
– Quaternary structure (interaction of different MMCs)

Better knowledge of the structure and function of biological macromolecules on one side, and increasing number of synthetic macromolecules and low molecular weight metal complexes/chelates/clusters on the other side, resulted in the recent years in an intensified research activity in the field of MMCs.

As seen from Fig. 1.1 the materials can be classified into type I, type II and type III MMCs. Figures 1.2–1.4 contain schematically possible combinations of macromolecules with metals and their derivatives.

The formation of an MMC is generally thermodynamically favoured by a negative free energy. The so-called polymer chelate effect (change of the free energy due to the addition of a metal derivative to a macromolecular ligand or stabilizing environment) gives comparing with the known "chelate effect" for the formation of low molecular weight metal complexes/chelates an additional contribution. Because of local, molecular and supramolecular organizations of macromolecules that may be changed by formation of an MMC, the situation is much more complicated in comparison with low molecular weight analogs. This includes formation constants, cooperative effects, reorganization of structural arrangements and other transformations. In Fig. 1.5 it

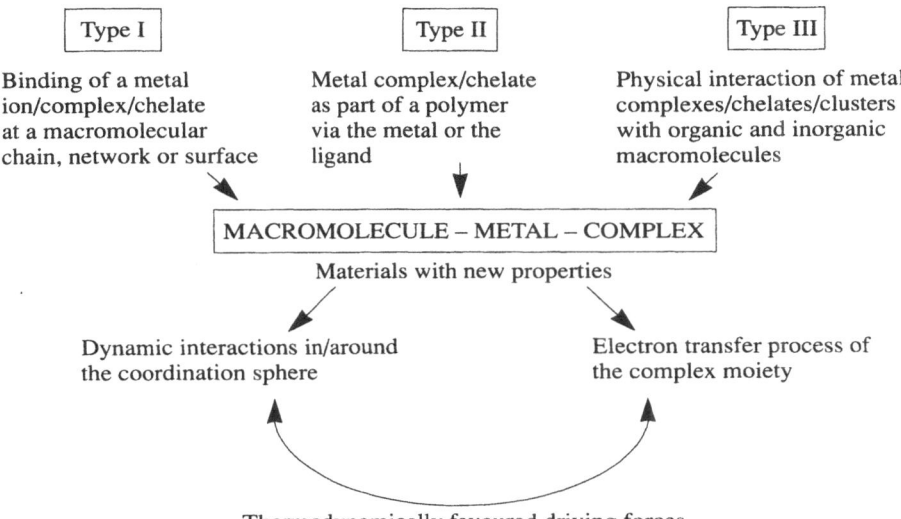

Fig. 1.1. Macromolecule–metal complexes (MMCs)

1. Electrostatically bound to a macromolecule.

2. Coordinatively bound to a macromolecule.

3. Covalently bound to a macromolecule.

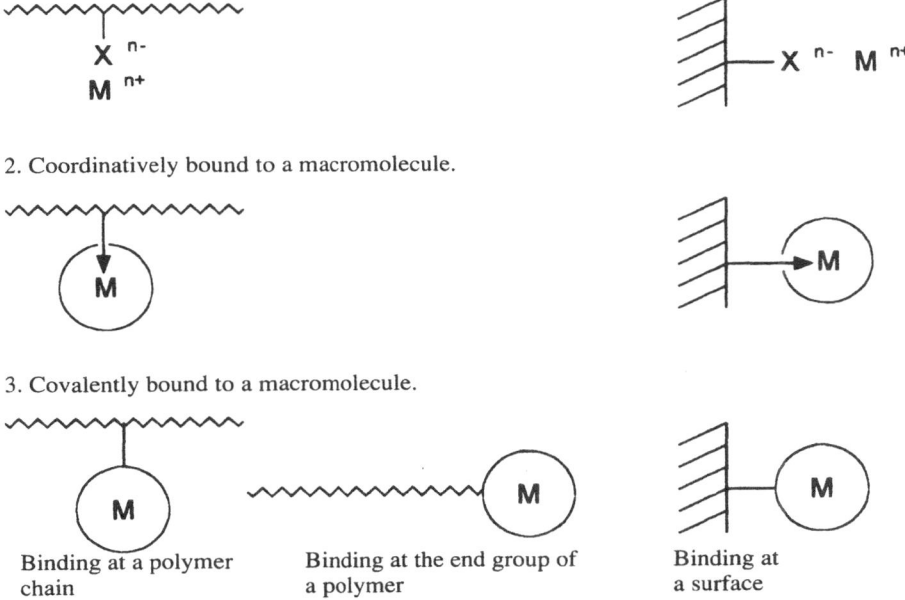

Binding at a polymer chain

Binding at the end group of a polymer

Binding at a surface

Fig. 1.2. Type I: Metal ions/complexes/chelates bound to a chain or a surface of organic or inorganic macromolecules

1. Part of a polymer chain or network via the ligand.

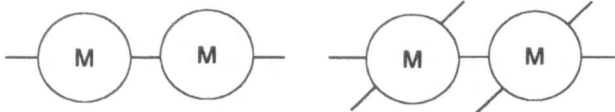

2. Part of a polymer chain via the metal ion.

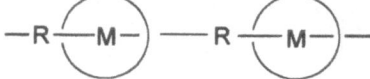

Fig. 1.3. Type II: Metal complexes as part of a polymer chain or network

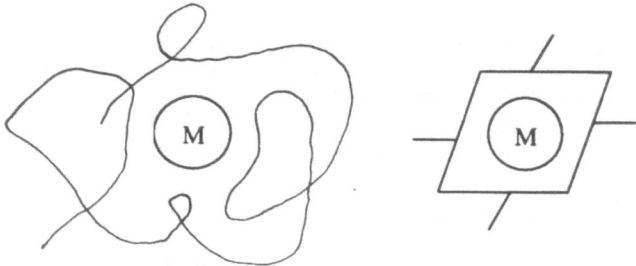

Fig. 1.4. Metal ions/complexes/chelates physically interacting with organic or inorganic macromolecules

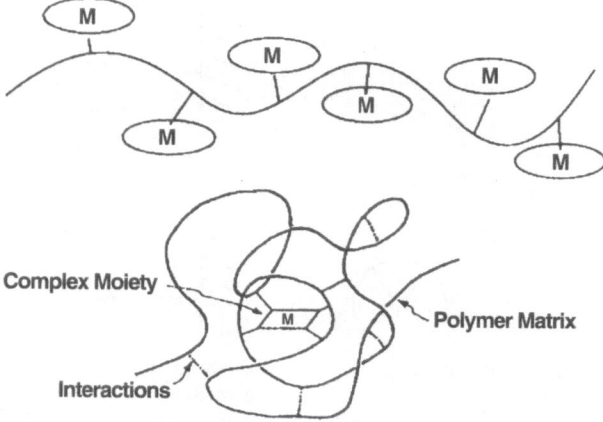

Fig. 1.5. Type I MMC: Schematic representation of a conformational change of the polymer metal complex

is exemplarily shown that multiple and dynamically interactions between a polymer matrix and a moiety strongly control the formation of a highly ordered structure.

The interactions originated by various secondary binding forces of the specific macromolecular environment are strong long-range coulombic or electrostatic forces, weak short-range dipole–dipole interactions and medium short-range hydrophobic interactions. Low molecular weight metal complexes exhibit a stable coordination structure and static configuration, whereas MMCs, depending on the kind of macromolecular environment, can show also labile, unsaturated and strained coordination structures. A well-known example are dynamic conformational changes during O_2 binding in polymer iron porphyrin complexes. In most cases metal complexes/chelates/clusters in types II and III MMC are stabilized by incorporation in or as part of macromolecules.

It is essential to study in detail the composition and structural arrangement of MMCs. Only on this basis can a relation between structure and property of the materials be discussed.

1.3 Properties and Application of MMC

In MMCs chemical reactivities and physicochemical properties are strongly affected by interactions with the macromolecular environment (types I and III MMC) or are fundamentally changed going from a low molecular weight to an MMC/chelate (type II MMC). In MMCs physicochemical properties and chemical reactivities of the complex moieties are often strongly affected by interactions with the polymer matrices, which surround the complex moieties. Interactions in MMCs consist mainly of various weak binding forces, such as coordination bonds, hydrogen bonds, charge-transfer interaction, hydrophobic interaction and so forth. These interactions are weak, but significant and act multiply and dynamically. Because they are plural, these binding forces cooperatively play an important role in MMCs. The electron transfer processes of the complex moieties of the MMCs are often affected by the dynamic conformational change of the molecular environment around the complexes. As a result, various kinds of electronic interactions are observed in MMCs (Fig. 1.6). Multiple interactions in and around the coordination sphere control

Multi-electron Transfer

| Electron Transfer Process of Complex Moiety | Soft Multiple | Weak Coordination / Dynamic Interactions in/around Coordination Sphere |

Molecular Conversion

Conformational Change

Fig. 1.6. Some dynamic interactions in MMCs

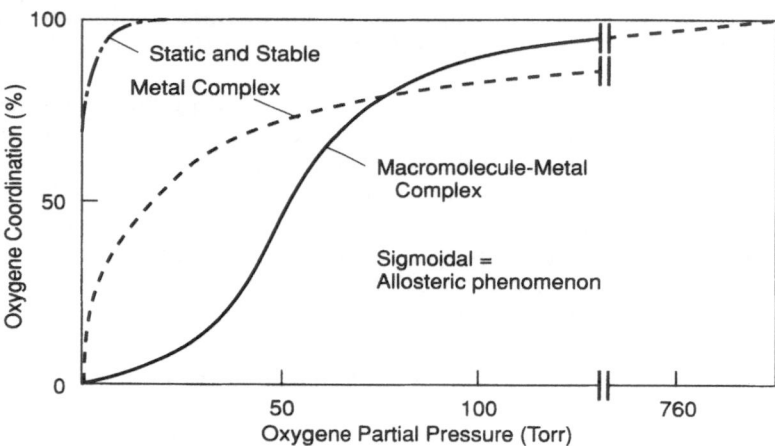

Fig. 1.7. Allosteric phenomenon in MMC

the function of an MMC. Characteristic aspects are binding equilibria, ligand substitution, multielectron transfer processes, molecular assemblies, molecular conversions and so forth. Molecular environments control the electronic configuration and mobility.

As in the case of metalloenzymes and metalloproteins, which are typical MMCs in a couple of decades the interactions and/or coordination behaviour between synthetic polymers and metal ions have been clarified gradually. Thus, in hemoglobin (Hb) it is well known that dynamic conformational change of the globin chain, induced by a small movement of the central iron ion from the porphyrin plane in the first dioxygen binding, reduces the activation energy of the second, third and fourth dioxygen binding to the hem complexes, surrounded by the neighbouring globin chains (Fig. 1.7). The profile of multistep coordination conjugated with the higher-order structural change of the macromolecules revealed the so-called allosteric phenomena usually observed in biological systems. The electron-transfer process of the complex moieties in the MMC are affected by a dynamic change in their configuration not observed for low MMCs with stable coordination structure.

Alignment of complex moieties along the macromolecular matrix or as macromolecule yields an integrated electronic process. Bulk and quantum effects, for example, formation of electron-transfer channels, charge separation processes induced by photoexcitation, tunnel effects and strong suppression of pertubation around the complex moiety as partly known for low MCs are expecially expected for MMCs (including finely dispersed metal clusters). Electronic interactions in MMC leads to the creation of higher functions. Thus, interfacial fixation of the metal complex moieties forms a sequential potential field (Fig. 1.8).

Properties and potential applications of MMCs are presently under investigation and are summarized exemplarily in Table 1.1.

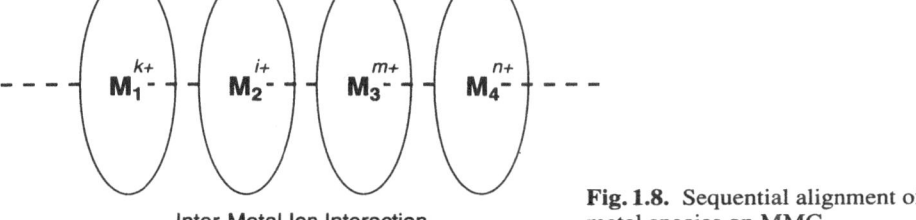

Inter-Metal Ion Interaction

Fig. 1.8. Sequential alignment of metal species on MMC

Table 1.1 Examples of properties and potential applications of macromolecule-metal complexes

Properties	Potential applications
Selectivity	Gas transport, separation, sensor
Mixed valence solution systems	Multielectron transfer, catalysis, photocatalysis, artificial photosynthesis, electrochemistry
Solid-state photon interactions	Photovoltaic cell, photoelectrochemistry, electro-luminescence, optical information storage
Electron, photoelectron conduction	Molecular devices, photoconductors, lasergraphy, electrochemistry
Ionic conduction	Electron-capture-detector devices, superconductors, polymer battery
Nonlinear optical effect	Modulator, integrated optics, high-power laser
Therapeutic effect	Drugs, photodrugs
Preceramics	Thermally stable compounds, quantum devices

1.4 Aim of the Book

The main objective of the present book is to provide the reader with a general overview of the present knowledge of MMC with reference to preparation, structure, properties and applications as new and versatile materials, without forgetting their role in biosystems.

First of all, Chap. 2 is a very broad section where an attempt is made to describe all possible synthetic methods of obtaining MMC and the relations between preparation method and structure. Chapter 2 Sect. 2.1 deals with the main structural principles governing the formation and characteristics of MMCs. Then three different situations are described in Sects. 2.2, 2.3 and 2.4 dedicated, respecitvely, to metal complexes bound to macromolecular carriers via the ligands or metal ions, metal complexes or metal as part of a polymer chain, and metal complexes of zero-valent metal or metal clusters physically connected to macromolecules.

Chapter 3 discusses first in Sects. 3.1, 3.2 and 3.3 the different structures of ligands and metal ions occurring in biological systems, and in Sect. 3.4 the applications of metal complexes in living systems.

Electronic processes involving MMC are presented in a broad sense in Chap. 4, where Sect. 4.1 deals with transport phenomena of metal ion in macromolecules, and Sects. 4.2–4.5 describe, respectively, applications for separation of small molecules, oxygen affinity, catalysis in conversion of small molecules and multielectron transfer.

Chapter 5 deals with photoinduced electron transport in solution (Sect. 5.1), at the solid/liquid interface (Sect. 5.2) and in solid macromolecular complexes (Sect. 5.3).

All these aspects are comparatively summarized, and views for future developments are reported in the last chapter (Chap. 6).

References

1. Jian YY (ed) (1988) MMC II. J Macromol Sci Chem A 25, vol 10 and 11
2. Tsuchida E, Toshima N (eds) (1989) MMC III. J Macromol Sci Chem A 26, vol 2 and 3; (1990) A 27, vol 9–11
3. Barbucci R, Ciardelli F (eds) (1992) MMC IV. Macromol Chem Macromol Symp 95
4. Wöhrle D (ed) (1994) MMC IV. Macromol Symp 80
5. Tsuchida E (ed) (1991) Macromolecular complexes, dynamic interactions and electronic processes. VCH Publishers, New York
5a. Tsuchida E, Nishide H (1977) Polymer-metal complexes and their catalytic activity. Springer Verlag, Berlin
5b. Tsuchida E, Abe K (1982) Interactions between macromolecules in solution and intermacromolecular complexes. Springer-Verlag, Berlin
6. Pomogailo AD, Uflyand IE (1991) Macromolecular metal chelates. Nauka, Moscow
7. Ray NH (1978) Inorganic polymers. Academic Press, New York
8. Stone FGA, Graham, WAG (eds) (1962) Inorganic polymers. Academic Press, New York
9. Carraher CE, Sheats JE, Pittmann CU (eds) (1978) Organometallic polymers. Academic Press, New York
10. Carraher CE, Sheats JE, Pittmann CU (eds) (1982) Advances in organometallic and inorganic polymer science. Marcel Dekker, New York
11. Sheats JE, Carraher CE, Pittmann CU (eds) (1985) Metal-containing polymeric systems. Plenum Press, New York
12. J Inorg Organomet Polym
13. Kepler BK (1993) Metal complexes in cancer chemotherapy. VCH Publishers, New York
14. Reedijk J (ed) (1993) Bioinorganic catalysis. Marcel Dekker, New York
15. Sigel H (ed) Metal ions in biological systems. Marcel Dekker, New York (over 25 volumes)
16. Spiro TG (ed) Metal ions in biology. Wiley, New York (7 volumes till 1985)
17. Pomogailo AD (1988) Polymeric immobilized metallocomplex catalysts. Nauka, Moscow
18. Hartley FR (1985) Supported metal complexes, a new generation of catalysts. Reidel, Dordrecht
19. Yermakov VI, Kuznetsov BN, Zakharov VA (1981) Catalysis by supported complexes. Elsevier, Amsterdam
20. Yermako VI, Likholobov V (1987) Homogeneous and heterogeneous catalysis. VNU Science Press, Utrecht

21. Jacobs PA, Jaeger NL, Kubelkova L, Wichterlova B (eds) (1991) Zeolite chemistry and catalysis. Studies in surface science and catalysis, vol 69. Elsevier, Amsterdam
22. Karge H, Weitkamp J (eds) (1989) Zeolite as catalysts, sorbents and detergent builders. Studies in surface science and catalysis, vol 46. Elsevier, Amsterdam
23. Ramamurthy V (ed) (1991) Photochemistry in organized and constrained media. VCH Publishers, New York
24. Ciardelli F, Braca G, Carlini C, Sbrana G, Valentini G (1982) J Mol Catal 14:1
25. Wöhrle D (1983) Adv Polym Sci 50:45
26. Korshak VV, Kozyreva NM (1983) Russ Chem Rev 54:1091
27. Hanack M, Lang M (1994) Adv Mater 6:819
28. Ozin GA, Gil C (1989) Chem Rev 89:1749
29. Sherrington DC (1988) Pure Appl Chem 60:401
30. Wöhrle D (1992) Polymers with metal in the backbone. In: Kricheldorf H (ed) Handbook of polymer synthesis, vol B. Marcel Dekker, New York, p. 1133
31. Wöhrle D (1989) Phthalocyanines in polymer phases. In: Leznoff CC Lever ABP (eds) Phthalocyanines, properties and applications, vol 1. VCH Publishers, New York, p. 55
32. Pomogailo AD (1992) Russ Chem Rev 61:133
33. Pomogailo AD, Uflyand IE (1990) Adv Polym Sci 97:61
34. Tsuchida E, Nishide H (1977) Adv Polym Sci 24:1
35. Kaneko M, Wöhrle D (1988) Adv Polym Sci 84:141
36. Hanack M, Deger S, Lange A (1988) Coord Chem Rev 83:115
37. Biswas M, Mukherjee, A (1994) Adv Polym Sci 115:89

2 Synthesis and Structure of Macromolecular Metal Complexes

A. D. Pomogailo and D. Wöhrle

2.1 Main Structural Principles of Formation and Characteristics of Macromolecular Metal Complexes

2.1.1 Classification, Main Requirements and Principles of Macroligands for the Formation of Macromolecular Metal Complexes

Macromolecular metal complexes (MMC) can be classified into three main groups (Chap. 1.) considering the "chemical" of binding of a metal compound (MX_n) to suitable macroligands or higher functional ligands: **Type I** MMC, described in Chap. 2.2, are those with the metal ion or metal complex binding in the side chain of an organic polymer or on the surface of an inorganic high molecular weight compound. The general route of type I synthesis is usually as follows: support $\rightarrow$ support functionalization $\rightarrow$ interaction with MX_n $\rightarrow$ separation of unbound reagents. Another possibility uses the polymerization of vinyl group substituted metal complexes (or ligand followed by metalation).

In **type II** MMC the bifunctional or higher functional ligand L and metal ion/metal are part of a polymer chain or network (Chap. 2.3). Different possibilities, such as the reaction of a bifunctional/higher functional ligand L with MX_n or the reaction of a bifunctional low molecular weight metal complex with another bifunctional reagent, exist.

In addition, **type III** MMC, included in this monography and described in Chap. 2.4, means the "physical" interaction of a metal complex or metal cluster with an organic polymer or inorganic high molecular weight compound. Preparations are realized by different methods such as impregnation, (co)reprecipitation, sorption, deposition by evaporation, sublimation and insertion into a structured support, dispersion, microcapsulation, etc. These methods as a rule demand a well-defined polymer or a developed support surface and cavity structure, respectively (specific surface and porosity). Sometimes the binding process is accompanied by chemical bond interactions between the metal ion and surface ligands. The nature of the generated bonds is often unknown.

In Scheme 2.1 the fundamentals for the formation of **types I – III** MMC are shown.

2.1.1.1 Main Requirements of Macroligands

For **type I** MMC, besides polymers different high disperse inorganic high molecular weight materials, such as oxide-type (silica gel, silica, natural zeo-

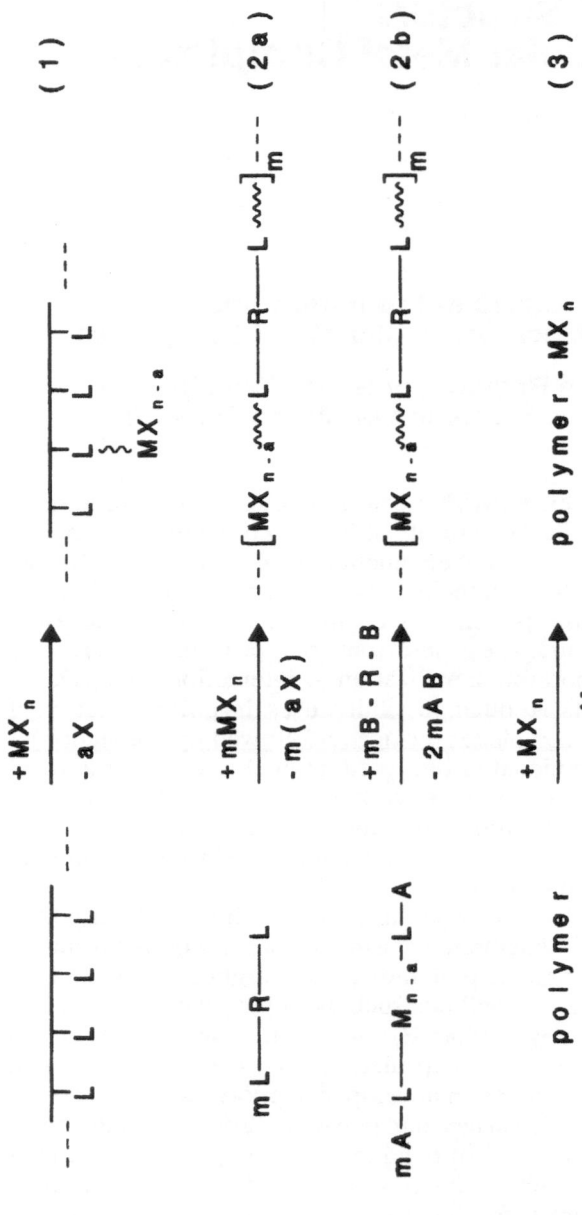

Scheme 2.1. Different possibilities of MMC formation ((1) type I MMC, (2a, b) type II MMC, (3) type III MMC)

lites, etc.), activated carbon, laminated materials (graphite, mica, sulfides, etc.), are used. The specific features of such supports are (a) rigidity of skeleton, which hinders the multiligand coordination with transition metal ions, (b) hardness by heat treatment which offers the use of such materials' exploitation at higher temperatures (above 150 °C), (c) easy-to-control surface parameters such as specific surface value, S_{sp}, and porosity, V_0.

The main features of organic polymer supports as macroligands are:

1. A great number of functional groups and comparable easy modification
2. A large number of supports with different S_{sp} and V_0
3. Controlled solubility or swelling in the reaction mixture
4. Controlled mobility of chain segments and functional groups
5. Low hygroscopicity and chemical stability on a large pH scale
6. The low reagents and/or substrate sorption ability

Depending on the goals of further MMC usage, macroligands may be produced in the form of powders, granules, fibers, textiles, films (including membranes), and swelling gels.

In a more general way one can postutlate the main requirements of the macroligands used for MMC formation:

1. Permeability
2. A low number of synthetic steps leading to functionalization and MMC formation
3. Presence of well-characterized reactive functional groups capable of binding a great amount of MX_n (sometimes up to 3 mmol/g)
4. A well-fitted structural geometric characteristic (the shape and size of particles, S_{sp}, porosity)
5. Easiness of unbound reagents' separation and good resistance under mechanical treatment

One more requirement for macroligands are the easiness to determine the degree of substitution with functional groups and their availabilty for MX_n interaction [1–3].

Formally, macroligands can be considered on one side as small porous subjects formed by particles of different shape and size communicating with each other through a porous-like system. On the other side, macroligands are spongy porous objects or continuous solid-phase networks consisting of cavities and channels. A number of effective methods for the preparation of porous materials has been elaborated in detail for copolymers of styrene with divinylbenzene. The pore radius, r, is a very important parameter, because pores are the channels for MX_n penetration to functional L of the polymer. According to their size, pores may be schematically classified as micropores ($r \leq 1.5$ nm, the total pores volume reaching value of 0.5 cm^3/g; S_{sp} in this case has no role, *meso* (moderate) pores ($r = 1.5–30$ nm; $S_{sp} = 700–900$ m^2/g; $V_0 = 0.8$ cm^3/g) and *macro* pores ($r \approx 30–6500$ nm). Pores may be isolated (from each other and have no opening to the surface), deadlock (ending at the surface but isolated from each other), and through-like (channels connected to each other and the surface of the polymer support as transport arteries for

MX_n transfer). Macroligands possess as a rule pores of different type and size. Reaction conditions lead to the formation of cross-linked polymers with both uniform cross-linked density. Large values of S_{sp} and V_0 are characteristic for such supports. Cross-linked isoporous copolymers of styrene with divinyl-benzene show pores of close size (5–40 nm; $S_{sp} \approx 690 \, m^2/g$, $V_0 \approx 0.6 \, cm^3/g$) [4]. In contrast, partially crystalline polymers (PE, PVA1, PVDC, etc.) are non-porous ones ($V_0 = 0$; $S_{sp} = 1 - 10 \, m^2/g$), although their amorphous phase may exhibit significant porosity. Especially, supports with high porosity are more sensitive to functionalization and MX_n binding than inorganic oxide supports. Macrolig and materials of particle size $d \approx 50 \, \mu m$, $S_{sp} = 5 - 600 \, m^2/g$, $r = 4 - 200 \, nm$, and $V_0 = 0.3 - 0.4 \, cm^3/g$ are the most convenient objects and commercial products. It is important to mention that properties such as basicity of polyligands, the rate of MX_n interaction with a polymer, and even material composition, strength, and distribution of formed complexes also depend on porosity parameters.

Type II MMC are generally materials with no porosity or low porosity if they are prepared according to Scheme 2.1, routes 2 a, b, because the reaction of bifunctional or higher functional ligands with MX_n leads to a high concentration of metal complexes per gram. In the case of **type III** materials, metal complexes or metal cluster are often included in a low concentration in the organic or inorganic carrier system, and this situation can be discussed analogously to **type I** materials.

2.1.1.2 Classification of Macroligands

For the classification it is necessary to take into account three levels of a structural organization of macroligands: The first level is the molecular level, which characterizes the chemical arrangement of polymer chains, functional groups' distribution along the polymer backbone, and stereochemical structure of the polymer chains. The second level is the supramolecular level, which displays intermolecular interactions as well as the degree of macromolecules' ordering and packaging. Finally, the topological level, which reveals the relations between some parameters of the polymer structure (molecular weight distribution (MWD), branching of three-dimensional structure characteristics, etc.) has to be considered. The structural organization is of prime importance for MMC formation.

For **type I** MMC the Flory principle postulating the independence of the reactivity of functional groups L from the chain length should be considered. Nevertheless, the microenvironment of L affects its reactivity. Ligands L can be part of linear soluble polymers, branched polymers, or copolymers incorporating other functional groups from the cross-linking comonomer. The L can be distributed in the whole polymer volume, localized only on the carrier surface, isolated by inert groups or associated (e. g., by hydrogen bonds). Connection of L directly with the polymer backbone or via a spacer is possible. All these possibilities for ligands will result in different reactivities for the interaction with MX_n bonding, which may be reduced, due to steric reasons.

Macroligands for **type I** MMC should be classified according to the pre-history of their formation (polymerization, copolymerization, grafting polymerization, polycondensation, polymer analogous reactions) and then by their functionality. Although such a two-parameter classification is schematic, it more or less definitively characterizes the macroligand. The formation of **type II** MMC can be described similarly via a polycondensation or polyaddition reaction. **Type III** materials are obtained via physical incorporation into the carrier.

According to their origin macroligands for **type I and II** complexes can be classified as protic (e.g., -OH, -COOH, >NH, -NH$_2$, >PH, RO(OH)$_2$), -SH, -C(S)OH, -C(S)SH, etc.) and aprotonic (groups or heteroatoms with unpaired free electron pairs capable of donor-acceptor or π-interaction with MX$_n$, such as > C=O, -COOR, -N=O, -NO$_2$, -N=N-, > C=S, - C≡N, > C=C<, -SCN, -PR$_2$, >As<, etc.) [78–80]. The ligand as a carrier unit may contain either one (monofunctional) or several (polyfunctional macroligands or ligands) similar or different groups. The last type includes chelated macroligands with polydentate functional groups capable of metal cycle formation with MX$_n$. Generally speaking, any interaction of MX$_n$ with macroligand or a ligand are attributed to a chelate reaction (in most cases multiple metal ion binding with ligands). Modified copolymers of styrene with divinylbenzene (1–20 mol.% DVB) are most widely used for **type I** MMC. The general way for the modification of the copolymer by chloromethylation was elaborated as early as 1953 [5].

2.1.1.3 Methods of Polymer Carrier Functionalization

Various organic reactions leading to **type I** MMC are used for the functionalization of polymers. One characteristic example is chloromethylated polystyrene [6]. Polymer analogous reactions for macroligand functionalization with oxygen-, nitrogen-, phosphorous-, and sulfur-containing groups are generalized in Schemes 2.2–2.5. These methods allow the yielding of three-dimensional macroligands with controlled surface, volume and pore size parameters and predetermined type and concentration of the corresponding L. Chelate nodes are obtained (see Scheme 2.6) when MX$_n$ interacts with both open-chain amines and cyclic amines [7]. Macroligands with chelate-like groups, such as dipyridyl, N,N-, N,O-, O−O-, and polymeric Schiff and Mannich bases, are widely used for MMC creation. Macroligands with chelate nodes are also often formed by polymer analogous reactions (Scheme 2.6). Water-soluble polymers with chelate properties are also formed from linear polymers such as polyethyleneimine (Scheme 2.7), polyvinylamine, and polyacrylic acid by introduction of chelate groups, e. g., pyridine-2-alde-hyde, iminodiacetic acid, 8-hydroxyquinoline, hydroxylamine, *N*-methyliso-cyanate, etc. [8].

Another possibility is polymerization or copolymerization of functional monomers themselves containing substituents for chelate formation of the chelates. Macroligands with hydroxy groups are obtained by copolymerization of allyl or propargyl alcohols, vinylphenols, and those with carboxylic groups from acrylic (AA) and methacrylic acid (MAA), as well as 4-vinyl-

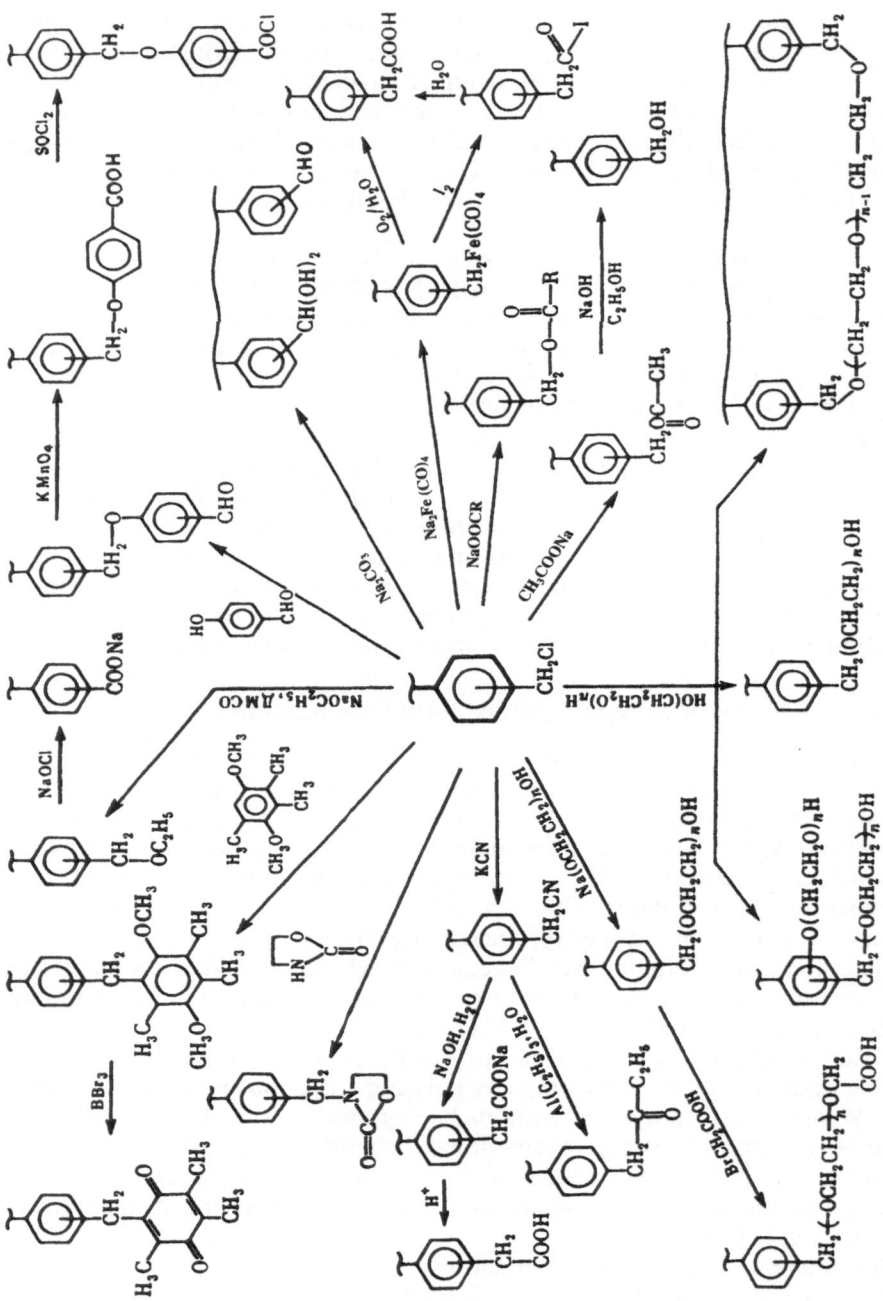

Scheme 2.2. More general polymer analogous reactions of chloromethylated polystyrene leading to oxygene-containing macroligands

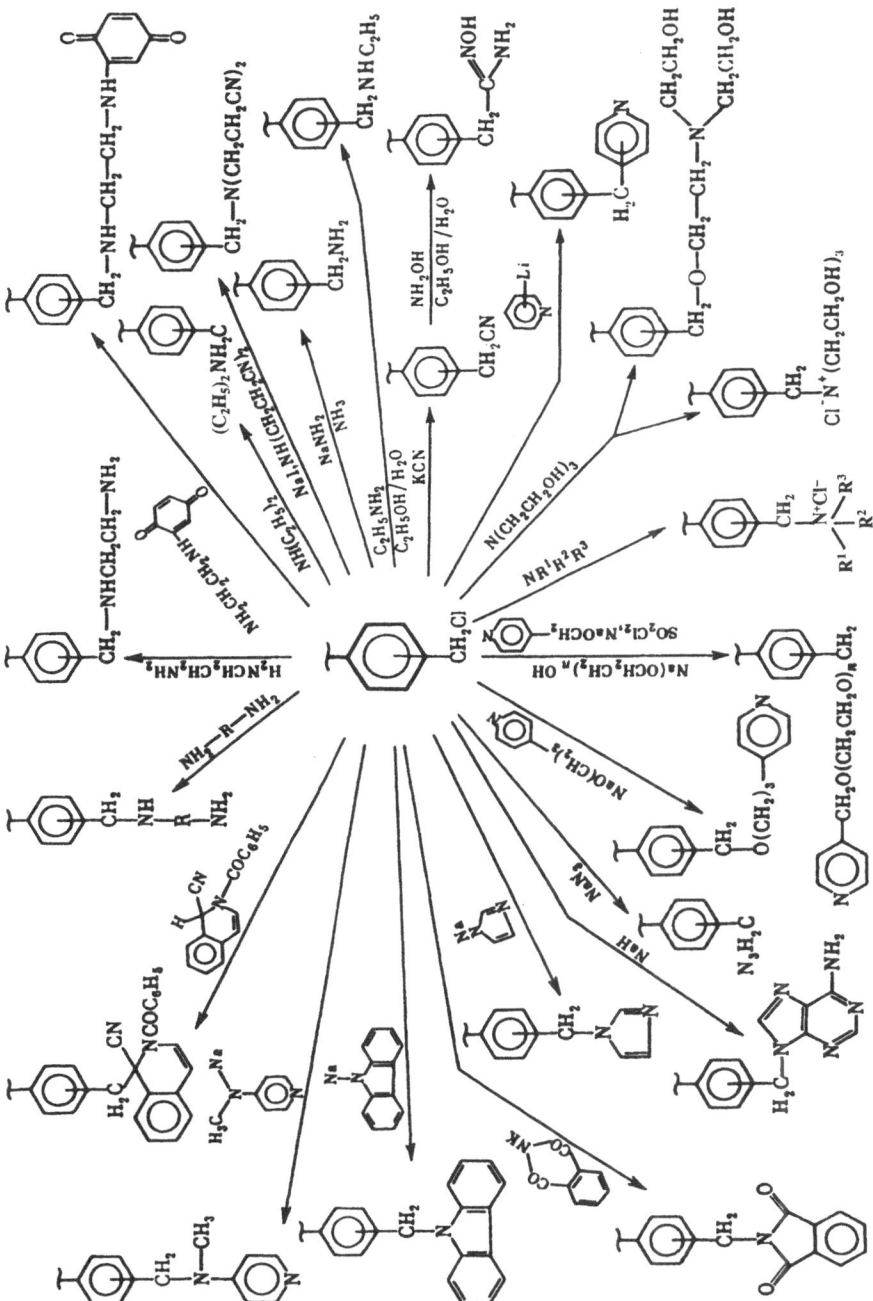

Scheme 2.3. Some principal ways of nitrogen-containing macroligands on the base of CMCS

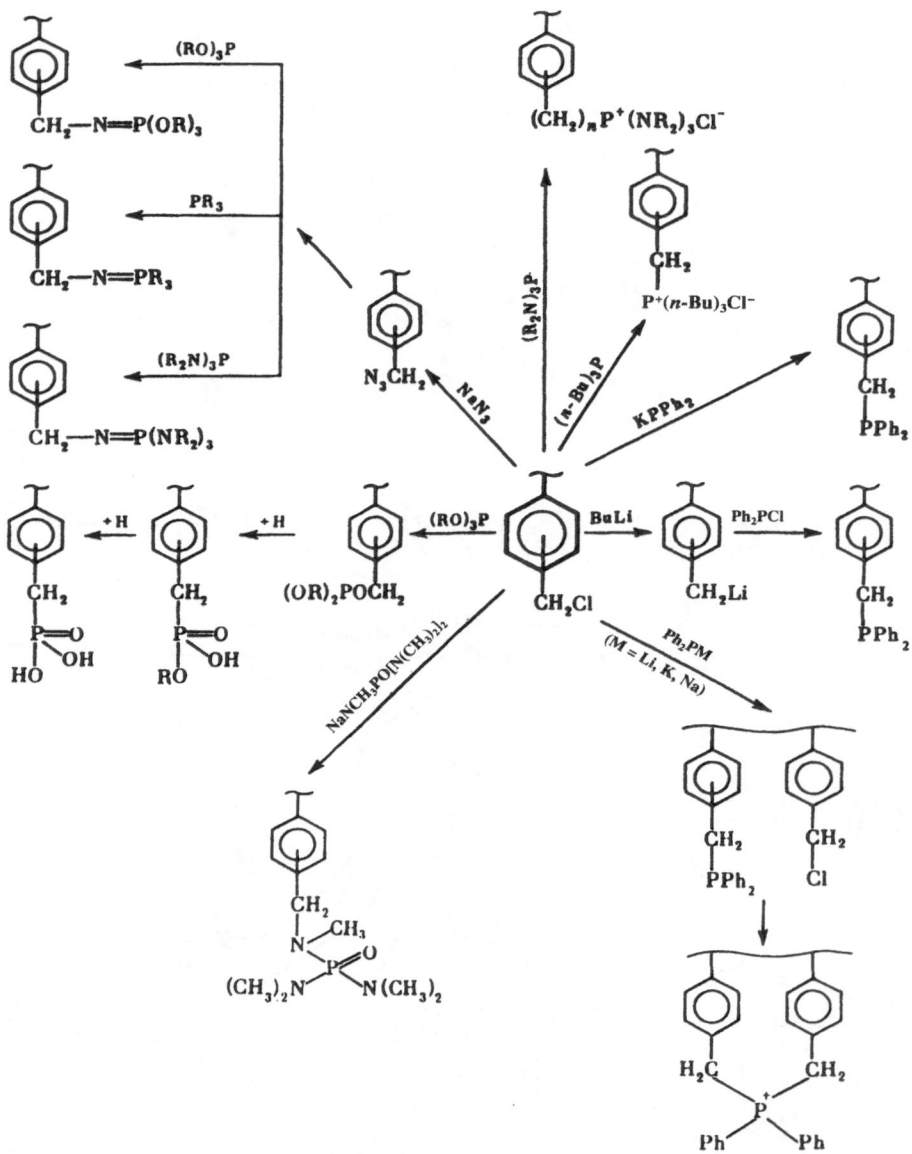

Scheme 2.4. The principal polymer-analogous reactions for phosphorous-containing macro-ligands (on the base of CMCS) formation

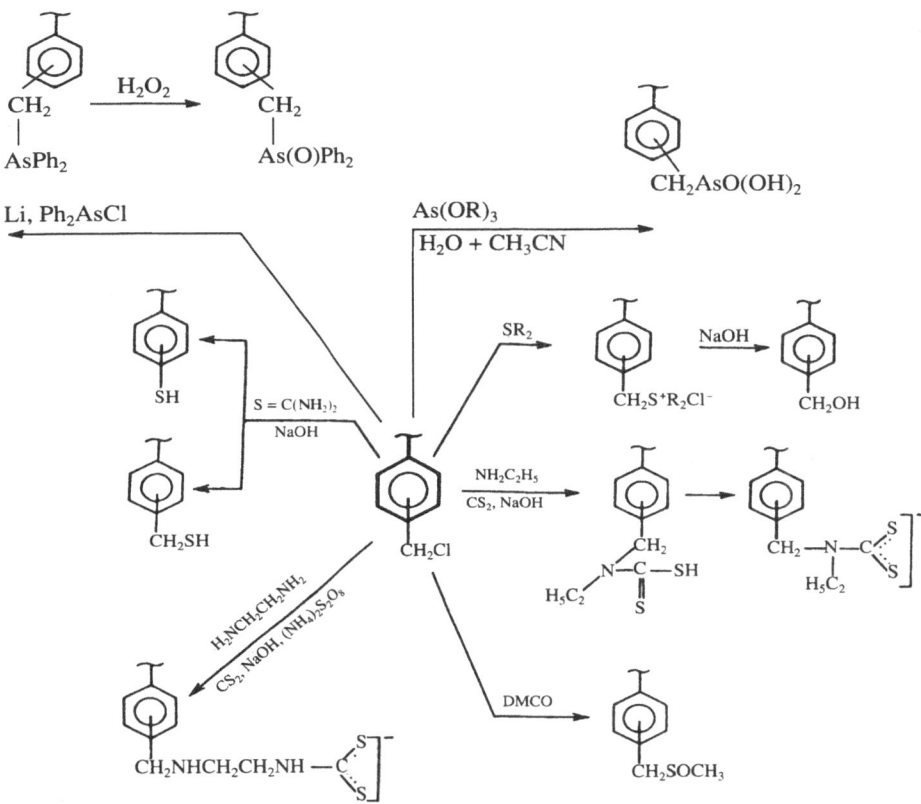

Scheme 2.5. The principal ways of S- and As-groups introduction into macroligands on the base of CMCS

benzoic and itaconic acids, maleic anhydride, etc. Macrolignads with donor-acceptor groups are produced by poly- and copolymerization of methylmethacrylate (MMA), vinylacetate (VA), acrylamide (AAm), N-vinylpyrrolidone, vinylpyridine (VP), vinylimidazole, acrylonitrile, vinyldiphenylphosphine, etc. Polymeric Schiff bases can also be formed by polymerization of respective monomers, such as 2-butyl-amino-4-vinylphenole or 1-hydroxy-5-vinylbenzaldehyde [9]. Almost all macroligands are also prepared by polycondensation or polyaddition reactions of suitable monomers.

For catalytic purposes it is necessary to create reactive sites on the surface of an organic polymer or a high molecular inorganic carrier. This requirement can be realized by a grafting polymerization of different monomers on a surface or presurface (10–30 nm) by mechanical, chemical or irradiated-chemical treatment of carriers in the presence of functional monomers (Eq. (1)) [10–12]. Gas-phase grafting polymerization initiated by irradiation

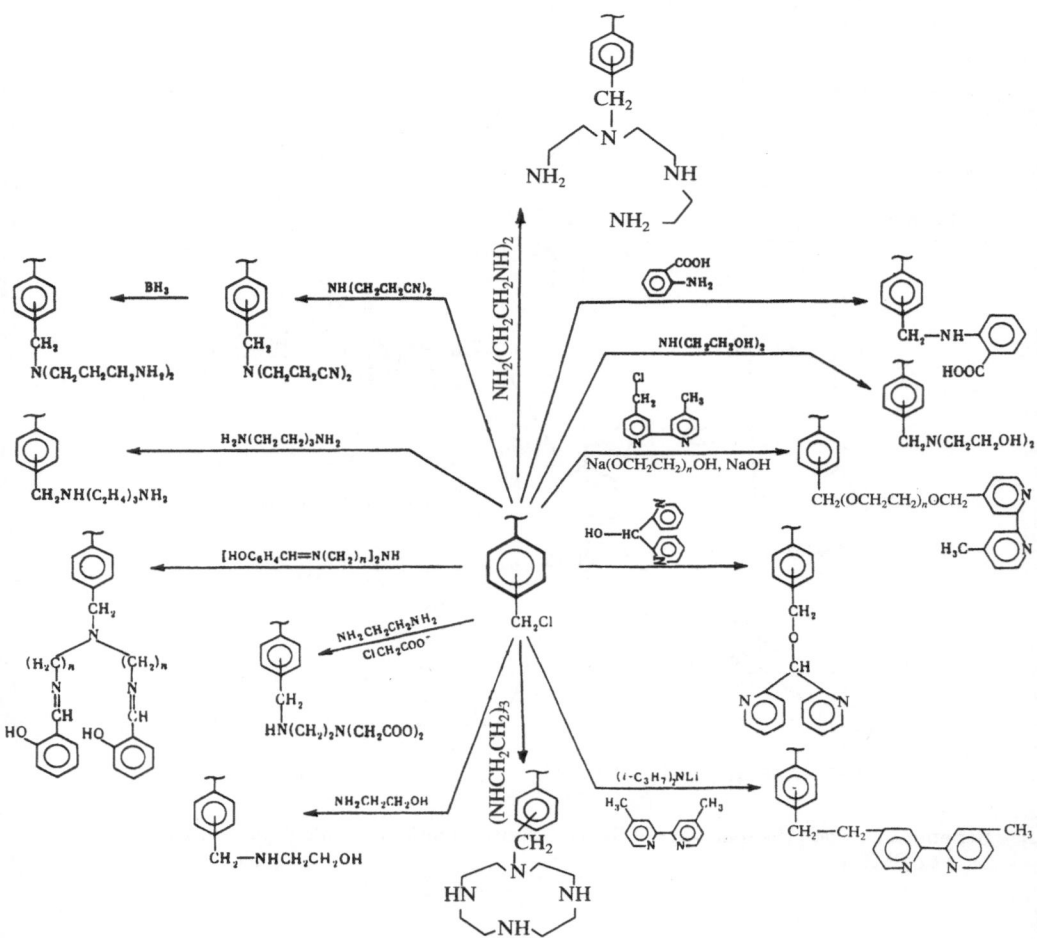

Scheme 2.6. Creation of chelate nodes on macroligands on the base of CMCS

(γ-irradiation of ^{60}Co, accelerated electrons, low pressure gas discharge, etc.)
is the most efficient way for this carrier modification [6, 10].

$$\left]\xrightarrow{\text{initiation}}\right]^{\bullet} + nCR_1R_2=CR_3L \longrightarrow \left]-(\overset{\overset{\displaystyle R_1}{|}}{\underset{\underset{\displaystyle R_2}{|}}{C}}-\overset{\overset{\displaystyle R_3}{|}}{\underset{\underset{\displaystyle L}{|}}{C}}-)n$$

the surface of support

$R_1\ R_2,\ R_3 = H,\ CH_3,\ C_2H_5,\ CH_2=CH,\ C_6H_5,\ Hal$

L-functional group or heteroatom   (1)

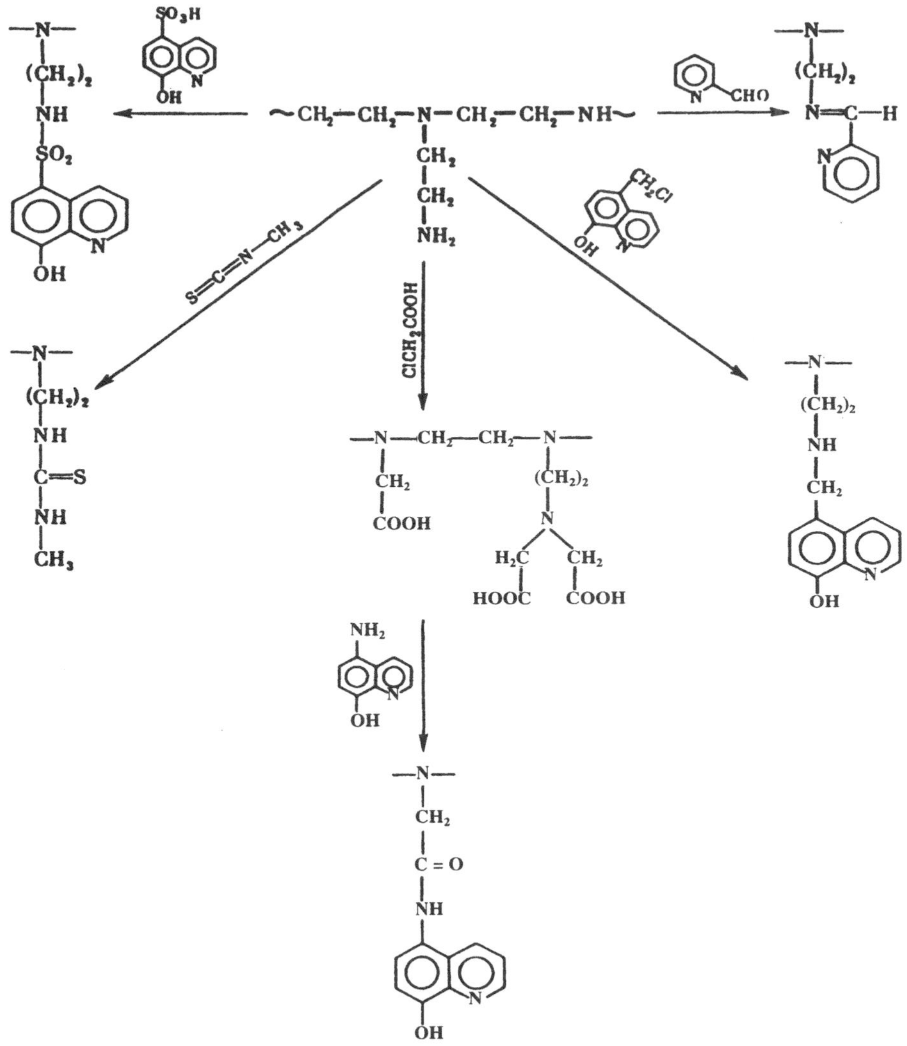

Scheme 2.7. Ways of complication of the construction of chelate nodes from primary amino groups of polyethyleneimine

The great advantage of this method is that nearly all polymers in the form of powders, fibers, and films can be modified. Polyethylene with grafted poly-allyl alcohol (PE-gr-PAA1), polyethylene grafted polyallyl- and polydiallyl-amine (PE-gr-PAA, PE-gr-PDAA), polypropylene with grafted MMA, poly-vinylacetate, polymethylvinyl ketone, polyvinylpyrrolidone (PE-gr-PMMA, PE-gr-PMVK, PP-gr-PVPd), polystyrene with grafted polyacrylonitrile, and poly-4-vinylpyridine (PE-gr-PAN, PE-gr-P4VP) are some examples. Differ-ent dispersive inorganic materials, such as silica gel are also used as matrixes. The more generally used method of supports functionalization is "anchoring" of functional compounds by surface hydroxylic groups, for example accord-ing to Eq. (2) [13]. γ-aminopropylsilane, triethoxysilane, trichlorosilanes, and other silicon organic compounds are used as anchoring agents. The L content is regulated both with the temperature of preliminary silica dehydration spe-cifying the concentration of the residual hydroxy groups and the temperature for this reaction. The principal ways of silica gel modification, including those leading to chelation, are shown in Scheme 2.8.

$$SiO_2(OH)_x + (EtO)_3Si(CH_2)_2PPh_2 \rightarrow$$
$$SiO_2(O)_x\text{-}Si(EtO)_{3-x}(CH_2)_2PPh_3 + x\ EtOH. \qquad (2)$$

Supports such as metal oxide matrixes modified with grafted or adsorbed poly-mers responsible for MMC formation have been elaborated recently. These composites allow us to combine the advantages of polymeric and inorganic supports. The methods of modifications are based on the reactions of poly- and copolymerization, grafting polymerization, polycondensation carried out in the presence of inorganic oxides (e.g., silica gel), and polymer analogous reactions [14]. Anchoring of macromolecules by one bond on the surface of inorganic supports maintains the advantages of existing polymer coil as homogenous microreactor and easiness of such MMC separation from the reaction mixture (Fig. 2.1).

Also, for the stabilization of **type III** MMC the modification of carriers is important. Physical interactions of modified organic polymers or high mole-cular weight inorganic carriers lead to specific physical interactions that can

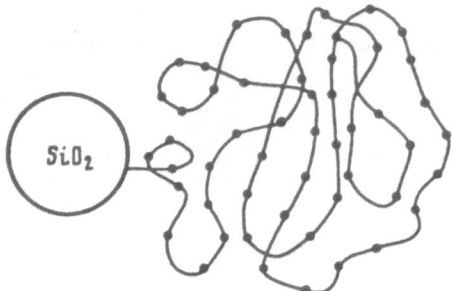

Fig. 2.1. Support of "mixed"-type macroligand with functional groups (points) attached to unporous silica gel surface

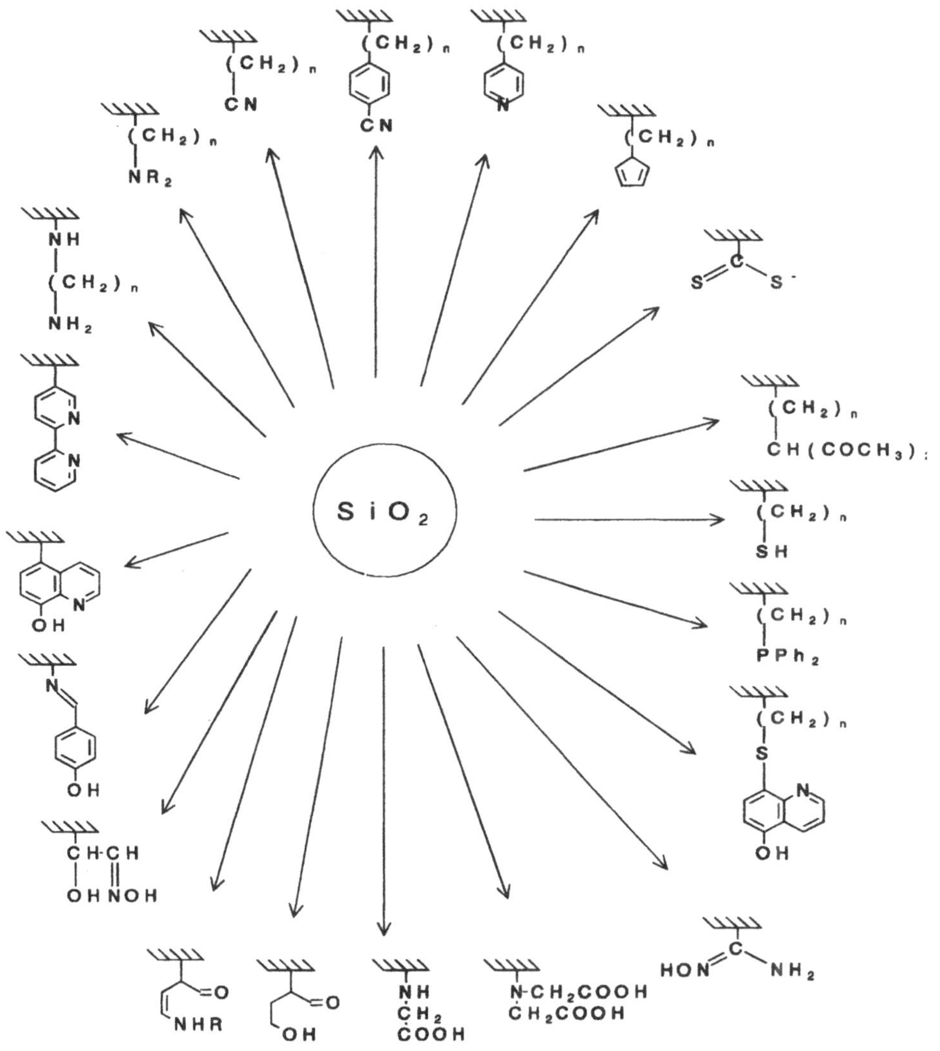

Scheme 2.8. Main ways of silica gel modification

improve properties such as conductivity, photoconductivity, optical information storage, and others. Metal clusters in these carriers are stabilized
similarly.

2.1.1.4 Natural Polymers as Macroligands

Natural polymers as macroligands lead to **type I** MMC. Recently, different
natural polymers and their derivatives have been used as metal binding agents.

A a rule, such macroligands are polyfunctional, but sometimes they have to be additionally functionalized. Polysaccharides are more widely used for the purpose especially cellulose. Cellulose is a comparatively inflexible polymer, its macromolecules possessing high asymmetry due to cyclic structure of element unit, and high polar OH groups lead to intensive intermolecular interactions. The complicated cellulose structure favors its mechanical strength being permeable to modifying reagents. Although cellulose itself does not show well properties in MX_n bonding, its derivatives (Scheme 2.9) are very effective [15]. Chitin and its desacetylated derivative, chitozane, are widely used for macroligand preparation (Scheme 2.10) [16]. Other polysaccharides, such as starch, dextranes, and alginic acids, as well as pectines, are comparatively rarely used for these purposes.

From other types of natural polymer some important biological systems, humic or fulvo acids, are mentioned, which contain different structural units. Humic acids possess, besides acid groups, significant amounts of ester, lactone, and amide ones. These supports are interesting as natural complexing compounds, e.g., to bind metal ions in soils. Gelatin, a comparatively simple collagen protein on the basis of amino acid residue of oxyproline including also different basic- and acid ionogenic groups, is also a very interesting natural macroligand. It is easily cross-linked by interaction with formaldehyde.

2.1.2 General Principles of MMC Formation

As mentioned in the beginning of Chap. 2.1.1, macromolecular metal complexes [81] were classified into three types (see also [17, 18]). In **type I** MMC (metal ions in the side chain) the metal ion can be removed without breaking the backbone of the organic polymer or the network of the high molecular inorganic carrier. Metal ion withdrawal in **type II** compounds (metal ion in the backbone) leads to decomposition of the polymer chain or network. In **type III** composites the components carrier/metal complex or cluster can be seperated easily from each other by solubilization processes.

Generally, for macroligands and their MMC the same classification can be applied as for low molecular weight ligands and their complexes. Therefore, interactions of MX_n with macroligands (**type I** MMC) or bifunctional and higher functional ligands (**type II** MMC), respectively, can be subdivided into donor–acceptor (coordinative), covalent, ionic, chelate type, and π-binding. The kind of MX_n binding depends on the nature of reagents, both macroligand/ ligand functional groups (L) and MX_n. Reaction conditions, such as kind of solvent (especially protic or aprotic), pH values, ionic strength, reaction temperature, presence of additives, etc., are also of great importance.

The type and strength of the formed M–L bonds depend on the ionization potential and electron affinity of the transition metal ion as well as on the distance between the charges. The main differences between formed bonds are displayed in long-range action. Particles (ion–ion, ion–molecule, mole-

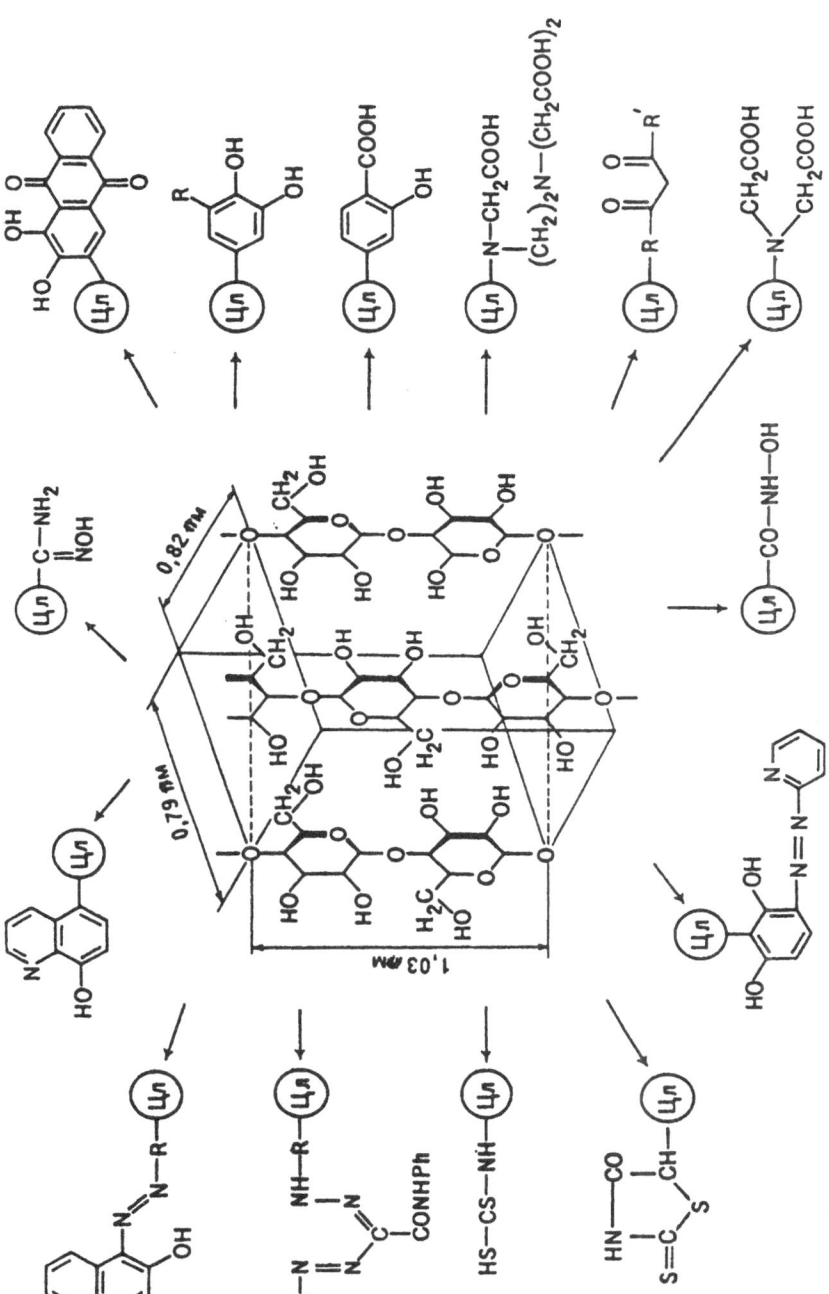

Scheme 2.9. The functionalization of cellulose (Ππ)

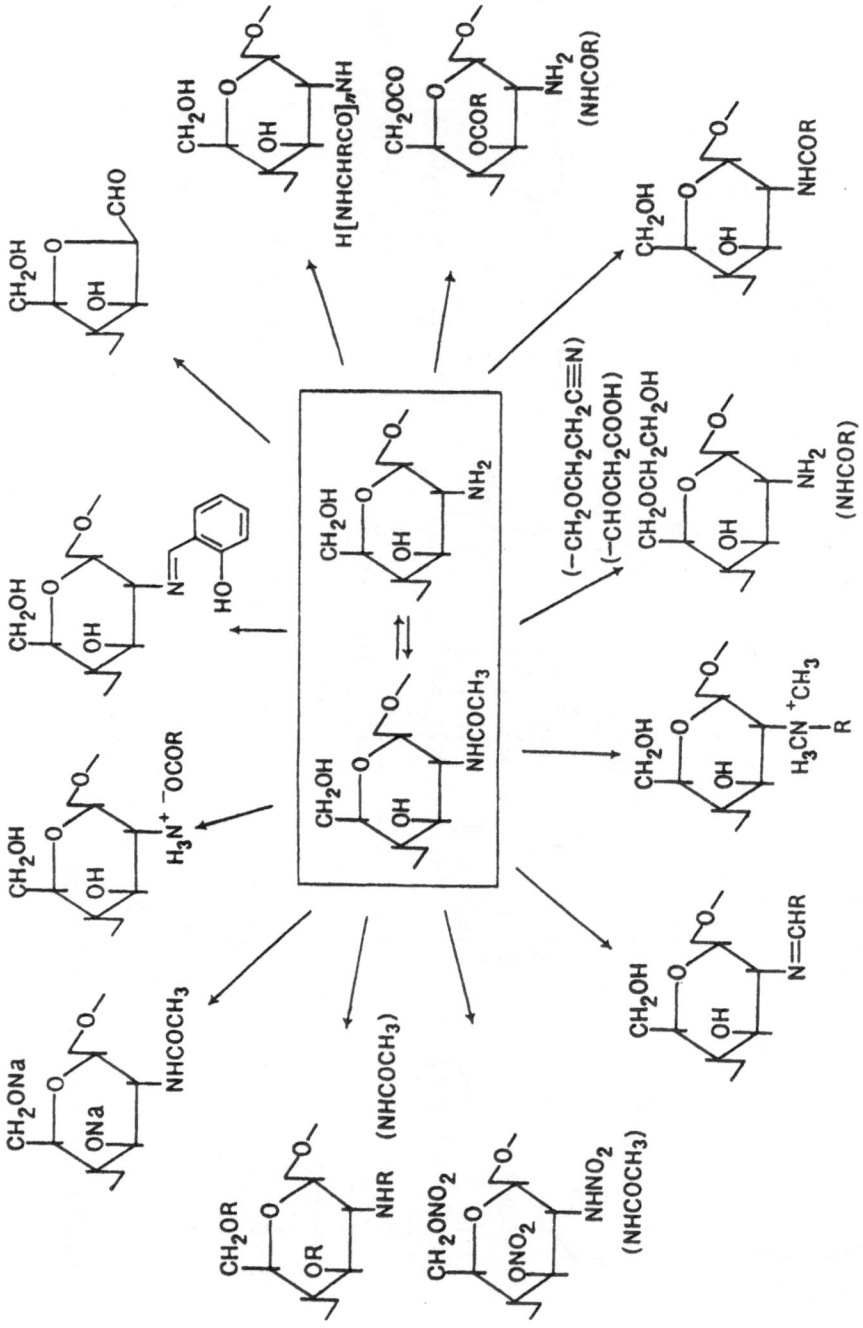

Scheme 2.10. The main way of chitin and chitozane modification

cule−molecule) can interact in long-range electrostatically and in more near distance approximately as the sum of covalent radii caused by covalent or coordinative bond formation. The total complex binding energy (combination of electrostatic and donor−acceptor) increases with increasing particle contact. The detailed investigation of all thermodynamic and kinetic regularities allows revelation of all reaction peculiarities as well as the structure of the formed polymer complexes.

Composition and structure of polymer complexes are affected not only by the nature of M and L, but also by specific polymer parameters such as degree of cross-linking, distribution of L, and, in the case of insoluble polymers, by the topography of the macroligand. The process of complex formation is accompanied by numerous complicated factors such as ion exchange equilibrium, ligand conformational changes, influence of change in electrostatic surface potential, etc. Therefore, the ratio of M/L (or in other words the amount of loading of the macroligand with metal ions) for polymer complexes is of more significance in comparison with the metal-to-ligand ratio for soluble low molecular weight metal complexes.

According to the kind of organic polymers in **type I** MMC, the following classification can be given: First, these are soluble MMC derived from linear unlinked polymers. They are obtained by reaction of reagent solution followed by precipitation of formed MMC in an appropriate solvent. The MMC of the second group are derived from weakly cross-linked (usually containing 1−2% of cross-linking agent) polymers of gel type. Swelling by a suitable solvent results in complete penetration for substrates. The MMC of the third group are metal complexes bound to strong cross-linked (10−20% of cross-linking agent) polymers in the form of hard beads ("popcorn" polymers or copolymers). These polymer supports show usually developed surface and unchanged porosity. Solvents slightly influence their properties. As a result of the high density of cross-linked porous structure, these supports interact only with functional groups of thin external or internal surface layer. Mass transfer is lower compared with dissolved systems.

Up to now no clear nomenclature exists for MMC. Usually rational nomenclature based on MX_n and macroligand are used for **type I** MMC. For instance, $NiCl_2$ complex with poly(4-vinylpyridine), sodium salt of polyacrylic acid, sodium polyacrylate, or zinc(II) complex of 5,10,15,20-tetrakis(4-aminophenyl)porphyrin covalently bonded at poly(methacrylic acid). Nomenclature of **type II** MMC are mentioned in the beginning of Chap. 2.3. For **type III** MMC for example zinc(II) complex of phthalocyanine in poly(N-vinylpyrrolidone) or platinum cluster in the cavities of faujasite X are used.

There are many unsolved problems in the field of physical chemistry of complex formation in the systems macroligand-MX_n of **type I** and especially for **type II** MMC. Mainly the absence of quantitative measurements of macrocomplex steadiness, composition, and reactivity is difficult to describe. No clear physical model considering specific features of such systems, for instance their heterogeneous nature, exists. No clear criteria for the quantitative calculation of complex formation equilibrium, etc., can be used. In the following some current approaches are discussed.

2.1.3 Main Approaches for Calculation of Rate Constants of MMC

For **type II** MMC, shown in Scheme 2.1, reaction 2b, where bifunctional low molecular weight metal complexes and another bifunctional reagent are employed, the usual rate constants, equilibria, and kinetics for polycondensation or polyaddition reactions can be used. No details are discussed here. Insolubility of these MMCs may lead to chain termination and oligomers (for examples see Chap. 2.3). A special situation is discussed for the thermal polycondensation of dihydroxy(metallo)phthalocyanines in the solid state (Eq. (3)) (Chap. 2.3.5) [19]. The polymeric product is crystalline with the chains in a fully extended (rigid rod) configuration. It was found that the polycondensation is topotactic and under topochemical control. This means that the reaction is guided by the lattice inside which it takes place with well-defined intermolecular distances and interactions. The kinetic studies show that the fraction of unreacte –OH and groups X over time t corresponds not to a first-order kinetic ($X = \exp(-k_1 \cdot t)$), but to a "power 2" Avrami-type expression ($X = \exp(-k_2 \cdot t^2)$).

$$n\ HO-M(Pc)-OH \rightarrow HO-(M(Pc)-O-)_n H + (n-1)\ H_2O$$

$$M = Si,\ Ge,\ Sn \tag{3}$$

$$n = 50 - 200\,.$$

For the reaction to **type I and II** MMC (Scheme 2.1, reactions 1 and 2a) the situation is different. One attempt for a quantitative description of the equilibrium of metal ion complex formation with macroligands in solution is based on the law of mass action in its general form for equilibrium metal ion complex formation with low molecular weight ligands [20, 21]:

$$M^{z+} + nL^- \xrightarrow{\ \bar{\beta}_n\ } ML_n^{(z-n)+} \tag{4}$$

$$\bar{\beta}_n = \frac{[ML_n^{(z-n)+}]}{[M^{z+}][L^-]^n} \tag{5}$$

$$\text{and } \bar{n} = \frac{k_1 [L^-] + 2k_2 [L^-]^2 + ik_i [L^-]^i}{1 + k_1 [L^-] + k_2 [L^-]^2 + ... + k_i [L^-]^i} \tag{6}$$

where $[M^{z+}]$ is equilibrium concentration of metal ion (mol/l), $[L^-]$ and $[L_n^{(z-n)+}]$ are corresponding concentrations of the ligand and formed complex (mol/l), $\bar{\beta}_n$ ist the total constante rate of complex formation, $\bar{n}$ ist the average number of bound ligand per one metal ion. For **type II** MMC of Scheme 2.1, reaction 2a, comparable equilibria can be assumed if instead of a monofunctional ligand now a bifunctional ligand (monomer with two functional ligand sites) with electronically separated ligand sites is employed in the reaction with MX_n.

A more complicated situation has to be discussed for **type I** MMC of organic polymers. They behave in diluted solution differently to low molecular weight analogues due to the presence of a variety of active sites and, particularly, with conformational changes of macromolecules in the course of the interaction with MX_n. The concentration of functional groups is determined by the number of polymer units carrying ligands. Such an approach is suitable for solutions of flexible linear macromolecules, because the units of such polymer molecules are separated from each other and react kinetically independent. On one side the process may be considered as the formation of isolated internal coordinative nodes with separate ligands (L) or like the formation of whole macrocomplexes. In this case two sets of constant rates of complex formation for the total process description are necessary: k_i, which characterize the Gibbs function of separate coordinative nodes, and $\bar{\beta}_n$, which is the total complex formation rate constant characterizing the whole thermodynamic potential of MMC. The past parameter describes the step-by-step coordinativ nodes addition to a macromolecule. Different models were used to describe the MX_n-macroligand interactions.

In the first considered model the polymer ligand L is the central particle and M is added in stepwise manner, then equilibrium constants will not depend on the molecular weight of the macroligand. Only the amount of added M (as it is for low molecular analogues) is considered:

$$LM_{i-1} + M \overset{k_i}{\leftrightarrow} LM_i \qquad (7)$$

(LM_{i-1} is the macrochain containing $(i-1)$ added M). Then it follows:

$$k_i = \frac{[LM_i]}{[LM_{i-1}][M]} . \qquad (8)$$

The total constant rate of MMC formation expressed in terms of actual values of $[L]$ and $[M]$ will be:

$$\bar{\beta}_n = \frac{[LM_i]}{[L][M]^i} = \prod_{j=1}^{j=i} k_i . \qquad (9)$$

If all centers are uniform and independent each complexation is described by the characteristic constant k_i. To define the values of k_i and N (the number of binding sites) it is necessary to know the degree of metal ions bound to the polymer (Θ) will be (index "0" means the initial concentration of corresponding reagents);

$$\Theta = \frac{[M]_0 - [M]}{[L]_0} = \frac{\sum_{j=1}^{j=N} i[LM_i]}{[LM_i]} . \qquad (10)$$

Parameter Θ is concerned with characteristic complex formation constant by the relation:

$$\Theta = \frac{Nk[M]}{1 + k[M]} . \qquad (11)$$

The model is uniform and independent centers are usually described in terms of $\Theta/[M] - \Theta$ coordinates (Scatchard's coordinates[1] [22]):

$$\frac{\Theta}{[M]} = kN - k\Theta. \tag{12}$$

The linear dependence of the experimental data on these coordinates demonstrates the independence of binding centers. Equation (12) allows definition of not only the value of equilibrium association constant and the total value of binding centers, but also the concentration of free metal ions corresponding to certain complexes concentration as well as the corresponding concentration of both free and bonded complex centers. The mass balance equation for $[L]_0$ and $[M]_0$ can be represented as follows:

$$L_0 = L + [L] \sum_{i=1}^{i=N} k_i [M]^i \tag{13}$$

(where N ist the largest number of ligands capable of addition of one metal ion), whereas the formation function is defined by:

$$\bar{n} = \frac{\sum_{i=1}^{i=N} ik_i [M]^i}{1 + \sum_{i=1}^{i=N} k_i [M]^i}. \tag{14}$$

The dependence of the degree of transformation on the initial concentration of ligand and metal ion are of S shape. The effective value of complex formation constant (K_{eff}) is defined by this relation. For the last case $\bar{n}$ values correspond to the number of ligands directly coordinated by metal ions. In the case of MMC, $\bar{n}$ means the average number of ligands that have been effectively excluded from the total amount of ligands for binding of a metal ion, regardless if metal ion has formed coordinative bonds with all $\bar{n}$ ligands.

The described simplest equilibria in "analytical" form are realized in aprotic solvents. The nature of solvents (dielectric constant, protic, donor – acceptor properties) can significantly affect the composition and steadiness of MMC, not only via the influence of metal ions in solution, but to a further extent by changes in the ionization state of ligand centers. The details of electrostatic influences are beyond the scope of this analysis; here it is only noticed that the electrostatic power during complex formation must increase with the increasing solvent polarity. In these cases the probability of macrocomplexes formation with unusual geometry and electronic structure differing from analogous complexes with low molecular ligands is significantly

---

[1] The calculation of equilibrium constants based on approach of independent interaction of each part of macromolecule with low molecular compound is often called Scatchard's method, whereas that which takes into account mutual influence of macromolecular parts (cooperative interactions) is Hill's method [23].

enhanced. Electrostatic interactions are usually considered in the case of charged particles. Dissociation of each LH group of macroligands (e.g., of polyacids) is impeded, due to the accumulation of the electrostatic potential ψ. Describing changes in binding constants for each isolated center in the polymer matrix [24] Eq. (11) is transformed into:

$$\Theta = \frac{Nk \, [M] \, e^{2\omega\Theta}}{1 + k \, [M] \, e^{2\omega\Theta}},$$ (15a)

where ω is an empirical parameter[2] with positive values in the case of electrostatic repulsions. For the definition of k and ω values this equation can be assigned as:

$$\log \frac{\dfrac{\Theta}{N}}{[M] \left(1 - \dfrac{\Theta}{N}\right)} = \log k - 2\omega\Theta$$ (15b)

with experimental data on a log $[(\Theta/N)/\{[M] - (1 - \Theta/N)\}] - \Theta$ plot. Experimental results of these parameters are presented in Chap. 2.2 under discussion of complex formation with some specific macroligands. The above model (the central particle–macromolecule) ist not applied to polymers where the free chain energy significantly changes under metal ion binding (Sect. 2.1.4). The concept of "site of bonding" is indefinite for polymers with cross-linking or charge-density changes along the chain (or parts of the chain), i.e., one metal ion may coordinate several arbitrary or functional groups [25].

Another model based on metal ion as central particle is described by the Flory concept of infinitely large chains. It assumes that the reactivity of binding centers is independent of their position in the polymer chain and considers macroligands as an assembly of independent binding centers. Such an approach allows description of the process of complex formation in terms of low molecular systems. Gradual ligands' addition to metal ion can be considered by the set of simultaneous equations:

$$M + L \underset{k_{-1}}{\overset{k_1}{\leftrightarrow}} ML$$ (16a)

$$ML + L \underset{k_{-2}}{\overset{k_2}{\leftrightarrow}} ML_2$$ (16b)

$$ML_{N-1} + L \underset{k_{-N\cdot}}{\overset{k_N}{\leftrightarrow}} \ldots$$ (16c)

---

[2] The value of ω is defined in the Debye-Hückel approximation for spherical particles with uniform charge surface density in nearest distance [26].

The total ML_n complex formation constant will be:

$$\bar{\beta}_n = \prod_{i=1}^{n} k_i = \frac{[ML_n]}{[M][L]^n} . \tag{17}$$

The average number of ligands bound with central metal ions is defined by the relation:

$$n = \frac{[L]_0 - [L]}{[M]_0} = \frac{\sum\limits_{n=1}^{n=N} n \, \beta_n \, [L]^n}{1 + \sum\limits_{n=1}^{n=N} \beta_n [L]^n} . \tag{18}$$

Because $\bar{n}$ depends only on $[L]$, it is possible to determine series of β_n ($n = 1, 2, \ldots, N$) from sets of N pairs. When it is difficult to measure $[L]$ and $[M]$ values, the method of competitive reactions is used [27]. If the third component is H^+, then:

$$[L] = \frac{[L]_0 - \bar{n} [M]_0}{\sum\limits_{j=0}^{j=i} \beta_j^H [H^+]^j} , \tag{19}$$

where β_j^H are constants of the protonated species. The dependence of $\bar{n}$ on log H^+ is denoted as formation function. The constants of step-by-step formation are defined from this dependence by use of half-integer meanings [28]. As was mentioned previously charges of polyanions significantly determines their conformation, and it is difficult to classify what parameter is now important for the polyligand complex ability.

The total complex formation constant was recently described as shown in Eq. (20) ([29] and references cited therein) (n: not-complex repeating units; c_p and c_s: initial concentration of repeating polymer units and metal salt; α: fraction of not-complexed metal salt by the macroligand)

$$\bar{\beta}_n = 1 - \alpha/[\alpha (c_p/n - c_s (1 - \alpha))] . \tag{20}$$

Detailed analysis show that Eq. (20) is not totally correct because $[L] \neq c_p/n - c_s (1 - \alpha)$ due to different sequences of vacant ligand units. Different $\bar{\beta}_{n1}$, $\bar{\beta}_{n2}$, $\bar{\beta}_{n3}$, etc. must be considered. Theoretical treatment of the equilibria were carried out [29] and the results were evaluated by the interaction of Na^+ (as $Na^+B(C_6H_5)_4^-$) with poly(oxyethylene) in methanol.

Metal complex binding by cross-linked macroligands can be considered as a surface adsorption that can be described by Eq. (10). Electrostatic factors significantly effect the composition, structure, and stability of macrocomplexes in the case of cross-linked macroligands. The meaning of $\bar{n}$ calculated by chemisorption isotherms is of more formal value and is little or not equal to the number of ligands coordinated by metal ions. The considered models are not applied to heterogeneous systems when L are mainly arranged on a macroligand surface and their concentration in solution is zero. For such systems calculated values of formation "constants" considerably depends on the macroligand loading with metal ions and others. For complex formation with cross-linked macroligands it is necessary to take into account diffusion (especially at above the glass temperature, T_g) and topological restrictions. Primarily, presurface functional groups interact with MX_n, and penetration of the transition metal compound inside of polymer block is restricted. The interaction between network nodes is often observed for systems leading to incomplete participation of ligands in complex formation. However, at low polymer loading one can neglect inactive bonding center concentrations. Then complex formation values can be evaluated by a Langmuir equation [30]:

$$\frac{[M]}{[M]_{bond}} = \frac{1}{K} + \frac{1}{f_{max}} [M] , \tag{21}$$

where f_{max} corresponds to extreme metal ion binding (maximum of adsorption) by a macroligand [82]. For example, adsorption parameters for Cu^{2+} binding by a cross-linked polymer containing bis(carboxymethyl)amino groups [31] are $\bar{K} = 3.5 \cdot 10^3$ l/mol and $f_{max} = 0.075$ mmol/g.

Support porosity also influences the composition and stability of formed products. Usually, when going from gel-like to porous structure, one can observe the drop of average values of formation constants. This is the result of a more compact package of polymer chains and less mobility of L leads to a significant loss in free energy of complex formation (see Sect. 2.1.4). Complex stability is also affected with surface complex lability, which depends on the number of bound L. More stable complexes are those with low [M]:[L] ratios and vice versa. Generally, the influence of the polymer chain on complex formation is stronger for cross-linked polymers than for dissolved ones. If for soluble polymers complex formation proceeds with high rates ($k = 10^6 - 10^9$ l·mol^{-1}·s^{-1}) than in networks this process is diffusion-controlled (MX_n diffusion to chelating units, the formation of outer sphere complexes, their transformation into inner sphere ones, etc.). For a nonporous solid matrix with identical ligands uniformly grafted over the surface, the complex stabilities are independent of the degree of surface coverage [32]. This was shown for the sorption of $CuCl_2$ and $PdCl_2$ from acetonitrile on Aerosil.

2.1.4 Cooperative Effects During MMC Formation

The cooperative effect is discussed for **type I** MMC at linear organic polymers. One of the specific features or reactions at polymer chains is the possibility of the formation of MMC at different positions of the polymer chain. When analyzing some quantitative parameters of MMC, the composition of not-uniform MMC distribution and macroligand free blocks should be considered. In addition, it is necessary to take into account different possibilities for MX_n binding with the macroligand [33]. If q molecules of MX_n will be bound to a polymer chain containing p active ligand sites, the number of possible complexes is equal to C_p^q (combinations of q molecules at p chain centers per time). The MX_n binding can proceed either via bond formation at a single chain or several macromolecules. In diluted solutions mainly intramolecular complexes are formed [34], whereas in concentrated solutions and polymer matrixes intermolecular complexes are possible. It is difficult to separate all consecutive processes if rates of complex formation are fast as usual. However, in some cases it was shown that k_i increases with higher degree of loading in contrast to low molecular weight analogues. This is the result of a cooperative "chain effect" which connected change in chain conformation in solution under complex formation. The MX_n addition is accompanied not only by chemical interactions, but also by changes in the local chain mobility. Binding during MX_n addition leads to an increase of the macroligand reactivity, e.g., polyethyleneglycol [35]:

$$\tag{22}$$

The complex formation constant can be represented as:

$$\bar{\beta}_n = \prod_{i=1}^{i=N} k_i = \sigma k_i , \tag{23}$$

where $\sigma = k_i/k_{i-1}$ is a factor of coordination, i.e., increase of formation constant in successive steps. It is postulated that all constants, except the first one, are similar because main entropy changes proceed during the first addition. All following steps can be seen as intramolecular cycle formations. The values of $\sigma = k_2/k_1$ are within the range of $10^{-4} - 10^{-8}$.

This "chain effect" is exhibited by a high local concentration of functional groups. Then the complex formation constant is determined not by the average values of reagent concentrations, but by their local content in polymer coils as microreactors. The chain effect leads to not uniform MX_n distributions among macromolecules. The first MX_n addition facilitates the following ones, and interactions with the polymer chain will be completed only when all potentially active centers are occupied. The presence of extremely loaded and nearly unloaded polymer chains with metal ions can be observed. The principle of "all or nothing" is realized in this case. For instance, this effect was demonstrated by rapid sedimentation of (Cu^{2+}-poly(4-vinylpyridine)) [36].

2.1.5 Thermodynamic Description of Macromolecular Metal Complex Formation

The thermodynamic situation is influenced not only by changes in conformation of macromolecules, in chain flexibility, and of electrical charge of the polymer during the reaction, but also by the energy content of the starting materials and the reaction products [37].

Thermodynamic characteristics of **type I** MMC formation can be evaluated by the temperature dependence of $\bar{\beta}_n = f(1/T)$. Usually, the greatest contribution of the Gibbs energy (ΔG).

$$\Delta G = -RT \ln \bar{\beta}_n = \Delta H - T\Delta S \tag{24}$$

comes from the entropic term (ΔS), and not from the nearly unchanged enthalpic one (ΔH). Both parameters of coordinative node formation (inner coordinative nodes, CNd formed by separate functional groups of the macroligand with MX_n) and of chain changes (similar to evaluation of MMC formation constants, Chap. 2.1.3) should be considered for the calculation of thermodynamic parameters. Addition of MX_n to a macroligand results in the formation of new covalent, coordinative, ionic, charge-transfer bonds at or between (cross-linking) macromolecules. This is shown for coordinative bonds formation:

$$\tag{25}$$

The differences in the free energies for the reactions a, b, ΔG_{CNd}, and for c, ΔG_{MMC}, are connected with the changes in macrochain mobility. If all free energy terms are additive, then:

$$\Delta G = \Delta G_{CNd} + \Delta G_{MMC} . \tag{26}$$

For low molecular ligand interactions it is well known that addition of a metal ion to one bidentate ligand is more favorable than to two monodentate ligands ("chelate effect", ChE) [38, 39]. The ChE, depending on the kind of solvent, is on the order of -5 to $-20\,kJ/mol$, and mainly determined by entropic terms (change of solvation and also transfer, inner rotations, symmetry, vibrations, and rotation entropies). The polymer chelate effect (PChE) can be described as the change of free energy ($\delta\Delta G_{pch}$) due to the addition of a metal ion to a macroligand (ΔG_{pch}) in comparison with the addition of a metal ion to a monodentate ligand (ΔG_{pm}) [40, 41]:

$$\delta\Delta G_{pch} = \Delta G_{pch} - \Delta G_{pm}. \tag{27}$$

In contrast to low molecular ligands, estimation of PChE is significantly more complicated, because the correct evaluation of all terms of the polymer effect is difficult. This includes local, molecular, and supramolecular organization of macromolcules. In supposition of additivity of all terms of free energy changes in the course of chelation with macroligand one can derive

$$\delta\Delta G_{pch} = \Delta G_{loc} + \Delta G_{mol} + \Delta G_{spm} - \Delta G_{pm} \tag{28}$$

(ΔG_{loc}, ΔG_{mol}, ΔG_{spm} are changes of free energy for local, molecular, and supramolecular levels, respectively). The difference $\Delta G_{loc} - \Delta G_{pm}$ characterized the local chelate effect (ΔG_{lch}). Two additional terms are related to the organization of molecular and supramolecular structures of MMC. Under certain conditions the changes of free energy for one or another level can be neglected. Such an approach allows to estimate the main contributions (ΔG) of each level. Therefore, in diluted polymer solutions the association of MMC ($\Delta G_{spm} \to 0$) is neglected:

$$\delta\Delta G_{pch} = \delta\Delta G_{lch} + \Delta G_{mol}. \tag{29}$$

In the case of infinitely long chains with a low degree of coordination, chelation will not significantly affect shape parameters and conformation of the macromolecule ($\Delta G_{mol} \to 0$). Under this supposition macroligands are comparable with low molecular weight analogues; both enthalpy and entropy terms of PChE are determined by the same contributions as in the case of low molecular weight chelates. Actually, similar values of ΔH have been observed [42] for chelation of Cu^{2+} with polyamines or low molecular weight amines, and PChE of MMC is mainly determined by entropy terms. This is confirmed by similar dependencies of both ΔG and ΔS values on the coordination degree (Θ) (Fig. 2.2).

As was mentioned previously, the contributions of transfer, rotation, symmetry, and isomerism are positive terms in the entropy part of chelate effect. The contribution of the transfer term are expected to be similar for low molecular weight complexes, and with MMC some differences are probably connected with rotation term, because both initial macroligand and formed MMC are nonrotatory particles. The groups of symmetry are considerably dissimilar for low- and high molecular compounds. As the number of possible isomers becomes significantly less for polymer chelates, the contribution of isomerism decreases. One of the known examples are Ni^{2+} chelates with salicylaldeimine ligands existing as two isomers: planar complex (green) and tetrahedral one (brown) [43]. Only the pseudotetrahedral form were observed for polymer analogues [44]. The contributions of vibration, inner rotation, and solvation to the entropy term of the chelate effect are negative. In correspondence to the stereochemistry of the chelate node either increase or decrease of solvation contributions will be observed for MMC. This discussion is valid for **type I** MMC (to a different extent for soluble macroligands, on one side, or cross-linked organic polymers and high molecular weight inorganic carriers, on the other side). For **type II** MMC; Scheme 2.1, reaction 2a, often insoluble amorphous or crystalline more or less solvated polychelates are formed and an additional negative contribution, especially going from monomers in solution and precipitating MMC, must be considered. **Type II** MMC, Scheme 2.1, reaction 2b, has to be discussed as usual for polyreactions in

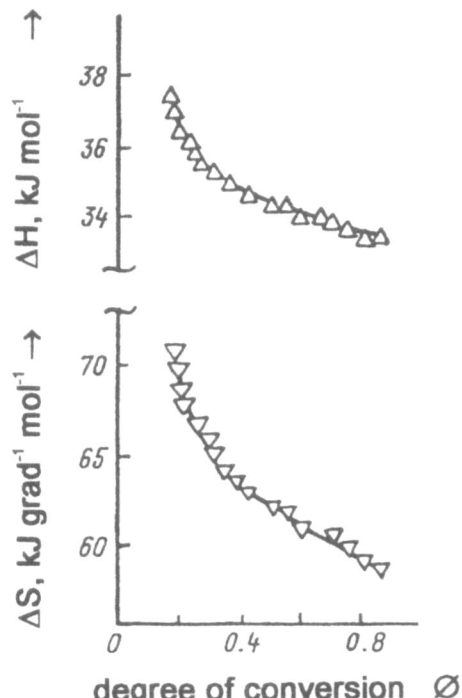

Fig. 2.2. Dependence of the thermodynamic terms of Cu^{2+} chelation by poly(amidoamines) from the degree of conversion

macromolecular chemistry. For all **type III** MMC a negative contribution to the entropy term for the stabilization of metal complexes or metal clusters by the surrounding high molecular weight matrix has to be considered. As well as for low molecular weight chelates, the polychelate effect also significantly depends on the dimension of the chelate cycle. Moreover, the well-known Tschugaeff's rule for cycles is also valid for MMC. The most stable are five- and six-membered cycles. But it must be mentioned now that detailed discussions considering the different contributions of the entropic term for MMC are available in the literature.

Besides some similar contributions to the enthalpy and entropy of chelation for low molecular weight and high molecular metal chelates, a difference can be assigned. The "alignment" of local fragments accompanied by the changes in local flexibility was mentioned previously. Although the energy of such "alignment" for soluble macroligands is significantly less than that of the chemical interaction, this effect significantly changes the stereochemistry of formed nodes (to a considerable degree of valent angles and to a smaller degree of bond length).

The effect of neighboring groups (both reacted and free) to the free energy of chelate formation, and consequently, on their structure, is also of great interest. The effect is caused mainly by steric factors and leads to metal center repulsions, their deformation hindering complex formation with the rise of coordination degree. In this case the length and the nature of "spacer" between the plane of polymer matrix and bound centers are of great importance. The optimization of "spacer" parameters (as a rule optimal length of around 10 atoms) provides binding fragments with space mobility for overcoming the diffusion hindrances in polymer media as well steric inhibition of polymer-bound ligands.

Intermolecular interactions between initial macroligands and MMC and **type I** complexes significantly rise when going from diluted polymer solutions to concentrated ones as well as to suspensions. In this case it is necessary to take into account the previously mentioned supramolecular MMC organization. Cross-linked metal polymers produced at high L concentrations (see Eq. (25)) change into more stable intramolecular complexes with time. Therefore, it is necessary to take into account the time of their relaxation, and consequently, the dynamics of change of properties. Because of the kind of chemical bond preserved in the process, $\Delta H = 0$ and these transformations are caused by entropy terms. Transformation of intermolecular chelates of Fe^{3+}-polyhydroxamic acid to intramolecular ones [45] were shown to be of first-order reaction with rates of $5.21 \cdot 10^{-6} s^{-1}$ at 303 K and $49.7 \cdot 10^{-6} s^{-1}$ at 333 K, with an activation energy of 63.3 kJ/mol and entropy factor of $+ 1$ e. u.

Two parameters, the ultimate polymer concentration (L^{ul}) and the temperature of liquid $\Leftrightarrow$ gel transformation (T_{tr}), are of great importance for intermolecular **type I** MMC formation. The L^{ul} means the polymer concentration above the intermolecular interactions become determined [46]. The L^{ul} is inversely proportional to the molecular weight of the macroligand. Above L^{ul} the T_{tr} value is practically independent of the concentration and molecular weight of the polymer and is described by:

$$T_{tr} = \frac{\Delta H^0}{\Delta S^0 - R \ln [(L - L^{ul})/L^{ul}]}, \tag{30}$$

The T_{tr} is therefore mainly the function of thermodynamic properties connected with formation of metal–polymer bonds. In the case of Fe^{3+}-polyhydroxamic acid infinity networks were formed when the probability of intermolecular metal binding was above 50%.

Often **type I** MMCs show a more perfect morphological structure than the employed organic macroligands. During the process of MX_n binding some structure reorganization may proceed, such as micronetwork rearrangement or macromolecule torsion [47]. All structural rearrangements follow steric requirements of the transitional metal ion (minimal loss of free energy). In all cases the tendency to torsion of chain molecules depends on chain rigidity and nature of the solvent. By variation of these factors, chain torsion and tension can be strengthened or weakened. Changes in supramolecular organization cause an increase of chain rigidity and should display also changes in relaxation transitions of chains and chain segments. Evidently, α-relaxation, which is attributed to micro-Brownian movements of long chain segments above glass temperature (T_g), is affected. At low macroligand loading, the total restriction of chain mobility is insignificant. Higher loading leads to growth of cross-linking and consequently ascent of T_g. The increase of MX_n content under cyclization also causes increase of T_g, due to a rise of rigidity of separated chain segments and drop of chain mobility. The β-relaxation processes are practically independent of the metal content.

In summary, it can be established that a correct evaluation of all terms of the polymer effect in thermodynamic description of complex formation with polyligands is difficult. More important in this connection are conformational, electrostatic, concentration-dependent (different local concentration of functional groups), and supramolecular (macromolecule association and aggregation during complex formation) effects.

2.1.6 Main Transformations of Macroligands and Transition Metal Compounds During Complex Formation

Both reacting components, macroligand of an organic polymer and MX_n, often (practically every time) suffer different transformations during **type I** MMC formation. For macroligands these are also the previously mentioned conformation changes, reorganization of structure, macrochain breakage, and also modification of functional groups. The most significant transformations are discussed in the following.

Two effects are most important for **type I** MMC of organic polymers. These are change of the configuration in different spatial orientations of neighboring groups ("neighbor effect") and the conformation displayed in the rearrangement of the whole macroligand during the reaction. A different number of monomer units can be involved (including coil-ball conformational changes). The presence of hydrogen bonds and steric reasons hinder bending and rotation, i.e., conformational optimization. This also may be the reason why a lot of functional groups are exluded from complex formation.

On the other side, complex formation often results in stabilization of untypical polymer conformations and tautomeric forms. For instance, if polyethyleneglycol (PEG) as helix-like molecule with alternating *trans-* and *gauche-* C–O, O–C and C–C bonds (conformation T_2G helix 7_2) [48] is treated with $HgCl_2$, complexes of regular composition $HgCl_2(OCH_2CH_2)_4$ with an elementary unit composed of four molecules of $HgCl_2$ and 16 units of (OCH_2CH_2) [49] is obtained (Fig. 2.3). The resulting chain carrying bound complexes is in $T_5GT_5\bar{G}$ conformation (G and $\bar{G}$ mean dextro- and levorotatory *gauche* forms) with increased part of turned-off configurations (*gauche*), each elementary unit of PEG changing from monoclinic to orthorhombic. These transformations cause change in the linear orientation of the $HgCl_2$ molecule (the angle ClHgCl is equal to 176°). The more noteworthy changes proceed in the period of identity of the macromolecule. It is composed of four monomer units instead of seven for the uncomplexed polymer. Structural studies show [50] that MMC properties in PEG-alkali metal ion complexes are better described in terms of the double-helix model. Other examples are: changes in the structure of polyethers by interaction with transition metal halides [51], conformational transformations of poly(2-vinylpyridine) during the complex formation with Co(AcAc)$_2$ [52], conformational modifications of linear poly(amidoamines) caused by their protonation and complex formation [53], etc.

Important also are intermolecular interactions with the result of a decrease in chain flexibility and more pronounced intramolecular interactions as mentioned previously. Any interaction which affects intermolecular processes usually also change MMC stability and morphology. For instance, an increase in the degree of cross-linking of polyvinylpyridine causes changes in Cu^{2+} complexes' structure from stable square planar to deformed tetrahedral [54]. Also, the molecular weight influences the kind of intermolecular interaction. Polymers with low molecular weight usually form stable intramolecular complexes, whereas those with high molecular mass show a tendency to intermolecular complex formation by chain cross-linking (see Eq. (25)). Complex formation of MX$_n$ with inflexible polymers (i.e., cellulose) leads to cross-linking of individual structural elements (microfibrils, fibrils, etc.) and of polymer chains inside of these structural elements [55]. Therefore, MX$_n$ affects not only processes of molecular, but also supramolecular, organization.

Another point to consider is the destruction of the polymer chain of the macroligand of **type I** MMC during MMC formation. This phenomenon is often observed even in diluted solution (including nonaqueous ones) under inert atmosphere. Viscosimetry, GPC, and rapid sedimentation techniques are usually used to reveal this aspect [56, 57]. The decomposition degree defined as the ratio of average-number molecular weight of the employed macroligand ($\bar{M}_n$) to that of the obtained macroligand ($\bar{M}_n$)° can reach great values. For instance, in the case of macrocomplexes of TiCl$_4$ with polymethylmethacrylate ($\bar{M}_n = 6.06 \cdot 10^5$), polyvinylacetate ($\bar{M}_n = 1.97 \cdot 10^5$), and PEG ($\bar{M}_n = 0.225 \cdot 10^5$) ratios of 30.3, 3.8, and 1.8, respectively, were determined [58, 59]. The

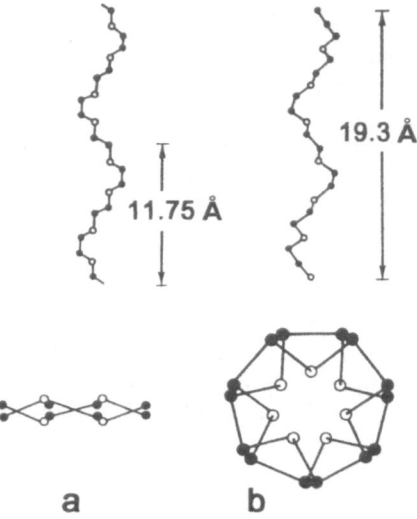

Fig. 2.3. Helix structure of PEG (**a**) and its complex with HgCl$_2$ (confirmation T$_5$GT$_5$ $\overline{G}$) (**b**)

a   b

destruction of polyvinylchloride (PVC) was observed [60] during the interaction of THF with iron salts and of PEG ($\bar{M}_n = 2 \cdot 10^4$) during the interaction with TiO_2 in CCl_4 [61]. The mechanisms of the destruction process are not clear. However, some suppositions about the radical character of the process have been made [57]. A mechanism of thermally activated bond cleavage is also possible, which is connected with the breakage of the polymer chain under additional participation of mechanical forces (σ) stretching the polymer chain during the complex formation (mechano-cracking). The stretch leads to a decrease of the C–C bond energy to a value $E = E_0 - \gamma\sigma$ (γ is the structural factor indicating distribution of stretching strength through a bond). In principle the destruction is possible for all polymers, and consequently for all bond types, between ions and macroligands. One mechanistic proposal for the destruction of polymethyl-methacrylate is given:

$$(31)$$

Changes in the chemical nature of functional groups in the systems MX_n-macroligand of **type I** MMC are now considered. The interaction is often accompanied by macroligand protonation, ionization, and alkylation, as well quaternization reactions, oxidation, reduction of functional groups, etc. In particular, interaction of polyvinylalcohol with V^{5+} is accompanied by partial transformation of hydroxylic groups into carboxylic groups [62]. Partial transformation of $-C\equiv N$ groups into $-C(Cl)=N-$ were observed during the interaction of $MoCl_5$ with PE-grafted-poly(acrylonitrile) [63]. Polymeric ketones were obtained by the reaction of $NaFe(CO)_4$ with chloromethylated styrene-divinyl benzene copolymer (CMCS) [64].

The local strength of metal-polymer bond in MMC is known to differ from those in low molecular weight complexes by the value of chain rigidity in **types I and II** MMC. Incorporation of functional groups into the polymer backbone leads to changes in their redox [36], donor–acceptor, and other properties. Therefore, the valus of $\Delta\upsilon_{c=o}$ as measure for the ligand basicity showing the carboxylic groups' vibrational shift during the complex formation decreases in the order: monomer > oligomer > polymer (e. g., in systems $TiCl_4$-MMA, oligo-MMA, and poly-PMMA) [65]). Molecular weights of polymers also considerably influence values of $\Delta\upsilon_{c=o}$. For $CoCl_2$ complexes with polyvinylpyrrolidone [$CoCl_2 \cdot 2L$] with polymerization degrees $n = 1100$, 200, 360, and 6300 $\Delta\upsilon_{c=o}$ values of 54, 49, 48, and 43 cm^{-1}, respectively, were found [66]. Some additional reactions can occur during the MX_n complex formation: redox processes, monomerization of dimeric and cluster complexes, the formation of complexes of different composition and structure, exchange interactions, cluster formation, etc.

Generally, complex formation with a macroligand favors stabilization of the complex. However, exceptions are known. The oxidation of zero-valent metals in complexes of the general formula $M(PPh_3)_4$ (M = Pd, Pt, Ni) to the two-valent state was observed during the interaction with halogenated poly-styrene (PS) [67]. A similar effect was seen by sputtering of metal vapors onto polymer matrixes [68], during metal solvatation in the presence of macro-ligand, during mechanochemical synthesis, etc.

However, transition metal ion reduction is more frequent. Pd^{2+} reduc-tion to Pd^0 is observed in reactions of K_2PdCl_4 with macroligands in aqueous-alcohol mixtures. Examples for the formation of metal cluster by this reduction (**type III** MMC) are given in Chap. 2.4. $RhCl_3 \cdot 3H_2O$ in methanol interacts with polymers yielding Rh^{2+} bound complexes. The high reducibility of Mo^{5+} to Mo^{4+} and V^{5+} to V^{4+} in reactions with macro-ligands is well known. Similar processes are also known for some other metals. Cu^{2+} chelation with cross-linked poly-N-(acryloylaminomethyl)mer-captoacetoamide follows with Cu^{2+} to Cu^{1+} reduction; Cu^{1+} is oxidated again to Cu^{2+} by air [69]. Cr^{2+} complexes bound to polyethers modified by amines under addition of NaCl are oxidized to Cr^{3+}, however, easily reduced at the medium potential of -1 V [70]. These metal polymers, made in the form of interweaved threads, work as electrical muscles, and are used also as special resistane systems. Iron sandwich complexes bound to PVC or CMCS with a reversible redox couple ($Fe^{3+} \leftrightarrow Fe^{2+}$) were used for the preparation of redox electrodes [60].

Monomerization of dimeric complexes is one of the processes accom-panying **type I** MMC formation. Various transition metal compounds (VCl_4, $MoCl_5$, $TaCl_5$, etc.) are known to exist as dimers, tetramers ($Ti(OR)_4$), and even polymers (VCl_3, $TiCl_3$, $HfCl_4$, etc.) with chlorine atoms as the bridging groups. Their interaction with macroligands usually causes monomerization. The strength of dimeric complexes and the thermodynamic advantages of monomerization influence the probability of this effect. If the last factor predominates, complete complex monomerization is observed. If both factors are present, mixtures of monomeric and dimeric complexes bound to macroligands are formed. In strong cross-linked macroligands coordina-tive unsaturated complexes can be formed as has been observed for bis(cyclooctadienylrhodiumdichloride) $[CODRhCl]_2$ bound to phosphory-lated PS cross-linked with divinylbenzene [71]. The nature of functional groups also affect the process. Thus, polymer amines and phosphines interact with $[Rh(CO)_2Cl]_2$ without monomerization, whereas polymer thiols results in monomerization [72]. Degree of cross-linking of the macroligand is also of great importance. Flexible polymer chains (low cross-linking degree) favor dimeric forms of MX_n. The behavior of cluster complexes are discussed in Sect. 2.4. Another feature of complex formation with macrochain molecules is the metal complex distributions at a single polymer chain and between chains (Eq. (25)). As discussed previously several types of complexes formed during the binding of MX_n are converted to MX_n bound to a single chain, which is thermodynamically more stable. However, in some cases (e.g., rigid framework) uniform distribution of MMC is obtained.

The MMC topography is predefined not only by the geometric structure of the employed macroligands (MX_n can either fill micropores of the support or assign a new microporous structure on the polymer surface or in thin presurface layer), but also by reaction conditions. Usually, local MX_n concentrations in a presurface layer are above those in the inner polymer volume. Nonuniformity in metal-complex distribution on the polymer framework displays the coexistence of individual complexes and exchanging associations. Concentration and stability of associations depend on the nature and concentration ratio of reagents. Even for soluble MMC, the local concentration of metal complexes (C_m) may be 50–100 times above their average concentration (C_0). An example for a topotactic reaction under topochemical control is the **type II** MMC formation of the solid-state polycondensation of $HO-M(Pc)-OH$ (see Chap. 2.1.3, Eq. (3)).

Because the topography of MMC is of primary importance for their reactivity and properties, different physical and chemical methods were used for their investigation. The most valuable results have been obtained by ESR (if suitable metal ions or metal complexes are used), because the tendency to form areas with a high local metal complex concentration are proved to be intrinsic for a number of paramagnetic ions (Cu^{2+}, V^{4+}, Mo^{5+}, etc.). The effect is established in ESR spectra by the superposition of singlet exchange peaks of neighboring metal ions and multiplet spectra of isolated complexes (Chap. 2.1.7). Increase of concentration of paramagnetic ions in MMC is accompanied by broadening of ESR bands due to dipol-dipol metal ion interaction according to:

$$\Delta H = \Delta H_0 + AC_M, \tag{32}$$

where ΔH is the width of individual ESR peak, ΔH_0 is this in the absence of dipol interaction, and A is the factor depending on both the spatial distribution of paramagnetic centers in MMC and the shape of individual peaks (calculated in [73]). By electron spin resonance (ESR) three different distributions of MX_n in MMC can be measured. At low macroligand loading (Θ) with metal ions ΔH and C_M are linear dependent on the MX_n concentration. At high loading the C_M contribution is so high that the resulting ESR spectrum looks like a singlet due to strong spin-exchange interaction between metal ions. This effect was analyzed in detail for ionite macrocomplexes [74]. Another type of metal complex distribution is revealed when at low macroligand loading all metal complexes are in the form of isolated centers showing good-resolution multiplet ESR spectrum. With then-increasing concentration the increase of C_M is accompanied by the formation of associates showing a singlet peak in ESR spectrum. This type was observed for V^{4+} complexes bound to PE-grafted-to-poly(4-vinylpyridine) (Fig. 2.4) [75]. The higher amount of bound V^{4+} is accompanied by an increase of aggregates and a decrease of the amount of isolated ions (Fig. 2.4b).

A more complicated situation exists when MMCs contains all three types of distributions with an average distance of $\bar{r}_{isol} \geq 2.2-1.5$ nm; aggregations of clusters with a distance between the metal ions $\bar{r}_{agr} \le 0.7$ nm and, in addition, strong dipol-dipol interactions and clusters with strong antiferromagnetic interaction (or "exchange-bound clusters"). This distribution has been revealed for Cu^{2+}, V^{4+} fixed on PE-grafted-poly(acrylic acid). By combination of ESR technique and magnetic susceptibility the contribution of each type was carried out, and C_M and $\bar{r}_M$ values at different Cu^{2+}, V^{4+} contents were determined.

2.1.7 Methods for Characterization of Composition and Structure of MMC

Different chemical and physical chemical methods valuable for **types I, II and III** MMC are widely used for their characterization. Some of the methods give information on only some of the MMC organizing level (local, molecular or supramolecular), and some of them offer detail information on several levels. Many particularly older papers describing the synthesis of MMC include unsatisfactory characterizations of the reaction products, which is fundamentally important to evaluate the relation between structure and properties. Someone being active in the field of MMC has to be cautious in the evaluation of reported data.

All methods are subdivided in three main groups: (a) chemical and physicochemical analysis, (b) spectroscopic, and (c) magnetic and magnetospectral.

Elemental analysis (group a) shows the metal ion coordination number allowing estimation in a first approximation of the stereochemistry of macrocomplex. The most informative analysis for the total metal content is usually

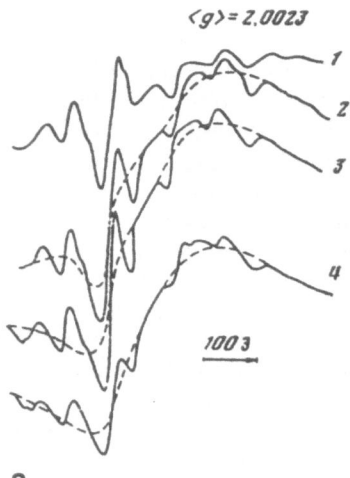
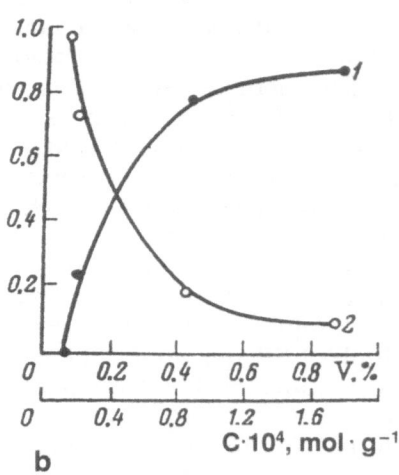

Fig. 2.4. a ESR spectra of V^{4+} bound to PE-grafted-poly(4-vinylpyridine); V^{4+} content in 10^{-4} V/g; 1:0.15, 2:0.20, 3:0.85, 4:1.76. **b** Dependence of aggregated V^{4+} (1) and the free ion V^{4+}(2) from the whole concentration of V^{4+} at PE-grafted-poly(4-vinylpyridine)

achieved by either destructive or nondestructive techniques. Traditional and atomic adsorption spectroscopy are widely used. Elemental analysis in combination with spectroscopic data and their comparison with those for low molecular complexes yield information on composition and structure of MMC coordination centers. Noninterger L/M ratio values often obtained for MMC indicate the presence of either complexes of several types or some part of free ligands poorly involved in interaction.

Gel-permeation chromatography (GPC), sedimentation techniques, viscometry, light scattering, and equilibirum dialysis are informative methods for investigations of MMC in solutions, macroligand conformation states, and their transformation during MX_n binding. Almost all studies on macromolecule conformation changes show contraction of polymers by complex formation (high-speed sedimentation, viscometry). This is caused by the metal ion cross-linking capability. Polyelectrolytical effects are observed in the case of charged macromolecules. Such MMC are swollen by aqueous and aqueous-methanol solvents similar to traditional polyelectrolytes. Under high ionic strength the electrostatic effect drops and the MMC structure gets more compact.

Microcalorimetric studies of MMC formation allows measurement of the enthalpy value of complex formation, but includes only poor information on MMC structure. Potentiometric and conductometric titration are widely used for characterization of water-soluble MMC due to easy experimental conditions. The formation constant $\bar{\beta}_n$ and actual values of metal ion coordination number can be calculated from the formation function $\bar{n}$ (change of average coordination number during complex formation determined by Bjerrum's, Gregor's, or Mendel's methods) evaluated by potentiometric titration data. Experimental evaluation of $\bar{\beta}_n$ and $\bar{n}$ is based on proton and metal ion competition for macroligand functional groups. The composition of metal ion coordinative sphere depends on [L]:[M] ratio, temperature, and pH value of the solution. Calculated by this method, $\bar{\beta}_n$ and $\bar{n}$ are more qualitative values.

Spectroscopic techniques [optical and luminescent spectroscopy, IR-, and Raman spectroscopy, acoustic vibrational spectroscopy, the method of dielectric loss, several types of X-ray and nuclear gamma resonance (Mössbauer) spectroscopy, X-ray photoelectron spectroscopy (ESCA), etc.] provide information about the inner coordinative sphere of metal ion and the localization of coordinative bonds. The shifts of some characteristic bands of IR spectra caused by MMC formation (sometimes considerable) are usually analyzed. The most interesting is the IR range where vibrations of metal–ligand bond are observed. Infrared spectroscopy being the most widely used technique for all type of MMC (both soluble and cross-linked) studies gives information about MMC composition, conformational peculiarities of macroligands, hydration and state of functional groups, and, which is usually most important, about the nature of coordinative centers and bond strength. This method, however, gives no opportunity to establish the precise composition and structure of individual coordinative centers as well as spatial organization of MMC. The application of IR spectroscopy to polyfunctional macroligands is also essentially restricted.

Electronic spectroscopy techniques (UV-, visible, and low-frequency infrared range) are also widely used for MMC studies. Three main types of characteristic bands can be detected in this connection. They are (a) inner molecular bands related to electron transfer from ligand-occupied orbitals to free ones, (b) bands of charge transfer characterized with high energy and intensity, and (c) d–d electron transitions between d-orbitals of central metal atom. When going from ion in gas state to coordinative compound the number of energy levels (terms) and accordingly number of peaks in the spectrum increases. Analysis of energetic diagrams often allows identification of coordinative node, and consequently, the covalent degree of metal–ligand bonds. Both methods of circular dichroism and dispersity of optical rotation are useful for the determination of central atom microenvironment and MMC structure.

Different diffraction methods, especially X-ray analysis, are rarely applied to **type I** MMC because of missing crystallinity. But X-ray and related methods are used more intensively for **types II and III** (metal clusters) MMC. EXAFS spectroscopy is less often used for MMC investigation, mainly for determination of atom-to-atom distances and the type of coordination. ESCA gives useful information about binding energies of electrons; the energy changes are connected with the state of valence electrons. The features that make ESCA so valuable are its ability to estimate elementary composition on the values of binding energies as well as to reveal the nature of macro–ligand, binding orbitals' symmetry, and valent state of transition metal ion. The method is restricted with its capability to reveal both conformational organization of MMC and unambiguous localization of ligands in the metal ion coordination sphere.

Although limited in the range of metal ions to which it is applicable (mainly isotopes of ^{57}Fe, ^{119}Sn), nuclear gamma resonance spectroscopy is widely used to determine the valent state of metal ions and the nature of bonds to ligands or groups, as well as for the investigation of dynamics of intramolecular movements. Thus, spectroscopical methods give important information about the nature of atoms or functional groups of macroligands bound to metal ions, binding energies, and the number and location of atoms in metal ion coordination sphere.

Magnetochemical methods are also often used for the investigation of MMC containing unpaired electrons (paramagnetic, ferromagnetic complexes without spin-lattice relaxation contribution). The magnetic moment value (μ_s) correlates with the metal ion valent state and its surroundings. For instance, values for Co^{2+} are within the range $\mu_s = 4.70-5.20\,B\mu$ for octahedral complexes and $\mu_s = 3,90-4.50\,B\mu$ for tetrahedral ones. The temperature dependence of the magnetic susceptibility of MMC gives data on antiferromagnetic and ferromagnetic metal ion interaction and indicates the presence of mono- and polynuclear (cluster) structures, as well as the character of interionic interactions in these cluster structures. However, magnetic susceptibility is a macro characteristic parameter. Therefore, it is difficult to distinguish the influence of ligands on metal ion states in the presence of complexes of different type.

Nuclear magnetic resonance (NMR) spectroscopy for MMC studies is connected with the fact that a more near distance of groups to paramagnetic center results in a greater chemical shift and band broadening. In most investigations H-NMR is used for solution spectra. However, NMR of ^{13}C, ^{31}P, ^{14}N, ^{23}Na, and H-NMR of high resolution on solid samples are applied.

The ESR technique is use most extensively among all magnetospectral methods. The technique is useful for MMC containing paramagnetic ions. The information provided by ESR may give detailed description of metal ion valence state, the nature of ligands and bonds and orientation of bound complexes. The ESR is more often used (see Chap 2.1.6), for studies of both soluble and cross-linked MMC with bound Cu^{2+}, VO^{2+}, Mo^{5+}, V^{4+}, Ti^{3+}, and Cr^{3+}, due to the high resolution of their spectra and for bound ions of Mn^{2+}, Co^{2+} and Fe^{2+} (at low temperatures). The ESR technique has proved to be very useful for different MMC characteristics such as kinetics of ion exchange, distribution of metal complexes (topochemistry) in MMC, exchange interactions, etc. Thus, each of the mentioned methods separately or in combination with each other are capable of giving comprehensive information about composition, structure, and spatial organization of **types I, II, and III** MMC.

A great variety of **types I, II, and III** MMC are available, and new ones will be synthesized and investigated in the future. Every MMC needs for a detailed structural analysis, besides the exemplarily mentioned analytical methods, special considerations. This is explained for **type II** MMCs.

Polymeric phthalocyanines (see Chap. 2.3.1.2) are prepared by two-dimensional propagation from various tetracarbonitriles as bifunctional monomers. An ideal polymeric phthalocyanine has a regular planar structure that can be treated in a two-dimensional Cartesian coordinate system with allowed positive integers (propagation directions of the polymers are denoted by the letters x and y) [76]. Figure 2.5 shows only two examples of possible structures. A model describing the structural features, such as degree of polymerization as well as size and shape of polymeric phthalocyanines, is described. Equation (33) correlates the number of macrocycles n (degree of polymerization) with the number of bridge monomers, e, and the number of end groups e:

$$n = b/2 + e/4. \tag{33}$$

Evaluation with experimental data (determination of number of nitrile end groups and groups of bridge monomers by quantitative IR spectra) leads to dependence on the kind of tetracarbonitrile and reaction conditions to values of $x = 4 - \infty$ and $y = 1 - \infty$. It was establised that the unique structure of polymeric phthalocyanines is exhibited by their fractal properties. They have a regular structure and four fractal dimensions for every size/shape/dilation combination [76]. The mathematical discussed model can serve as polymer model for discussing basic fractals.

Another approach describes the number of isomers $i_{r,t}$ for cofacial stacked polymeric phthalocyanines (see Chap. 2.1.3, Eq. (3), and Chap. 2.3.5) [77].

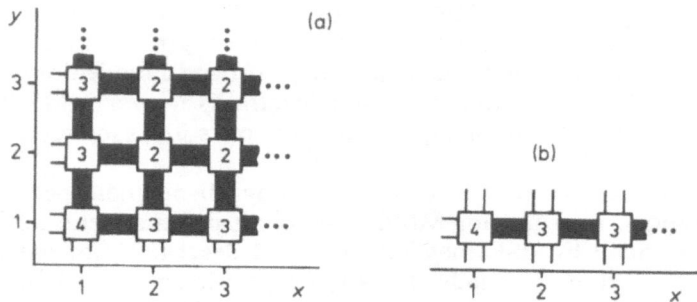

Fig. 2.5. Examples of polymeric phthalocyanine structures in a Cartesian coordinate system. Squares symbolize the macrocycle. Each line indicates an end group, lines between squares being the bridging parts between the macrocycles and are shaded; numbers inscribed into the squares show the number of monomers, m, required to form a macrocycle. Polymeric phthalocyanines with **a** a two-dimensional plain and **b** a linear chain

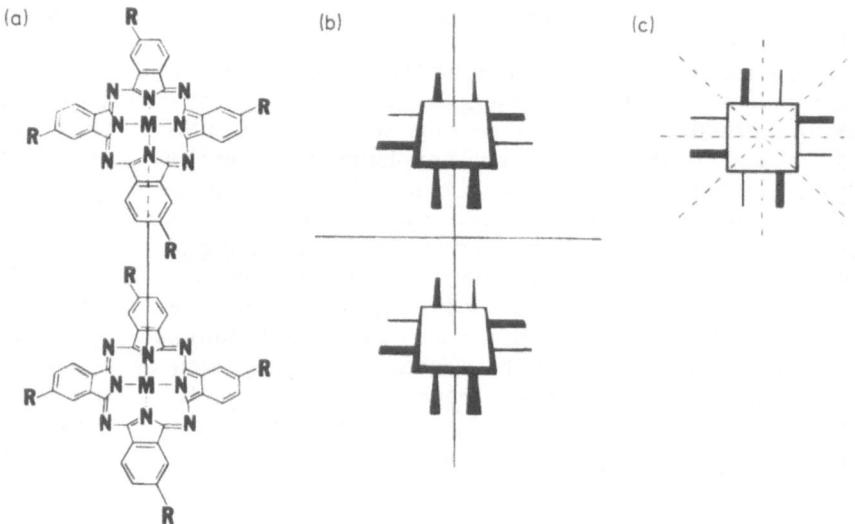

Fig. 2.6. a Stacked polymer phthalocyanines with tetrasubstituted rings. **b** Schematic description of this arrangement; the *thick lines* protruding from the square symbolize positions of substituents, *thin lines* denote unoccupied positions; the axis of the molecules is a C_4 and also a S_4 axis; the perpendicular line to the axis of the phthalocyanines is a C_2 axis. **c** The stacked polymer when viewed from the top

Polymers with tetrasubstituted phthalocyanines (Fig. 2.6) possess a number isomers according to:

$$\lim_{x \to \infty} i_{r,t} = 2^m/16 \,. \tag{34}$$

Depending on eclipsed or staggered positions and the substitution pattern, they show D_{2h}, D_{2d}, C_{4h}, C_4, or C_s symmetry, or can be described by a sole symmetry operation.

2.2 Type I: Metal Complexes Bound to Macromolecular Carriers via Ligands or Metal Ions

A. D. Pomogailo

The detailed analysis of the processes leading to MX_n bonding with a macroligand, as well as of the structure of formed products is of primary interest because it is very helpful for the study of specificity of complex formation and for optimizing MMC usage. The interactions in the sytem macroligand–MX_n can be specified according to the nature of reagents as coordinative, covalent, ionic, chelate type, and π-binding.

One can start either with attachment of binding of a ligand to a polymer with following attachment of MX_n, or to prepare the "procurement" by introduction at first of a ligand to a metal complex with further immobilization of the polymer. It is also possible to bring together the stages of synthesis of metal complex with desirable ligand and its fixation on the polymer support (the assembly method). Polymerization and copolymerization of metal-containing monomers are also perspective routes for MMC synthesis. Let us generalize the approaches for preparation of MMC of several types by their classification according to formed metal–macroligand bond nature.

2.2.1 Metal Complexes Coordinatively Bound to both Synthetic Organic Polymers and Inorganic Macromolecular Compounds

Macroligands with N-, P-, O-, and S-containing functional groups are usually used for the above-mentioned purpose (see Sect. 2.1.1). Polymers and copolymers based on vinylpyridines and vinylimidazoles (more rarely those with other nitrogen-containing heterocycles) as well as polyethyleneimine and polynitriles, are the most widely used among other N-containing macroligands.

2.2.1.1 MX_n Binding with Polyvinylpyridines

Both polymers and copolymers of several vinylpyridines have long been known as desalinization membranes of water and have been used for demineralization of organic solvents, oils, and as anticorrosive protective materials, etc. However, despite their application value, the mechanism of complex formation with polyvinylpyridine became intensively studied only recently, due to usage of formed MMC in catalysis.

The donor properties of nitrogen-containing heterocycles are specified by the presence of unshared electron pair at the nitrogen atom localized at sp^2-orbital and conjugated with π-electron system of the aromatic ring. Polyvinylpyridines are therefore the weaker bases compared with aliphatic polyamines, even weaker than their monomeric analogue (e. g., pK_a values for poly-2-ethylpyridine, 4-ethylpyridine are equal to 5.89 and 5.87, whereas those for poly-2-vinylpyridine (P2VP) and poly-4-vinylpyridine (P4VP) are equal to 3.50 and 3.95, respectively). Here we concentrate mainly on the synthesis and studies of MMC on the base of linear polyvinylpyridines with Cu^{2+}, Ni^{2+}, and Zn^{2+}. We also point to metal complexes of IVA-VIIA groups and those of platinum group, however, in less detail.

The MMC of the composition MX_2D_2 (where D is a monomeric unit; M = Co, Ni, and Fe, X = Cl, I, OAc, AcAc, etc.) have been obtained and isolated by mixing in inert atmosphere of deluted ethanol solutions of atactic P4VP [83]. Under interaction with MI_2 and $FeCl_3$ the complexes MI_2D_3 and $FeCl_3D_3$ have been obtained, complexes composition independent neither on the reaction temperature nor on macroligand concentration (within the range 0.01–10%). Meanwhile, interaction with low molecular analogue is dependent on the concentration ratios and is accompanied with coordination of either two or four ligand molecules. Insolubility of MMC in convenient solvents is caused by MX_n coordination with unshared electron pairs at nitrogen atoms of different polymer chains and yields to polymer cross-linking. At the same time the viscosity of diluted solutions of P4VP in methanol decreases by introduction of metal ions [84] specifying the contraction of polymer globules caused by intramolecular chelate formation.

The structure of MMC derived from $ZnCl_2$ and P2VP was assumed to be $[ZnCl_2D_2]D$ with two monomer units placed in the inner coordinative sphere of transition metal ion and one in the outer. The origin of the constancy of this complex composition under different synthesis conditions is the result of effects described in Sect. 2.1. Quasi-bound units are capable of coordination with partners possessing apropos steric and electronic properties, particularly with elementary iodine [83].

Stereostructure of macroligand is of great importance for complex formation. Thus, the complex of composition $NiCl_2D_2$ (**A**) possessing tetrahedron structure has been obtained by simultaneous precipitation of $NiCl_2$ and atactic P2VP [85]. At the same time isotactic P2VP in ethanol did not produce the complex in acetone-yielding MMC of tetrahedron structure (**B**) with bridged chlorine atoms (Scheme 2.11).

The number of interesting regularities has been obtained by the studying of P4VP interactions with Cu^{2+} complexes. For instance, by high-speed sedimentation, ultraviolet (UV)-spectroscopy, and viscosimetry techniques the equilibrium coexistence of both associated of macromolecules carrying the maximal value of bound metal ions and polymer chains completely free of metal ions has been observed [86], even in the lack of Cu^{2+} ions in water-methanol (1:1) polymer solutions. In other words, the principle "all or

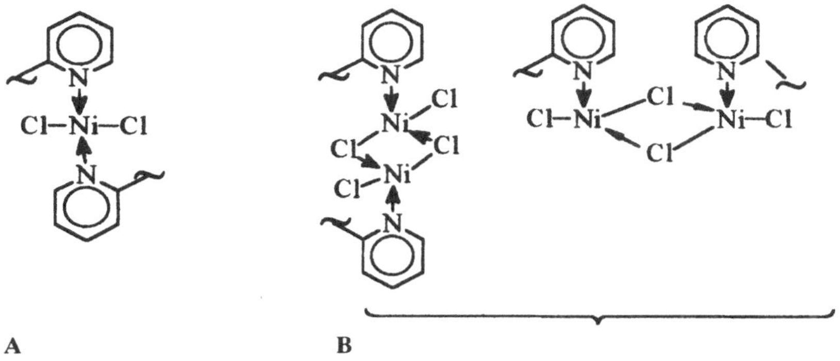

A   B

Scheme 2.11

nothing" (see Sect. 2.1.4) is sometimes realized for examined systems. The exchange kinetics between filled and free macromolecules are complicated [87]. If one takes into consideration the real concentration of interacting polymer units, the constant formation can be calculated according to the equation [88]:

$$\beta_n = \frac{[Cu^{2+}D_4]}{\{\gamma \frac{[D_0]}{4} - [Cu^{2+}D_4]\}\{[Cu]_0 - [Cu^{2+}D_4]\}}$$ (35)

where factor γ specifies the maximal value of P4VP units yielding $Cu^{2+}D_4$ complexes. The calculations showed that β_n value for P4VP was 40 times higher than that for complexes of Cu^{2+} with pyridine 124 and 3.2 l/mol, respectively.

The reaction constant rates of Ni^{2+} complex formation with polymeric and monomeric ligands are equal to 4500 and 3500 mol/l · s, whereas those for dissociation are equal to 6.6 and 2.3 s^{-1}, respectively. The addition of electrolyte (NaCl) significantly affects the rate of complex formation $k = 10$ mol/l · s [89].

The complex formation of $RuCl_3$ with linear P4VP or P2VP leading to either soluble or insoluble in CH_3OH six-coordinated complexes is also to be mentioned [90, 91].

The processes of complex formation also considerably affect the state of macrochains. Thus, the changes in the structure of atactic P4PV under complex formation with Cu^{2+} have been testified by ^{13}C-NMR [92]. The rise of syndio-triads content has been observed after the total removal of Cu^{2+} ions from MMC [92]. This fact indicates that P4VP molecules takes the configuration facilitating copper ion binding with several polymer chains.

In principle the same regularities have also been observed in the course of MX_n binding by polymers with grafted polymer coverings, such as PE-grafted-poly(4-vinylpyridine) (PE-gr-P4VP) [93]. The grafted complexes of Ni^{2+} have octahedron configuration, whereas Co^{2+} tetrahedron has one. The interaction with $MoCl_5$ is accompanied by its partial reduction to Mo^{4+} [94], and the same effect has also been observed for Pd^{2+} complexes [95]. The degree of functional groups loading with metal complexes in the case of grafted functional layer is usually significantly above those values for homo-, co-polymers, and the more for cross-linked macroligands including nitrogen-containing heterocycles.

The most important regularities of MX_n binding with examined polymers are caused by functional dependence of both composition and strength of formed MMC on the cross-linking degree. The main reason for that is the decrease of unit lability complicating their favorable orientation with regard to coordinative metal ion. At the same time the concentration of ligands in both cross-linked and twisted polymers is high and yields the rise of constant stability for Cu^{2+} complexes with P4VP cross-linked or quaternized with some amount of 1,4-dibromobutane [84, 96].

The rise of polymer carcass rigidity also leads to the increase of defects in coordinative centers by changing of their structure (for instance plain square Cu^{2+} complexes transform into tetrahedron structure), and are accompanied by the rise of coordinative unsaturation of transition metal ion [97]. The strength of $M \leftarrow N$ bond in the row of metal ions decreased in the order: $Cu^{2+} > Cd^{2+} > Zn^{2+} > Co^{2+} > Ni^{2+}$. The rows of steadiness of the cross-linked macrocomplexes of this type are usually in the coincident with famous Williams-Irving stability row for low molecular ligands.

2.2.1.2 Interaction of MX_n with Polyethyleneimine and its Derivatives

Both linear and branched polyethyleneimines (PEI) show high coordinative ability. The reason lies in the fact that repeating $-NH-CH_2-CH_2-$ group behavior is in a first aproximation similar to ethylenediamine (although triethylene tetramine, *trien*, is usually accepted as a monomer analogue. Theoretically the coordinative ability of PEI is high (3.88 mmol/g if one assumes the transition metal ion coordinative number, CN, to be equal to 4) [98, 99]. The mechanism of complex formation with PEI has been studied in detail for the cases of copper and cobalt salt interactions [100, 101].

Four-coordinative complexes $Co^{2+}-$ PEI have been formed (Scheme 2.12); the high coordinative ability of the units of macroligand is also confirmed by

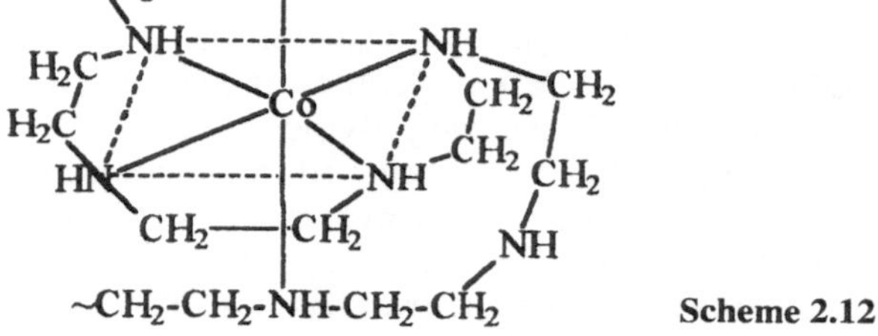

Scheme 2.12

the fact that all ligand groups of metal complexes have been replaced by PEI units within all the examined reagents concentrations [102].

The process of Co^{2+} or Ni^{2+} complex formation is assumed to be step-by-step and consists of progressive complexes (of the ratio $D/M^{2+} = 2, 4, 5,$ and 6) formation. It would look unlikely if the formation of $CoCl_2D_2$ complex (composite in a manner similar to ethylene diamine chelate cycles) had been confirmed by oncoming synthesis by interaction of $CoCl_2D_4$ with $CoCl_2$. The completion of these high-ligand complexes is realized by the use of mono- and bidentate units of different polymer chains [103]:

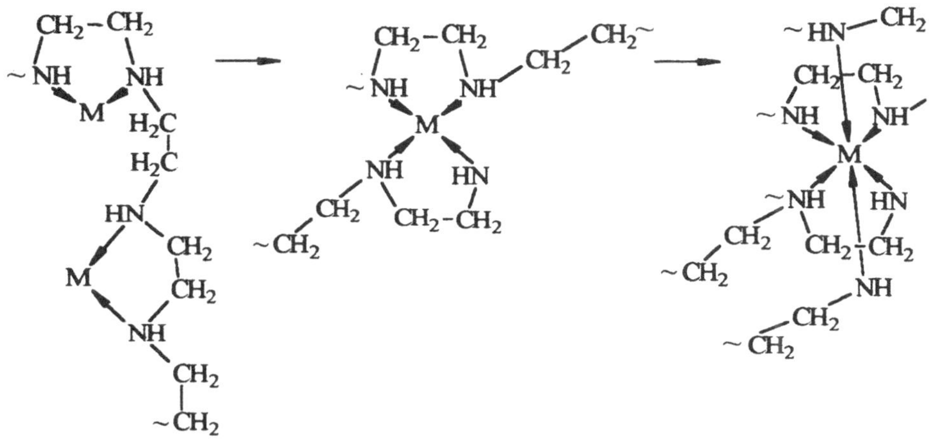

Scheme 2.13

The complexes of two types distinguished by the number of nitrogen atoms in Rh^{3+} coordinative sphere have been revealed in the system $RhCl_3$-PEI. These were "yellow" complexes of the composition of $RhCl_3D_3$ and "red" ones, $RhCl_3D_2$; complexes of the ratio $Rh/D = 1:6$ have not been formed, probably due to the steric reasons. The interaction of MX_n with polyamine anionites containing *en, diene,* and *triene* fragments principally follows the discussed regularities [104].

2.2.1.3 MMC Formed by the Use of Oxygen-Containing Donor Centers

Polymer glycols [polyethylene glycol (PEG), polypropylene glycol (PPG) of several molecular weights], polyvinyl alcohol (PVAl), polymethylmethacrylate (PMMA), polyvinylacetate (PVAc), polymethylvinylketone (PMVK), and polyvinylpyrrolidone (PVPd) are the most significant among other oxygen-containing macroligands.

Both PEG and PPG show considerably higher chain flexibility in contrast to PEI, due to facilitating of rotation about CO bonds. It should be noticed that oxygen of the ether residues according Pearson possesses strong basic properties [105]. One of the fundamental studies on the subject has been devoted to investigate the structure of MMC derived from high molecular weight polyoxyethylene and $HgCl_2$ (see Sect. 2.1.6).

A considerable amount of heat (8.9 kJ/mol) has been released by dissolving $CoCl_2$ in methanol, whereas a significantly less amount has been evolved after its adding to PPG solution in methanol specifying that interaction of $CoCl_2$ with PPG is accompanied by the rise of enthalpy factor [106]. The decrease of free energy is caused by an increase of entropy due to a decrease of Co^{2+} CN from 6 to 4 with evolving of two methanol molecules.

The formation of stable complexes of intrachain type with high stoichiometry [107] caused by increased flexibility of polymer chains and donor ability of oxygen atoms are the specific features of transition metal ions complex formation with PEGs.

It is interesting to notice the cases when the main contribution into complex formation is made be terminal hydroxyls of discussed ligands. Thus, the formation of oligomeric metal polyethers has been shown to be the first stage in the course of interaction of $TiCl_4$, VCl_4, $VOCl_3$, etc., with PEG (at $MX_n/PEG = 2$) [108, 109]:

$$HO(CH_2CH_2O)_mH + 2MX_n \rightarrow MX_{n-1}O(CH_2CH_2O)_m-MX_{n-1} + 2HX.$$

Oxyethylene residues are involved in complex formation under high reagent ratios, each two coordinated functional groups falling at one free due to passivation ("the neighbor effect").

A similar composition also show complexes of MX_n with PMMA or its derivatives. The complex formation of $TiCl_4$, VCl_4, and $MoCl_5$ with macroligands containing carbonyl groups proceeds effectively in nonaqueous medium [110]. The MX_n is bound to PVPd also at a carboxyl group of lactam type [111, 112]. The comparison of complex ability of PVPd and polymethacryloylpipyridine:

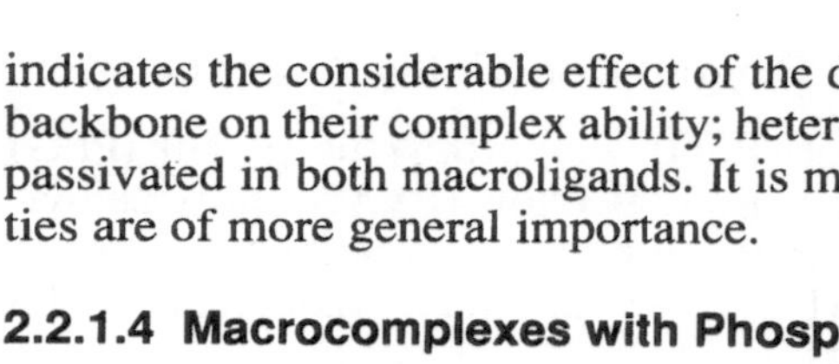

Scheme 2.14

indicates the considerable effect of the distance from donor atom to polymer backbone on their complex ability; heteroatoms adjoining to the backbone are passivated in both macroligands. It is most probably that examined regularities are of more general importance.

2.2.1.4 Macrocomplexes with Phosphorous-Containing Ligands

The interaction with phosphorous-containing macroligands is most widely used for MMC synthesis, although the choice of macroligands holding phosphorous is scarce with regard to N- or O-containing polymers. These are crosslinked polymers modified with phosphorous groups (especially phosphynated Merrifield's resins), as well as polymers with grafted phosphorous-containing functional cover.

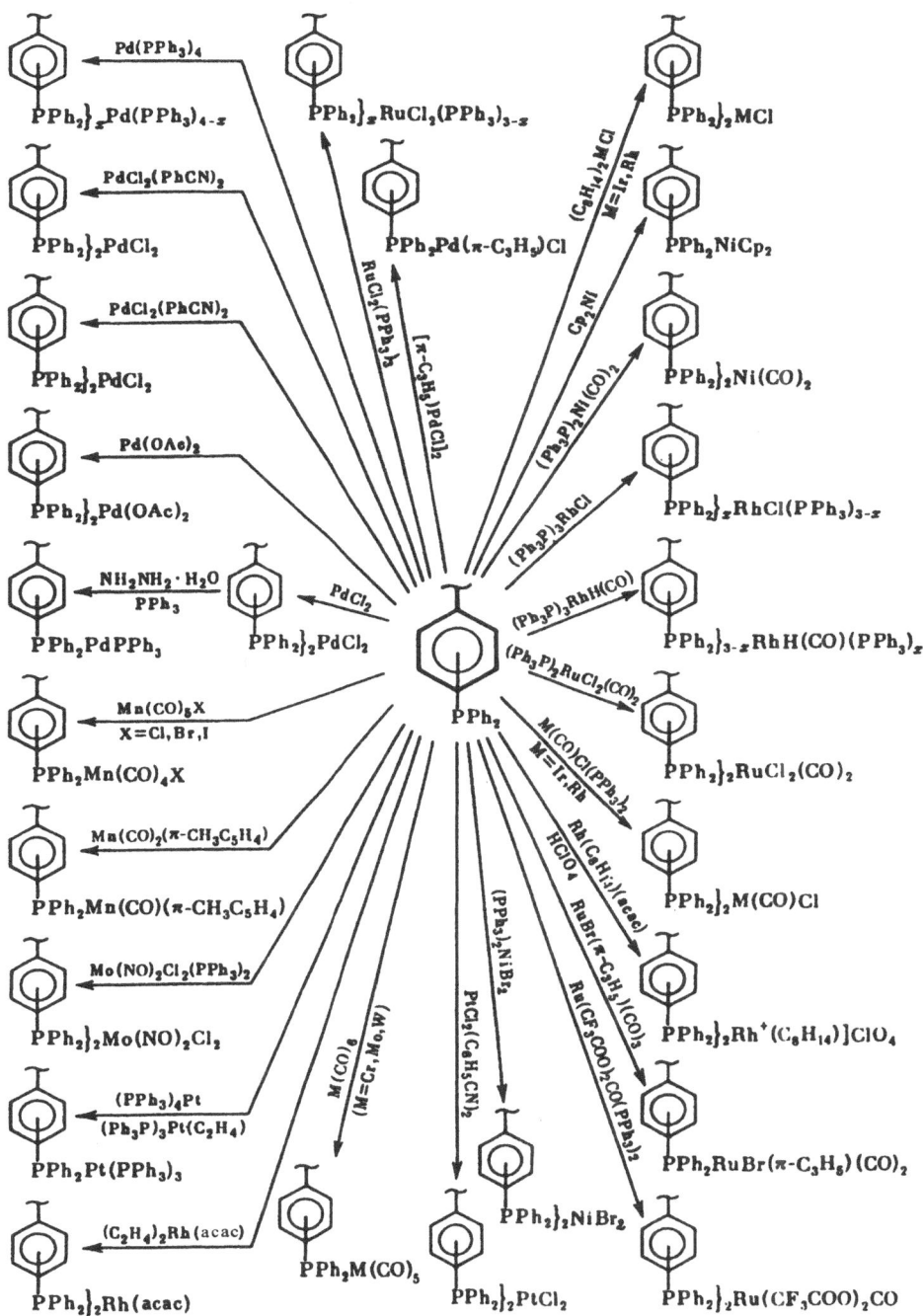

Scheme 2.15. MMCs based on poly(p-styryldiphenylphosphyne)

The method of competitive ligands (when polymer phosphorous-containing groups replace one or several groups of reactant) is mainly used for the synthesis of such MMCs. The replaced groups are usually PPH_3, $P(C_2H_5)_3$, and more rarely, AcAc, CO, norbornadienyl, etc. (Scheme 2.15). One, two, and even three functional groups of macroligands can replace PPH_3 residues of fixed metal complexes depending on the distances between diphenylphosphyne centers. This situation has been analyzed in detail using the examples of interaction of rhodium compounds with CMCS [113]. Phosphyne ligands are flexible. This is the reason that not only isolated diphosphyne complexes, e.g., of Pd^{2+}, but also bi-, and even polynuclear ones [113, 114], have been witnessed in produced MMCs.

Several polymers with grafted phosphorous-containing cover (Sect. 2.1.1) have been used for synthesis of MMCs with complexes of rhodium, iridium, and some others. Rh^{1+} bound to PP-gr-poly(p-styryldiphenylphosphyne) yielded MMC, which were well characterized by IR-, PMR- and NMR (^{31}P) spectroscopy [115]. p-styryldiphenylphosphyne has been grafted on the surface of PP, PVCl, and cross-linked PS. These macroligands were treated with $RhCl_3(PPh_3)_3$, $RhCl(CO)(PPh_3)_3$, $Rh(AcAc)(CO)_2$, and $IrClCO(PPh_3)_3$ [116]. The complexes of $RhCl_3$ have been formed on PP-gr-vinyldiphenylphosphyne [117].

However, phosphorous-containing MMC for production of chiral catalysts are of the most important among others. Also with optically active macroligands, the methods of such MMC synthesis are similar to those presented in Scheme 2.15. Thus, the reaction with 4,5-bis(diphenylphosphyne)methyl-1,3-dioxalane or dibenzophospholane groups follows a simple ligand exchange [118, 119] yielding the structures:

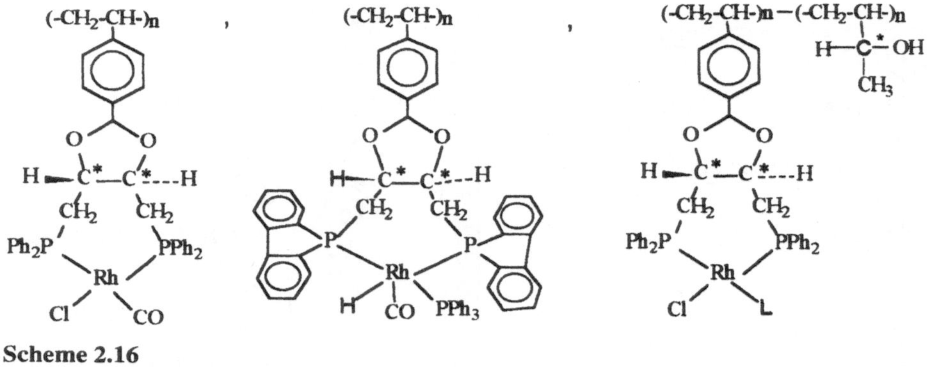

Scheme 2.16

In contrast to three-valent phosphorous, the five-valent one does not show donor ability, so the coordination with MMC containing $\equiv P=X$ (C=O, S, Se) groups proceeds mainly at X-atoms. For instance, donor ability of the oxygen atom at phosphoryl group is even higher than that of >C=O group [120, 121].

2.2.1.5 Sulfur-Containing Macroligands in Coordination Reactions

This type of polymer is merely used for metal complex binding, because sulfuric ligands are weaker electron donors with regard to oxygen-containing macroligands, due to lower electronegativity of the sulfur atom. Among others, thioethers and thiocarbamates are more widely used. Rapid and selective binding of Hg^{2+} complexes with thioethers is a well-known reaction (e. g., see [122]). Such high selectivity is caused by the fact that Hg^{2+} centers show by contrast to other transition metal ions, mild acidic properties, thioethers being mild bases according to classification of strong weak acids and bases.

It is interesting that interaction of $PtCl_2(PPh_3)_2$ with PS modified with thioether groups gives two geometric isomers, whereas this with PS modified with sulfoxides yields structural isomers [123].

Two, three, and even six dithiocarbamate groups (bi- or monodentate type) per one metal ion may be involved in complex formation depending on the nature of MX_n and reaction conditions [124]. This effect may be widely exemplified.

2.2.1.6 Metal Complexes Coordinatively Bound to Synthetic Inorganic Macromolecular Compounds

Two main approaches, such as MX_n binding with the surface of modified oxides (see Sect. 2.1.1, Scheme 2.8) and employment of mixed-type supports (mineral–polymeric) are usually used to obtain these MMCs. It seems that the phosphorylation of silica gel surface with the following binding of complexes of Rh, Ir, Ru, Pt, Co, etc., by coordinative mechanism [125] was the first example of examined MMC production:

$$\equiv SiOH + (EtO)_3SiCH_2CH_2PPh_2 \rightarrow \equiv Si\text{-}O\text{-}\overset{|}{\underset{|}{Si}}\text{-}CH_2CH_2Ph_2 \overset{Rh(AcAc)(CO)_2}{\text{------}\blacktriangleright}$$

$$\rightarrow \equiv Si\text{-}O\text{-}\overset{|}{\underset{|}{Si}}\text{-}CH_2CH_2PPh_2Rh(AcAc)CO.$$

Apart from the phosphyne group, some other nitrogen-containing "anchor" residues possessing donor properties are also used for metal complex binding [126]. These are

$$-(CH_2)_2CN, \; -(CH_2)_3NEt_2, \; -(CH_2)_3N\bigcirc, \; -(CH_2)_3\text{-}\langle\overset{N}{\underset{}{\bigcirc}}\rangle, \; etc.$$

The length of spacer between the oxide surface and functional group considerably affects the structure and stability of formed complexes. If it will be large, the behavior of the ligand in MX_n binding might be similar to a low molecular analogue.

The second approach is based on MMC formation with macroligands of mixed type (see Sect. 2.1.2, Fig. 2.1). The interaction of Cu^{2+} with SiO_2-grafted-P4VP [127] and $PdCl_2$ with SiO_2-grafted-polyvinylcarbazole [128] should be noted as examples. The PEI absorbed on SiO_2 binds Pd^{2+} by coordinative mechanism [129]. The binding of Pd^{2+} on PVPd absorbed on chromosorbe from methanol solution proceeds by coordination of an amide group of polymer fragments [130]. Macroligands of mixed type modified with phosphorous-containing ligands are also known. Thus, $CoCl_2$ is effectively bound by polyphosphyne ligands grafted to macroporous glasses [131]; $RhCl(CO)_2$ has been anchored to phosphyne groups of modified PS covering silica gel [132].

Thus, coordinative binding of MX_n with macroligands of different types gives wide possibilities for construction of coordinative centers. The macroligand in many cases facilitates the characterization and identification of formed complex structure, due to easiness of aim-product separation from the reaction mixture.

2.2.2 Metal Complexes Covalently and Ionically Bound to Synthetic and Inorganic Macromolecular Compounds

This type of metal complex binding can be realized both in nonaqueous and aqueous solutions. In the first case the high activity of M-X bonds in "protono-lysis" reactions can be achieved. The specific feature of such MMC consists of the presence of strong covalent bond (σ-bond) between metal ion and macroligand. Thus, the average value of Ti$-$O bond (formed in the course of covalent binding of $TiCl_4$ with alcohols) dissociation energy is within the range $420-460 \, kJ/mol$ [133] being close to that of the Ti$-$Cl bond in $TiCl_4$ ($430 \, kJ/mol$).

We have already mentioned previously (Sect. 2.2.1) that different mechanisms, depending on many factors, can be realized in the course of interaction between MX_n and macroligand. Somewhat similar situations have already been examined in the examples of PEG and PEI capable of both coordinative and covalent MX_n binding. However, under real conditions each of these limited cases is rarely separately realized. More often both of these mechanisms provide their own contribution in the process of MX_n binding.

Now we concentrate briefly on MMC formation with macroligands, including "protolytical" functional groups, such as with protonated PVAl, polyacids, polyaminoacids, etc.

2.2.2.1 Specific Features of MX_n Binding with Polyvinyl Alcohol

This problem has been widely studied by numerous investigators, especially concerning the interaction with Cu^{2+}. The general scheme of the reaction can be represented as follows:

$$LH + M^{n+} \rightleftarrows ML_n + nH^+.$$

It has been shown that each copper atom coordinates four hydroxyl groups of PVAl, two of them being protonated. This interaction leads to binding of two polymer chains:

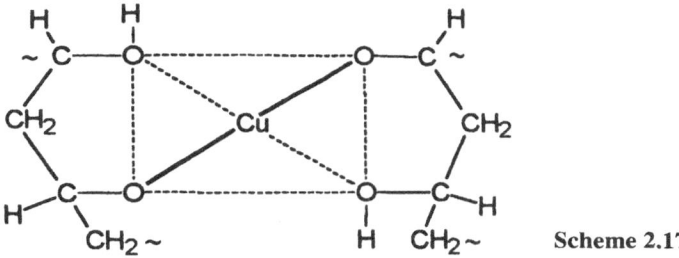

Scheme 2.17

The ESR studies have revealed the complicity of these processes [134] and showed their dependence on the basicity of reaction medium. Depending on the ratio ($F = HO^-/Cu^{2+}$, at least five ESR signals attributed to different Cu^{2+} complexes have been detected. They are the peak corresponding to isolated copper ions coordinated with two water molecules and two oxygen atoms of PVAl ($F = 0$); Cu^{2+} ions bridged with OH^- groups ($F = 1$); signal corresponding to dimeric Cu^{2+} complexes bound with either one or two bridging OH groups ($F = 2$). Also Cu^{2+}-ion-coordinated hydroxyl groups, but not water molecules, have been detected in strong basic solutions ($F = 7$). The chain of PVAl binding with Cu^{2+} ions was shown to carry a negative charge ($pK_a = 10.64$) [135, 136]. It was also shown that in PVAl obtained by PVAc hydrolysis the rest units of vinylacetate (~ 1–2%) do not take part in interaction with MX_n. The Cu^{2+} ions, however, influence the values of complex formation constants by their decreasing with the rise of length of vinylacetate blocks in the chain of copolymer. This example is the illustration of the importance of functional uniformity of macroligand. Both the values of constant formation and thermodynamic parameters of the system increase in accordance with Irving-Williams activity row (Table 2.1): $Co^{2+} < Ni^{2+} < Zn^{2+} < Cu^{2+}$.

A similar reaction an also proceed in nonaqueous reaction medium, excluding hydroxyl group protonolysis, for instance, in the interaction of PE-gr-poly-(allyl alcohol) with $TiCl_4$, VCl_4, $MoCl_5$ in CCl_4 [137]. It has been shown in the example of macroligands traced with ^3HO-groups that each molecule of VCl_4 is bound in average with 1.3 hydroxyl groups.

Table 2.1. Values of constants formation of M^{2+} complexes with PVAl and thermodynamic parameters of the reaction

M^{2+}	$logK_1$	$logK_2$	$log\overline{\beta_n}$	ΔG, kJ/mol		
				$-\Delta G_1$	$-\Delta G_2$	$-\Delta \overline{G}$
Co	5.67	5.32	11.06	34.2	26.8	66.9
Ni	6.21	5.60	11.82	34.9	33.0	66.9
Zn	6.94	6.63	13.57	42.4	39.8	80.4
Cu	8.07	7.86	15.93	48.2	46.4	94.1

2.2.2.2 MMC Derived from Polyacids

This type of MMC is very likely the most studied and most widely used type of the macrocomplexes. A great number of theoretical studies have been carried out namely for these macroligands (e.g., see Sect. 2.1). Polymers and copolymers of acrylic, methacrylic, and sulfuric acids are widely used as macroligands, rarely the polymers of maleic, itaconic, etc., acids. Usually the preionization of polymer acids is carried out, because it is more difficult to exchange hydrogen ions with transition metal ion than alkali metal cations.

The formed products may contain either terminal (**A**) or bridged (**B**) carboxylic groups:

A   B

In the first case carboxylic groups act as monodentate ligand, and as bidentate in the second case. The binding of the type $R-C\begin{smallmatrix} O-M \\ O-M \end{smallmatrix}$

seems to be hypothetical [138]. The decrease of pH value has been observed under addition of two-valent metal ion salt to the solution of polyacid indicating partial ionization of carboxyls:

$$R-COOH + M^{2+} \overset{K_1}{\rightleftarrows} R-COOM^+ + H^+$$
$$R-COOM^+ + RCOOH \overset{K_2}{\rightleftarrows} (R-COO)_2M + H^+.$$

The effect appears in the potentiometric titration curves by two sharp leaps (e.g., for Cu^{2+} the first one has been observed between pH values 6.0 and 8.0, the second closer to 10.5). However, the discussed processes usually follows the more complicated routes including the dissociation of polyacids and hydrolytic equilibirum via formation of intermediate hydrocomplexes $M(OH)_n$. Each Cu^{2+} ion interacts with either one, two, or four carboxyls. The products of the composition $2:1$ are presumably formed. This may be achieved either by interaction with two carboxyls of the same polymer chain (intramolecular interaction) or of a different one (intermolecular interaction) [139, 140]. Depending on the pH value ionized PMAAc and PAAc may form complexes of D_{2h} or D_{4h} (dimers) symmetry [139]:

D$_{2h}$ Scheme 2.18   D$_{4h}$

At high pH values PAAc chain has a drawn shape due to electrostatic repulsion of charged COO$^-$ groups; metal ions are then binding either with one or two neighbor groups. At a pH value equal to 4.5 macromolecular globule contracts and metal ion became capable for coordination of 2−4 carboxyls.

Metals of platinum group easily interact with polyacids, formed bonds considerably depending on the metal's nature [141, 142]. Bridging binding is more characteristic of ruthenium compounds, whereas rhodium (1+) mainly yields unbridged complexes. Both the mechanism and structure of ionic-bound rare earth carboxyls have been studied in detail [143, 144]. Different variations of f-element binding has also been investigated (e.g., the example of UO$_2^{2+}$ [145], (Scheme 2.19).

The data on MMC derived from stereoregular polyacids are scare. However, from general reasons one can believe that both configuration and flexicility of polyacids are to affect the structure of formed metal complexes as has already been discussed for PVP. It is known that isotactic PMAAc is 1.5 times as active with regard to Cu^{2+} ions as syndiotactic is [146, 147], although the contrast dependence has been observed with regard to Mg and Na ions.

The interaction of metal ions with polyacids is an isothermal process. Therefore formed complexes have to be stabilized due to respectively high changes in entropy factor. With the exception of MMC based on Cu^{2+} metal ions, ΔH is always of higher value for isotactic PMAAc than that of syndiotactic [148]: This means that the molecules of isotactic MMC are strengthened to a larger

Table 2.2. Thermodynamic parameters of Cu^{2+} and Mg^{2+} complex formation with syndio- and isotactic PMAAc [134, 148]

MMC	$K \cdot 10^{-7}$, (l/mol)	$-\Delta G$, (kJ/mol)	ΔH, (kJ/mol)	ΔS, (kJ/mol · grad)
Cu^{2+} − syndiotactic PMAAc	400.0	55.2	21.3	25.5
Mg^{2+} − syndiotactic PMAAc	1.0	40.1	0.6	135.8
Cu^{2+} − isotactic PMAAc	1200.0	58.1	15.9	246.6
Mg^{2+} − isotactic PMAAc	0.4	38.0	3.3	137.9

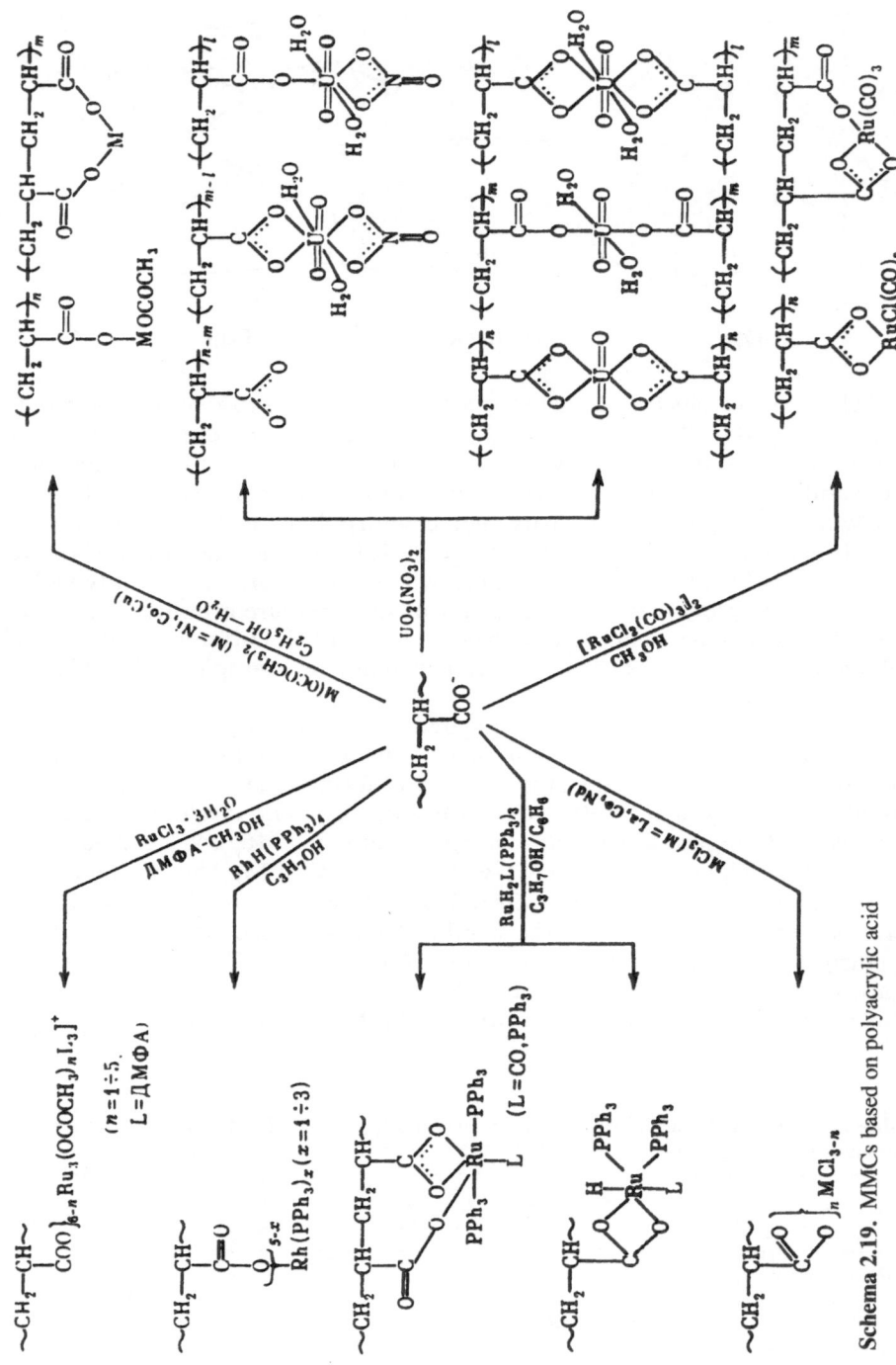

Schema 2.19. MMCs based on polyacrylic acid

extent. However, the differences between copper and magnesium complexes (Table 2.2) are probably related to their nature. The Cu^{2+} ions form mainly covalent complexes, whereas Mg^{2+} ions less strengthen and are characterized by smaller values of ΔH. The high values of ΔH and ΔS for macrocomplexes of Cu^{2+} are connected with H_2O molecules released in the course of complex formation; two bonds, water-metal are cleavaged in this case leading to the increase of entropy. (The case is somewhat simi-lar to the interaction of Co^{2+} with PEG in methanol, already mentioned previously.)

The grafted fragments' behavior in the case of transition metal carboxyls in water-alcohol mediums is similar to homopolymers [149, 150]. Some additional information on the composition of formed products as function of reaction conditions can be obtained by binding of traced transition metal salts $(M(OCOC^3H_3)_2)$.

The specific features of metal ions binding with cross-linked polyacids is also of great interest, due to their broad employment as carboxylic sorbents, both cationites and ampholytes. The composition of coordinative center depends on the conditions of polymer pretreatment as well as on the ionization degree of carboxyls. The structures with maximal value of unionized groups are predominate at $\alpha \to 0$. The ionic type of bonds became prevailing at higher α values, due to considerable electric field induced at the surface of the macroligand [151]. Coordinative node can include both protonated and nonprotonated groups. For instance, the complex $Cu^{2+}(RCOO^-)_2(RCOOH)_2(H_2O)_2$ contains four carboxyls in the plain of a square and two water molecules at the tops of a octahedron [152]. Cross-linking can considerably change composition, structure, and strength of formed complexes sometimes leading to reversal of stability rows. For instance, the row of stability constants for complexes with non-cross-linked PMAAc $(Cu^{2+} > Zn^{2+} > Ni^{2+} > Co^{2+})$ alters for cross-linked PMAAc $(Cu^{2+} > Ni^{2+} > Zn^{2+})$ [153]. A mixed-diffusive mechanism of binding is presumably realized for these systems. The contribution of inner or outer diffusion can be altered by changing of cross-linking degree and reaction conditions.

Phosphorous acidic groups of three-dimensional polyligands at different pH values can bind metal ions simultaneously by ionic and covalent mechanism, yielding four-membered cycles [154]. As well as in the case of carboxylates such binding leads to the leveling of electron density at oxygen atoms:

Ions of Cu^{2+} and UO_2^{2+} form chelates with these macroligands, whereas Ca^{2+}, Co^{2+}, Ni^{2+}, and Hg^{2+} are bound to polymers by covalent mechanism. The bond-covalence degree decreases for the row of metal ions [155]: $UO_2^{2+} >> Fe^{3+} >> Cu^{2+} > Zn^{2+} > Ni^{2+} > Cd^{2+} > Ca^{2+}$. Macroligands with sulfo-hydro-(-SH) as well as polymers and copolymers of p-styrylsulfonic acid should also be noted as useful macroligands.

2.2.2.3 Transition Metal Ions Binding with Polyamino Acids

With the exception of special cases that are accompanied by metal complex chelation, polyamino acids usually react as polycarboxylic acids. Thus, poly-L-glutamic acid and PAAc of molecular weights of 50000 and helical configuration (neutralization degree $\alpha < 0.2$) do not practically bind Cu^{2+} ions; however, the binding ability considerably rises with configuration transformation from helix to globule ($\alpha \approx 0.6$) [156]. The dentatity is equal to 2 for both compared polyacids. However, unlike PAAc or PMAAc, there is a second binding center in polyamino acids. It is amino group of the side chain that can take part in metal ions binding only under certain conditions. It has been shown that in the system Cu^{2+}–polyglutamic acid (molecular weight 60 000; $\alpha = 0.19-0.90$) complexes of two types have been formed. These are bound by carboxyls, and others are formed by two carboxyls and peptide nitrogen atoms of one or two polymer units [157]. This mechanism has been studied in detail in the example of Cu^{2+} binding with poly-N-methacryloyl-L-alanine [158]. The products similar to those formed in the course of interaction with polycarboxylic acids were formed at the first stage of the reaction. They were stable within pH values $3-5.5$, and most probably have the structures:

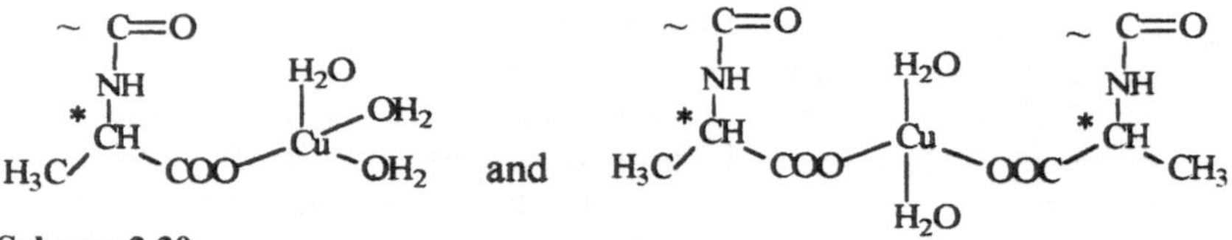

Scheme 2.20

The products of the structure:

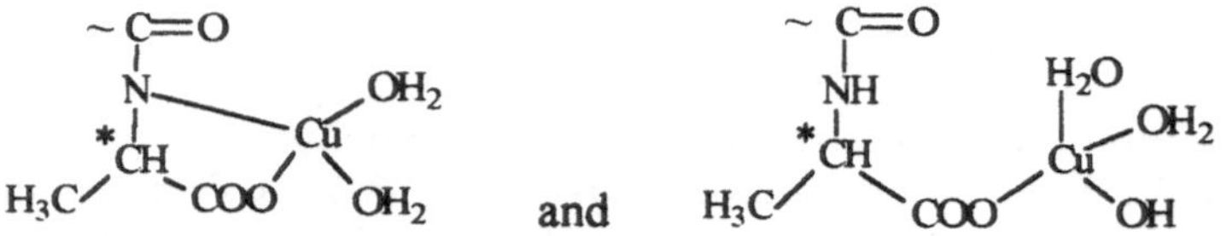

Scheme 2.21

have been formed at pH $6-9$. Stable five-membered chelate cycles have been further formed by nitrogen atom protonation (pH ≥ 5). The third stage (pH ≈ 10.5) has been accompanied by metal ion coordination with deprotonated nitrogen atoms and HO^- groups as well as COO^-:

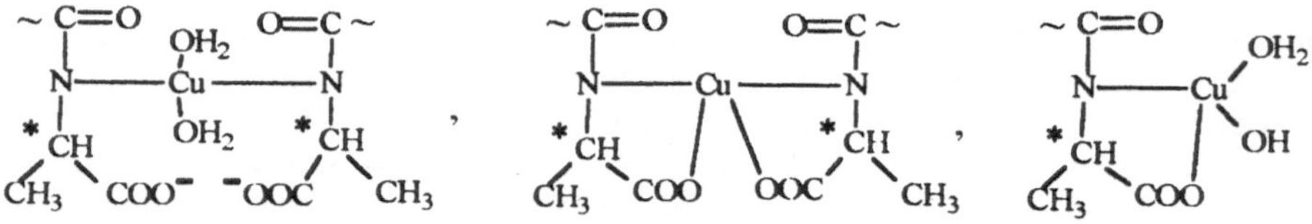

Scheme 2.22

It is important that chelates in the form of five-membered cycles, including protonated carboxyls and amino groups, show optical activity, whereas for complexes with two amino groups this was not observed. The constants formation of the products (in water at 25°C) with nonprotonated carboxyls depend on proportion of reagents and are within the range $(7.2-9.5) \cdot 10^3$ being close to those for other polyacids such as PAAc $(K = 4.2 \cdot 10^{-3} - 2.8 \cdot 10^{-4})$ or PMAAc $(K = 2.5 \cdot 10^{-4})$. However, the values for protonated carboxyls are very high $(K = 1.8 - 2.4 \cdot 10^9)$. A similar behavior has been detected also for some other systems [159, 160].

The data on metal ion binding with polymeric amido- and iminoacids are scarce.

2.2.2.4 Covalent Binding with Inorganic Ligands

Covalent and/or ionic binding of MX_n at the surface of inorganic carriers modified with protolytic functional groups is not so widespread as coordinative binding. Here we point only two examples: heterogenization of rhodium-pyridine complexes by silica gel modified with ethylene- and trimethylenediamine [161] and $[Rh(CO)_2]_2$ binding with silica gel modified with $(CH_2)_2SH$ groups [162]. Macroligands of mixed type are also rarely used, with the exception of $PdCl_2$ and H_2PtClt_6 anchored to SiO_2 with grafted (radical copolymerization in the presence of SiO_2) copolymer of acrylic acid and m- or p-divinylbenzene [163].

The conducted analysis shows that the variety of metal ion binding by covalent or ionic types gives possibilities for the synthesis of MMC with well-characterized composition and structure of coordinative nodes as well as of the certain bond nature.

2.2.3 Macromolecular Metal Chelates

The chelation of metal ions with polymers containing poydentate ligands is widely used for membrane filtration of solutions containing trace amounts of metal ions for the removal of toxic or radioactive contamination, etc. Polydentate binding (more often bidentate binding) provides strong bonding of metal ions in MMC. Macromolecular metal chelates (MMCh) can be classified as molecular, intramolecular, macrocyclic, and polynuclear (Scheme 2.23) [164, 165]:

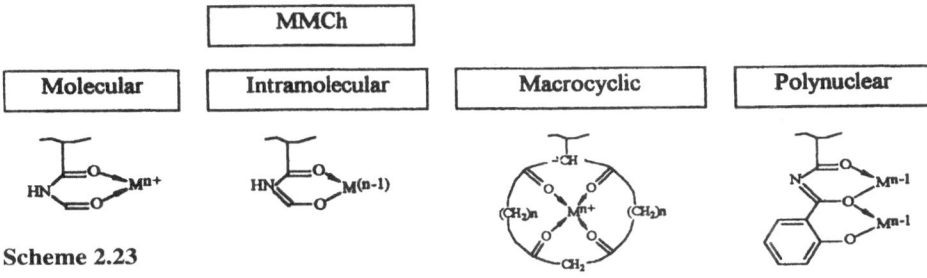

Scheme 2.23

Within the types MMCh can be divided according to the nature of coordinated atoms: O,O-, N,O-, N,N-, N,P-, P,P-, S,S-chelates, etc.

Metal complexes in which the metal ion is bound with donor atoms of chelating fragments only by coordinative bonds can be attributed to MMCh of molecular type. These complexes are formed either with the fragments containing donor centers in the position that predetermines the certain metal ions fixation or with flexible groups containing one or more donor centers. The MMCh of intramolecular type are those in which the chelating center of the polymer ligand is bound with metal ion via σ- or nv-bond. Polymers with porphyrins, phthalocyanins, cyclic ketones, and crown-esters mainly yield macrochelates of cyclic type. In MMCh of polynuclear type each chelating center of macromolecule is usually bound with two or more metal ions.

Regarding the main principles of these MMC syntheses as well as thermodynamic and kinetical aspects of their formation, this has been discussed in Sect. 2.1 briefly (because there are exhaustive reviews on the subject, e.g., [166−170], so let us concentrate only on specific features of synthesis and structure of the most widely used macrochelates.

The main approaches for synthesis of MMCh can be represented as follows:

The methods of MMCh synthesis		
Synthesis on the base of chelating macroligands	Polymer analogous transformations	the method of "assembly"
The interaction metal ion–macroligand. Metal ions exchange. Ligand exchange. Synthesis from zero-valent metals	Metal chelate binding in the polymer via ligand. Transformations of coordinative polymers into MMCh	Metal chelate fixation on the polymers via metal ion. Polymerization of metal-containing monomers. Polycondensation

Scheme 2.24

2.2.3.1 Molecular Metal Chelates

Polymers with polyamines (e.g., PEI) or 2,2'-dipyridyl fragments are typical examples of MMCh with N,N-chelating ligands (see also Chap. 5). The last one's behavior is similar to low molecular analogues. Thus, they show high affinity to transition metal ions, and they are also less effective in binding of nontransition metal ions and do not form complexes with alkali metal ions. The structure of coordinative node of soluble MMCh is also close to low molecular models: PS modified with dipyridyl groups yield trispyridyl complexes [171, 172].

It should be noted that MMCh are very stable macrocomplexes. Thus, reduction and reoxidation of metal ions bound with polymers containing dipyridyl ligands proceeds without metal cycle rupture. In going from dipyridyle to tripyridyl complexes the stability of complexes rises considerably [173]. This effect also corresponds to the behavior of low molecular analogues.

Previously mentioned macrocomplexes with polyamidoamines, as well as with polymers containing aminoalkylamide units in side chains, are more widely studied among other MMCh of N,O-type. The composition of Cu^{2+} complexes with the last mentioned macroligand depends considerably on pH values. Molecular chelates with N,O-coordination are presumably formed at low pH values, where as at high pH intramolecular N,N-chelates are mainly obtained [174]:

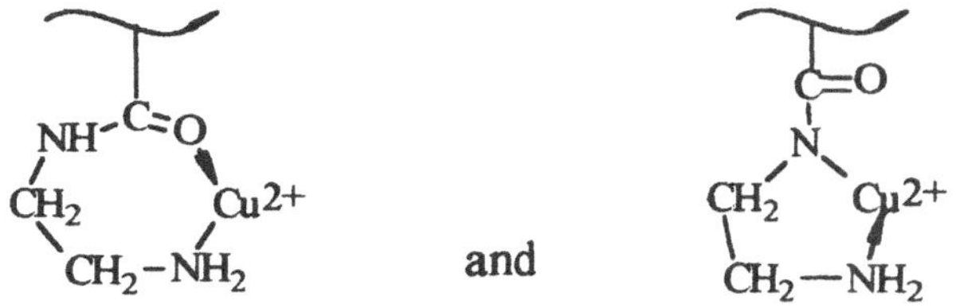

and   Scheme 2.25

It is evident that competitive coordination of nitrogen or oxygen atoms, depending on the pH values, takes place in this case. Rh^{1+} yields P,P-chelates with CMCS modified with 1,2-bis-(diphenylphosphyno)ethane, while a mixture of chelate and molecular complexes formed in the course of interaction with $Fe(CO)_5$ [175].

2.2.3.2 Intracomplex Compounds

This is the most numerous group of polymer metal chelates showing comparatively high chemical and thermal stability. Another specific feature of these compounds is the total saturation of the coordinative sphere of transition metal ion. Not going into detail, we only name the main types of examined chelate nodes.

Poly- β-diketonates are known to be the first representatives of intracomplex O,O-MMCh (e. g., see [169]. Stability constants of polymeric metal chelates are usually higher than those of low molecular analogue. This has been well demonstrated in the examples of polymethacryloylacetone (PMA) and pivaloyl acetone [176]. The increase of stability constants with the binding degree is caused by the "chain effect" appearing in the changes in macromolecule conformation in solution in the course of complex formation. This effect can be demonstrated in the example of PMA:

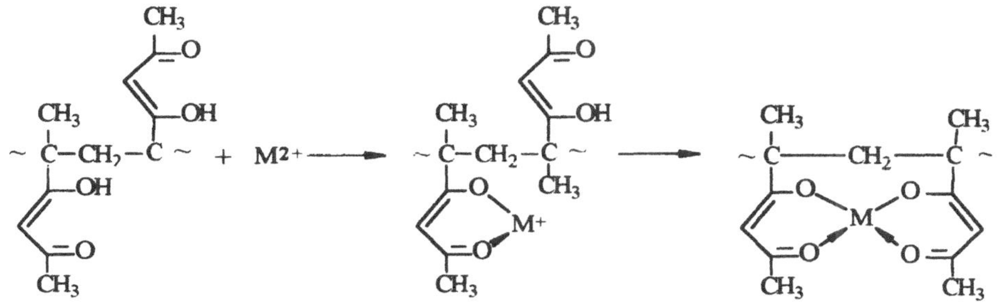

Scheme 2.26

In other words, chelation causes the bending of the polymer chain making favorable conformations for further interactions. The differences in the behavior of low- and high-molecular chelates practically disappear with the increase of binding bridge length between chelating unit and polymer backbone. Similar regularities have been observed in the course of complex formation with macroligands carrying the groups of salicyloaldehyde or salicylic acid. Plane square copper chelate complexes, octahedron nickel and manganese, and tetrahedron cobalt chelate complexes have been detected by magnetochemistry measurements and electron spectroscopy [177]. The order of stability of examined metal chelates within the tempeature range $548-913$ K was as follows: $Fe^{3+} > Co^{2+} > Cu^{2+} > Ni^{2+} > UO_2^{2+} > Zn^{2+} \approx Mn^{2+} >$ macroligand. In other words, complex formation with chelating polymer increases its thermal stability.

Polymers containing hydroxamine groups and polyhydroxamic acids are macroligands of O, O-chelating type. The later containing bridges $(CH_2)_8$ preferably to form complexes of composition $1:2$. At shorter chains $(CH_2)_6$ the formation of octahedron complexes between two units is impossible due to considerable stresses [178]:

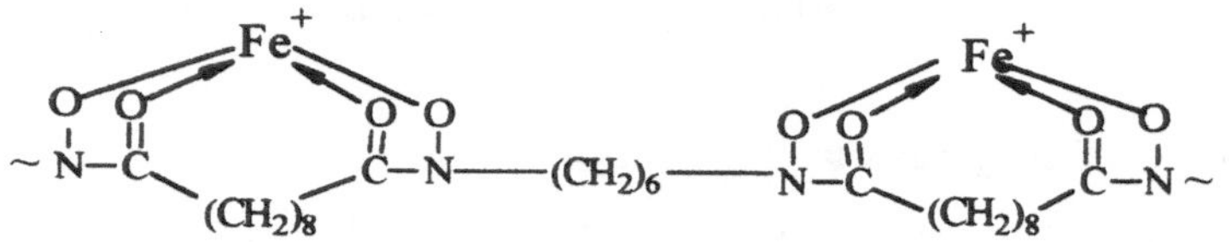

Scheme 2.27

Intracomplex MMCh with N,O-chelating nodes are the most widely used in MMCh practice. These are, for instance, chelating polymers containing fragments of iminodiacetic acid or Schiff bases as complexing groups. The broad interest in the first group of chelating polymers is based on the fact that they are commercially manufactured products (ES 466, Dowex A-1, Chelex-100, Wofatit MC-50, Amberlite, IRC-718, XE-318, Ligandex, IMAC SYN101, Unicellex, etc.). Usually, ligand surrounding of transition metal ion in such MMC is similar to low molecular analogues [179]. The increase of dentatity of chelating units (say, by introduction into the polymer of additional styrylsulfocarbonates, methylenphosphonic, etc., groups) leading to the rise of MMC stability constants.

At present, a number of metal chelates with Schiff polybases have been synthesized (Sect. 2.1.1, Scheme 2.38). The MMCh of this type are mostly tricyclic with binding bridge between nitrogen atoms not obligatorily bound to polymer backbone (a) (MMC of this type are discussed in Sect. 2.3.1.1), bound with side chain (b), or belonging to the polymer chain (c):

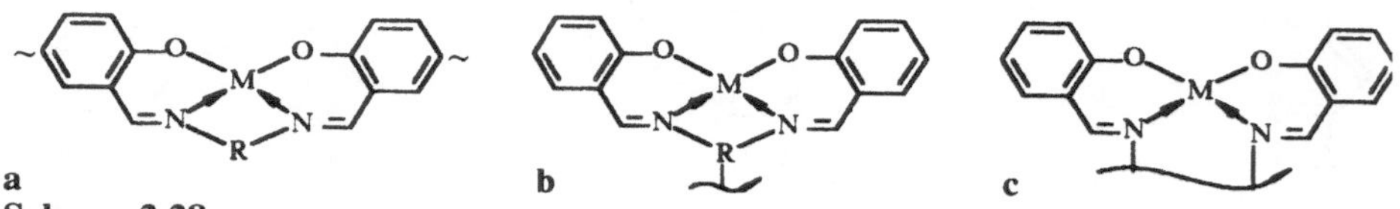

a   b   c
Scheme 2.28

Three main factors affect the structure of such MMCh. These are the nature of transition metal ion, the length of binding bridge between nitrogen atoms, and the nature of polymer chain. In the case of (a) only two of the factors are meaningful, the influence of the polymer chain being negligible. Numerous experimental data show that regularities established for the dependence of metal chelate structure on different factors for low molecular chelates are also valid for stereochemistry of MMCh. Thus, in the case of short (ethylene-like) bridge between nitrogen atoms the formed chelate node shows plane structure independent on the metal ion nature if only there is no additional coordination of solvent molecules at the axial position. The probability of appearace of other configurations, such as tetrahedron or octahedron, increases with the rise of bridge length, the real structure also depending both on the metal ion nature and the length of binding bridge. Thus, tetrahedron configuration is characteristic of cobalt (2^+) ions, while octahedron configuration with additional coordination of two solvent molecules, say of water, is typical for nickel (2^+) ions. So far as the field strength of azomethine ligand is considerably different from that produced by solvent molecules, tetragonal distorted complexes are formed in this case, and complex parameters can be obtained by the use of electron spectroscopy data.

All three previously mentioned factors should be taken into consideration for studies of complexes of type (c) (Scheme 2.28). This mainly concerns cross-linked macroligands considerably affecting the stereochemistry of coordinative node. (The effect is usually confirmed by the magnitudes of magnetic moments of the complexes.)

Polymers containing the groups of 8-hydroxyquinoline or amidoxime units are also effective N,O-chelating ligands [180]. Intracomplex metal chelates originating from polymers with N,N-, N,S-, O,S-, S,S-, and P,O-chelating nodes are considerably less studied. The studies of polynuclear MMCh (Scheme 2.23) are in their beginning.

2.2.3.3 MMCh as Macrocyclic Complexes

Polymeric crown-ethers (pseudo-crown-ethers, macrocyclic ketones, etc.) as well as polymeric diaza-crown-ethers (polycryptandes) as chelating ligands with regard to alkali metals and earth-alkali metals are well studied, although the data on transition metal ion binding are somewhat scarce.

Principally the interaction of MX_n with polymeric macrocycles can follow one of three mechanisms. The first one is binding with the macrocycle cavity, which can be considered as polydentate chelate node. Stoichiometry of these complexes is 1:1, formation constants depending on the cation nature, size of macrocycle cavity, and cross-linking degree of the polymer. In other words, this process can be assumed to be topochemically controlled. Complex stability is strongly dependent on the correspondence between the size of macrocycle cavity and those of M^{n+} (the principle of "keylock"). Another type of binding yields complexes of "sandwich" structure and composition of 2:1. This route is mainly realized when the cavity size is too small to form complexes of 1:1 composition:

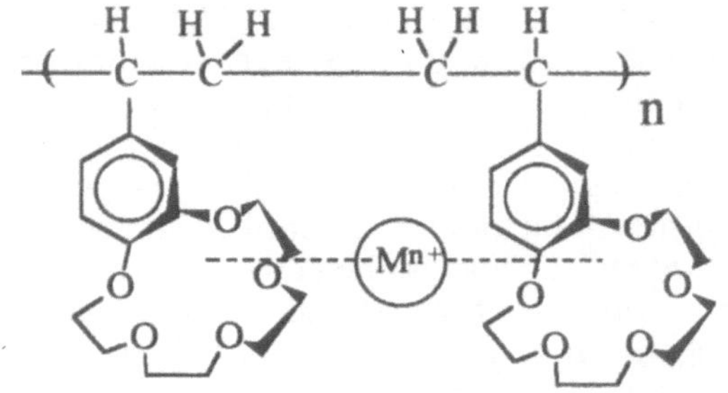

Scheme 2.29

Finally, the third type of binding leading to exocyclic coordination is realized in the cases when in macrocycles there are some other groups capable of coordination apart from main ones (ester, nitrogen-containing).

The data on polymeric macrocyclic metal chelates are concentrated in monographs [93, 164]. Here we discuss the problem of MX_n binding with polymeric porphyrins and phthalocyanines, which is presently the mostly developed direction. These are tetradentate polyligands capable of yielding very stable complexes of almost all transition metal ions by their incorporation into "windows" of macrocycles. The main approaches and polyligands used are reviewed in some studies [181, 182]. The method based on polymer analogous transformations is nevertheless the main one for synthesis of metalloporphyrins. The main stage is their binding (usually by Friedel-Crafts reaction) with polymers at reactive side groups, mostly carboxyls and amino or sulfo groups. Besides the pronounced tendency of metalloporphyrins for extra coordination of ligands occupying the fifth or six position in the inner coordinative sphere has been realized by binding of hems and hemins with P4VP, poly-N-vinyl-2-methylimidazole, as well as with copolymers derived from 1-vinylpyrrolidone, etc. The formation constant for polymeric complexes is sometimes 10^2–10^4 times higher than that for low molecular analogues which is probably caused by increased appeared value of reactive-center concentrations in the polymer domains.

It is obvious that similar reactions are considerably larger in value and of interest not only for synthesis of MMCs. In particular, sterically screened from one side by polymer metal porphyrines, they can be assumed to be models of hemproteins, such as hemoglobin, myoglobin, artificial analogues of hemoglobin, etc. (see Chap. 4).

2.2.3.4 Interpolymer Metal Chelates

Metal chelate formation may proceed via binding with functional groups of two different macromolecules (usually one of them provides "acidic" functional groups, and the second "basic" groups). These studies are not numerous in number, although metal binding with mixed biopolymers, the process that is of great importance for biological reactions, is intensively investigated.

Both polymer–polymer compositions and interpolymer complexes are used as chelating agents for synthesis of such chelates. It has been shown by studying of the systems PEI-PAAc-M^{2+} (M = Cu, Co, Ni) that Cu^{2+} ions give polycomplexes of two types [183, 184]. The first ones contain four coordinated amino groups of PEI and cooperatively bind polyanions of PAAc (at pH values above 10); other polycomplexes have two amino and two carboxylic groups in Cu^{2+} coordinative sphere (pH = 7−9). Stabilization of these complexes is provided by formation of both salt-like and coordinative bonds. Interpolymer complexes are usually formed within a short range of pH values. Thus, in the system PAAc-P4VP-Cu^{2+} at equimolar ratio of pyridine and carboxyclic groups only complexes of Cu^{2+} with P4VP are formed in acidic medium, whereas at pH 4.0−5.0 "mixed" complexes with chelate node $Cu(Py)_3(COO^-)$ are formed:

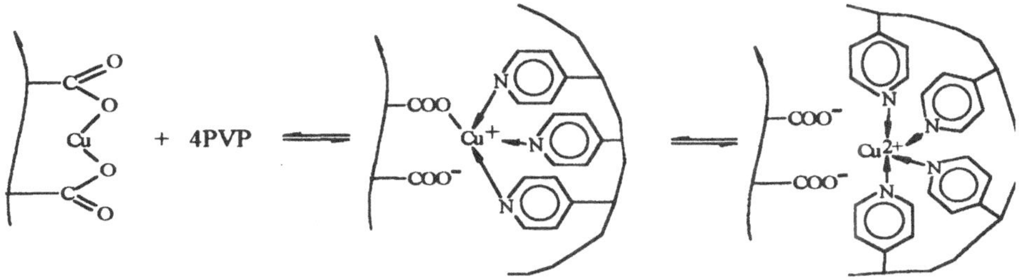

pH > 10   pH 4−5   pH 3.0–3.5

Scheme 2.30

The effect of stability restrictions of "triple" polycomplexes with regard to pH values can be used to carry out selective extraction of several metal ions (e.g., of Cu^{2+} ions by the systems PVPd-PAAc [185] or by PEI-polyepichlorohydrin [186]). Mixed-ligand MMCh derived from a combination of synthetic and biopolymers are of special interest. These polycomplexes are very stable compounds. For instance, interpolymer complexes of Cu^{2+} with P4VP-ox serum albumin are stable in acidic medium within the broad range of reagent ratios. The origin probably lies in the fact that both synthetic and biopolymer molecules carry the similar electrical charge and cannot react with each other without some intermediates. Such a complex is very compact. Its subunits are composed of 6−7 polycations that are associated with enzyme globules via chelate nodes [187]:

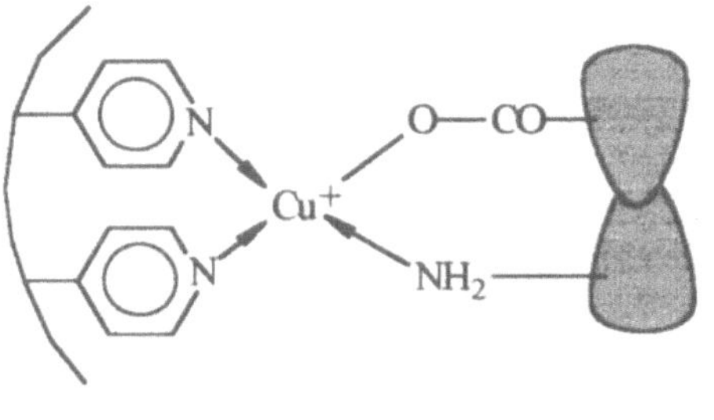

Scheme 2.31

Immobilization of enzyme variety (glucose oxidase, carboanhydrase, trypsin, α-chymotrypsin, etc.) on synthetic polymers can be carried out by chelation with the use of Cu^{2+}, Ni^{2+}, Co^{2+}, etc., ions [188]. These metal ions can also serve as bridges in complexes of protein with several polysaccharides [189].

2.2.3.5 Metal Chelates on the Surface of Oxides and Mixed-Type Polymers

Chelation of transition metal ions with polyamine ligands (*en, dien, trien*, PEI, etc.) anchored to the surface of silica have been widely studied. In many cases the similarity in both composition and properties of these complexes of molecular type with low molecular analogues are noted. The chelation ability was shown to be dependent on both the size of polyamine ligand and pH value of solution. The most high rate of Cu^{2+} ions chelation was observed for shortest groups of ethylenediamine at pH = 9, whereas at pH = 5 the rate of the reaction increased with the rise of polyamine chain length reaching the maximal value for grafted PEI [190]. The complexes of various composition (1:1 and 1:2) are mostly formed on the surface of modified mineral supports. These complexes do not transform each other under changing of the reaction conditions. Chelation ability of discussed ligands depends also on the length of the binding bridge ("spacer"). With the rise of flexible hydrocarbon chain-length chelating fragments will be drawn apart from the surface, so their behavior will be similar to soluble analogues. Such examples are numerous and usually indicate the higher stability of five-membered cycles over six-membered cycles.

The more-studied intracomplexes are compounds of MX_n anchored to the surface of silica particularly via acetylacetonate groups (e.g., [125, 191, 192]). As well as in the case of MMC with synthetic polymers, the interest to these systems is promoted by well-studied low molecular analogues and well-elaborated methods of β-deketone fixation on the surface of silica. Immobilization does not considerably affect complex ability; however, it affects stability constants being strongly dependent on the ionic strength of solution sometimes even distorting the structure of the chelate node. The composition and even mechanism of metal chelate formation can also be affected by pH values, solvent nature, the nature of anion in MX_n, etc. For instance, in basic medium the ligand chelates metal at enol form while at keto form in acidic medium [193].

Another effective way for the synthesis of chelate complexes on the surface of oxides is interaction with preformed low molecular metal chelates with monofunctional surface ligands (binding via ligand). By this way disalicyl-al-1,2-phenylenediamine Cu^{2+} complexes were immobilized via pyridine fragments on the surface of silica [194] as well as chelates N,*N*-disalicyliden-4-carboxy-1,2-phenylenediiminatocobalt (2^+) [195].

In the absence of steric restriction the complex abilities of macrocycles, crown-ethers, and cryptands stay practically unchanged after their fixation on inorganic supports. This fact is confirmed by close values of stability constants of metal complexes with 15-crown-5, 18-crown-6, and diaza-crown bound to silica surface and their low molecular analogues [196]. Metal porphyrins and metal phthalocyanins can be easily bound to SiO_2, Al_2O_3, etc., containing aminopropyls or imidazolic functional groups with the formation of metal – nitrogen covalent or coordinative bonds [197, 198]. This direction is presently extensively elaborated and probably should be analyzed separately.

Thus, even brief analysis of both main routes of synthesis and the structures of formed products show that chelation of MX_n with macromolecular ligands provides really unrestricted abilities for synthesis of MMC. The formation of chelate cycles of desirable structure as well as selective and multipoint binding of metal ions can also be easily reached. These facts are very important for the purpose of creation of new constructional materials, immobilized catalysts, selective sorbents, etc.

2.2.4 Polymeric π-Complexes of Transition Metal Ions

If MMC of coordinative type are formed via heteroatoms of macroligands providing unshared electron pair for metal complex binding, then MMC of π-type are formed via aromatic groups or unsaturated fragments of macrochain acting as π-donor and can be conditionally classified as π-complexes of aromatic or diene type. The former are the products of interaction of MX_n with such macroligands as PS (polystyrene), polyvinyltoluene, polyanthracene, copolymer of styrene with α-methylstyrene, β-vinylnaphthalene, etc. The stability of these MMC can be evaluated by the data of IR- or UV spectroscopy, the measurements of dipole moments, and heats of formation. Thus, interaction of PS with $TiCl_4$, $VOCl_3$ yields π-donor complexes of the composition $D:MX_n = 1:1$ (D is a monomeric unit) with charge transfer (electron affinity of above MX_n is equal to 1.57 and 2.03 eV, respectively; IR-adsorption bands correspond to 385 and 485 nm) [199]. It is of interest than π-complexes are sometimes formed regardless of the presence of other functional groups:

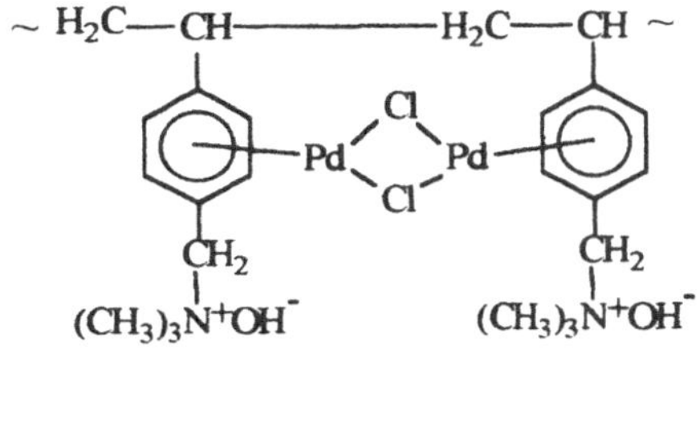

Scheme 2.32

The π-complexes with polyaromatics, particularly with graphite, can also be attributed to this type. Graphite can be considered an aromatic-type macromolecule (the number of rings is approximately 1000, and the distance between parallel planes is approximately 0.35 nm). The lack of chemical bonds between parallel layers of graphite provides ability for MX_n incorporation yielding flaky (or laminal) graphite compounds [200]. The general methods of discussed complex formation consist of interaction of graphite with metal vapors or with volatile metal chlorides (e.g., $TiCl_4$), cationic complexes of the type ML^+_n, all the routes leading to incorporation of transition metal ion between the layers of graphite. The interaction is often accompanied by π-electron density transfer from graphite to the incorporated layer of metal ions. Thus, a carbon network of graphite can be considered a polymeric ligand (e.g., for compounds of Mo, W, and Cr).

We now concentrate on MX_n immobilization via cyclopentadienyl rings that are characteristic groups of such macroligands. This route of MX_n formation can be generalized as follows [201] (Scheme 2.33).

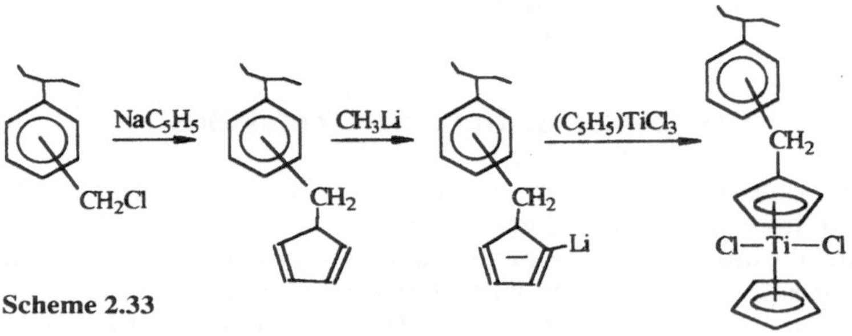

Scheme 2.33

This approach is also effective in synthesis of MMC based on niobocene dichloride, cyclopentadienyldicarbonyls of rhodium, cobalt, etc.

There are also known numerous π-complexes of transition metal ions with polydienes, copolymers of styrene, and dienes. Thus, thermolysis of $Fe(CO)_5$ in solution of *cis*-polybutadiene in xylene yields poly-1,3-octadienylirontricarbonyl [202]:

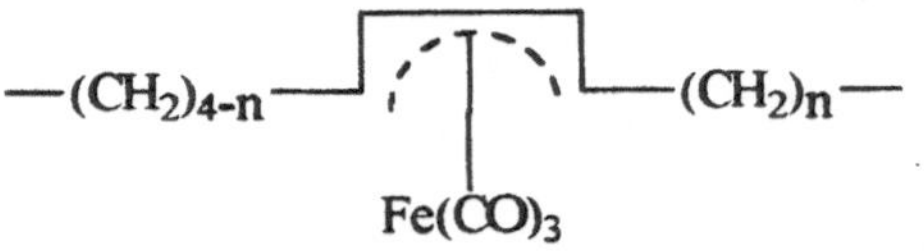

Fe(CO)₃   **Scheme 2.34**

The product consits of η^4-(butadienyl)irontricarbonyl untis with *trans-trans* and *cis-trans*-tetramethylene fragments. At the same time interaction of polybutadiene with $(C_5H_5)_2Zr(Cl)H$ (Schwartz reagent) is accompanied [203] by hydrozirconation yielding polymer products with σ-bound $Zr(C_5H_5)_2$ fragments at the side chain.

This perspective seems to be the direction in the near future for creation of new types of MMCs.

2.2.5 Macrocomplexes with Natural Polymers

Natural polymers being multifunctional systems provide difficulties for determination of active functional groups interacting with MX_n. Polysaccharides, especially cellulose and its derivatives, are the most widely used macroligands for MX_n binding. We have also noticed in Sect. 2.1.1 the special properties of cellulose that make it valuable as a macroligand. Here we only compare the complex ability of carboxymethylcellulose (CMC), phosphatecellulose, and cellulose modified with groups of hydroxamic acid. The CMC in the form of fine powder in H-form is a uniform macroligand containing no other groups apart from carboxyls $((5-6) \cdot 10^{-4} \, mol/g)$, which can only take part in ionic exchange at modest pH values (up to 10). Each of ten units is modified under these conditions. Besides, because of cellulose macromolecule rigidity, the possibility of coordination of two grafted carboxyls with one metal ion (e. g., Cu^{2+} or Mo^{6+}) is totally excluded [204]. Thus, the Cu^{2+} ion became bound via one coordinative bond and electrostatic interaction with the chain. At the same time Mo^{6+} ions are bound without proton elimination yielding binding complex acid:

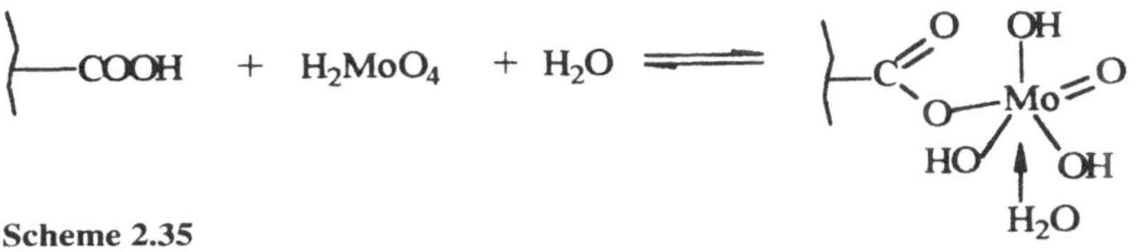

$$\Big\}\!\!-COOH \; + \; H_2MoO_4 \; + \; H_2O \; \rightleftharpoons \; \Big\}\!\!-C\underset{O}{\overset{O}{\lessgtr}} \underset{HO}{\overset{OH}{\underset{|}{\overset{|}{Mo}}}} \!\!\overset{O}{\underset{OH}{}}$$

$$H_2O$$

Scheme 2.35

In a similar way sorption of Mo^{6+} on phosphate cellulose in acidic solutions proceeds. However, the nature of functional groups significantly affects the strength of bound Mo^{6+} complexes, β_n for phosphate cellulose being two orders of magnitude higher than this value for CMC. The same regularities have been observed in the course of chelation with alginic acids.

Ions of Zn^{2+}, Pb^{2+}, Cu^{2+}, and Co^{2+} are effectively bound with pectin (polygalacturonic acid which carboxyls are partially etherificated with methyl alcohol, and showing chain composition similar to cellulose) [205]. The stability constant of Zn^{2+} MMC sharply decreases with the rise of etherification degree, a fact that can be used for regulation of Zn^{2+} ion content in organism. Carboxymethyl ether of dextrin shows well binding properties with regard to Cu^{2+}, Ni^{2+}, and Co^{2+} ions [206]. The high values of formation constants ($\beta_n = 1,5 \cdot 10^{10}$) for V^{5+} ions with cellulose modified with hydroxamic acid functional groups [207] provides abilities for this reagent usage for analytical purposes.

Cyclodextrins and dextrans are close to cellulose in their complex abilities, although CMC yield more stable MMC according to some data [208]. The possibility of formation of two different complexes depending on binding-group position (at inner or outer side of cyclodextrins) has also been noticed [209]. In the first case there are considerable steric limitations for interaction with MX_n, and as a result, less stable complexes are formed. In contrast of both α- and β-forms γ-cyclodextrins has such a big cavity that practically any metal chelates can be incorporated inside. The stability of these complexes is also increased. The same effect can be reached by cross-linking of dextrin chains, for example, by $(C_5H_5)_2TiCl_2$ [210]. The formed products have the structure (Scheme 2.36).

The binding centers of chitin and chitosan are both nitrogen atoms of amino or acetamido groups and hydroxyl groups. Chelation with soluble polymers proceeds readily and yields intramolecular complexes of the type [211] (Scheme 2.37).

As a rule metal ion binding is a heat-absorbing process, e. g., the changes in enthalpy are equal to -17.9, -9.38, and -5.15 kJ/mol for the reactions of uranyl ions with chitin phosphate [212], chitosan modified with salicylic aldehyde, and glutaric aldehyde, respectively [213]. Both chitin and chitosan show high selectivity with regard to transition metal ions and heavy metal ions, being merely reactive with regard to alkali and earth-alkali metal ions. However, in going from soluble to cross-linked polyligands the increase of chelate ability with regard to both types of metal ions has been observed.

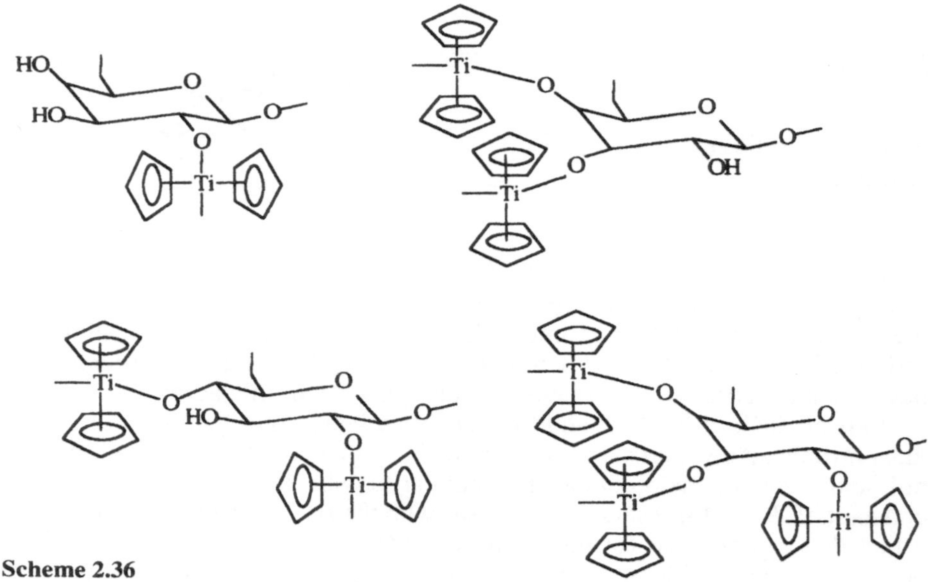

Scheme 2.36

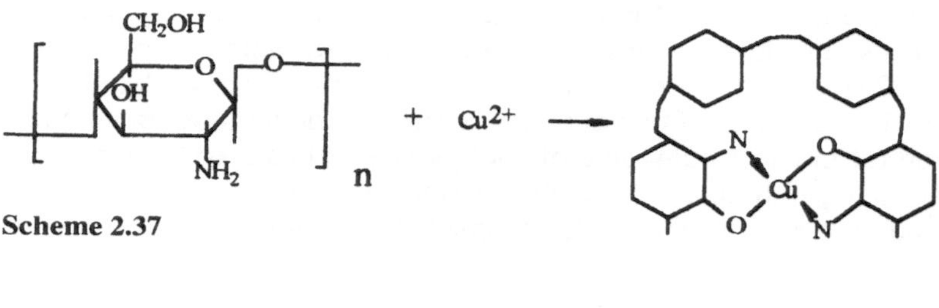

Scheme 2.37

Solutions and gels of gelatin (polyfunctional ligand) effectively bind Co^{2+} ions at $pH < 6$ by carboxyl groups, Cu^{2+} ions by chelation with carboxyl and amino groups, and Zn^{2+} by imidazolic and other functional groups [214]. Which functional groups will be involved in the reaction depends on pH values of the reaction medium [215].

The increased interest in organo-mineral derivatives of humic and fulvic acids is primarily based on their significance for several natural processes, particularly for their role in the transformation of biogenic metals into the forms assimilably by alive-organisms. These polyligands contain two types of donor atoms. These are oxygen atoms of carboxylic, alcohol, phenolic, quinone functional groups of humic acids, and nitrogen atoms of amine, amide, imidazole, and peptide fragments. Three types of complexes are formed in the course of interaction with MX_n: heteropolar salts (ionic bonds), complex heteropolar salts (coordinative bonds), and adsorbed complexes (intermolecular bonds) [216]. Complex salt metal ions are incorporated into the anionic part of the molecule. Because this complex compound has free carboxy and phenolic groups, it can further interact with formation of simple heteropolar bonds:

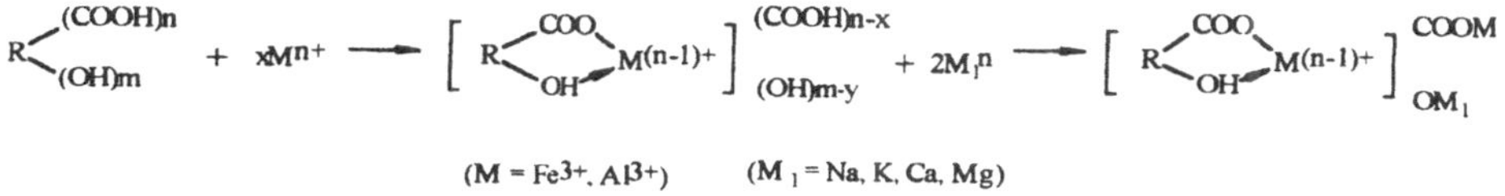

$$(M = Fe^{3+}, Al^{3+}) \qquad (M_1 = Na, K, Ca, Mg)$$

Scheme 2.38

The formed products can be considered as MMC with isolated coordinative nodes formed at the positions of localization of metal ions. The value of the binding constant of metal ions on the variety of reactive centers of macroligand is thus the mean statistical value, and can be analyzed in terms of equilibrium function [217] for separate reactive centers.

It follows from the above analysis that formation of MMC with natural polymers is possible as for synthetic macroligands. However, it is often difficult to determine the bond nature and the structure of coordinative nodes, due to polyfunctionality of these polymers.

2.2.6 Polymerization and Copolymerization of Metal-Containing Monomers as a Way for MMC Synthesis

Metal-containing monomers (MCM) are compounds that include both multiple bond capable of polymerization and chemically bound metal ions. According to their chemical nature these compounds can be classified [218] by types similar to those discussed in Sect. 2.2.2–2.2.5 (Scheme 2.39):

Multiple bonds in these compounds may be of different type, such as double, triple, allene, diene, or their combination. Synthesis and polymerization of MCM are discussed in detail in other works [219, 220]. However,

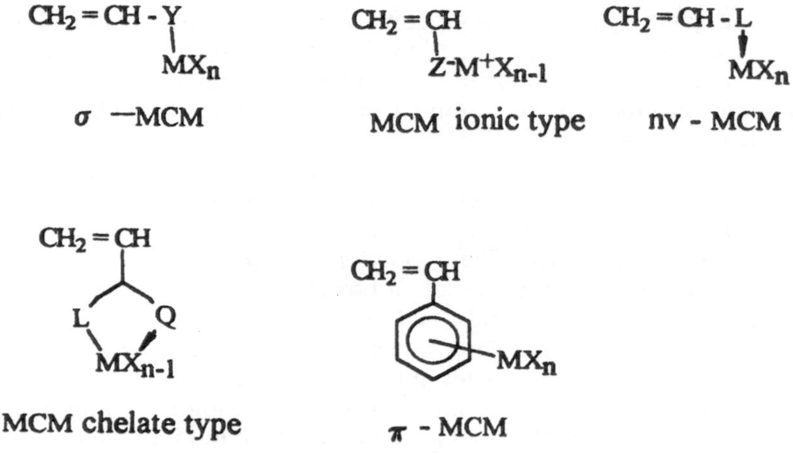

$CH_2 = CH - Y$
MX_n

σ —MCM

$CH_2 = CH$
$Z^-M^+X_{n-1}$

MCM ionic type

$CH_2 = CH - L$
MX_n

nv - MCM

$CH_2 = CH$
$L \quad Q$
MX_{n-1}

MCM chelate type

$CH_2 = CH$

⌬—MX_n

π - MCM

Y, Z, L, Q - the functional groups   Scheme 2.39

MCM of this type considerably differ from previously discussed types, even if they have no unbound functional groups. This is not always true: In the case of polymerization of coordinatively bound MCM (e.g., vinylpyridine complexes) dissociation of MCM with following copolymerization of released ligand with coordinatively bound MCM can take place. Certainly, there are a lot of difficulties in this way of synthesis that have not yet been overcome. It primarily concerns the influence of metal ion on all elementary stages of both polymerization process (initiation, chain propagation, chain termination, and restriction) and reactivity of comonomers in copolymerization. Radical polymerization of these MCM is often accompanied by redox reactions with metal ion participation as well as metal ion complex formation with primary or growing radicals, processes which can be competitive with polymerization. At the same time, polymerization of MCM is a widely investigated field of polymerization science [9, 17, 18, 81, 166, 181, 220, 243, 252, 253, 264, 279, 353] with great potential for new types of MMC production.

2.3 Type II: Metal Complexes and Metals as Part of a Polymer Chain or Network

D. Wöhrle

In classical polymers the backbone of a chain or network consists of only six elements (C, N, O, S, P and Si). But altogether among the approximately 110 elements of the periodic system only the six noble gases and 12 elements of groups IA and VIIB (including H), which are univalent, are incapable of producing stable polymers (but some of them in the solid state, and IA elements in a ligand environment). The remaining nearly 90 elements in varying combinations are principally capable as chain forming or as participating element giving rise to an unlimited number of materials.

This chapter selects only polymeric materials where a metal complex or a metal (partly semimetals are briefly mentioned) is part of a polymer chain or network. Various combinations are possible, and this chapter is divided accordingly. The IUPAC nomenclature for these "regular single-strand" and "quasi-single-strand" inorganic and coordination polymers are described in [221, 222] and is not discussed here. The following subchapters select only some characteristic examples for the preparation and the structure of these polymers. Additional literature are cited in the mentioned references. Properties of the materials are not discussed in detail and are part of subsequent chapters that also consider additional examples.

Very often the polymers described in Chap. 2.3 are insoluble and difficult to characterize with regard to structural purity and additional presence of side products. Someone interested in these polymers should carefully check the reported experimental parts including the analytical characterization (Chap. 2.1.7). Only on the basis of reproducible synthetic procedures and a good relation between analytical data and structure are investigations and discussions of properties of these materials substantiated.

2.3.1 Ligand of a Metal Complex as Part of a Polymer Chain or Network (Polymeric Metal Complexes)

In this case a polymeric ligand with donating groups L can bind a metal ion M as is usual for low molecular ligands:

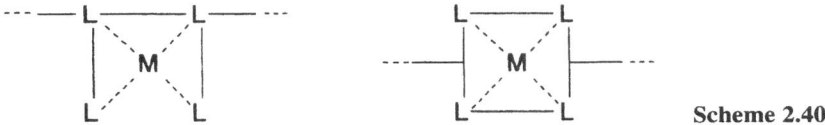

Scheme 2.40

If a multidentate ligand is part of a polymer chain, the Lewis basic donor atoms can easily bind metal ions to obtain a polymeric metal complex. At first it is essential to know the complexing behaviour of metal ions of the low molecular weight analogous ligands. The experiences in synthesis and characterization can then be used to study the complex formation and properties of a polymer ligand. Because of entropic effects, the complex binding ability of a polymer ligand resulting in a polymeric metal complex of higher order enhances the formation of the products.

Advantageous for the preparation is first to synthesize the polymeric ligand either by (a) the reaction of a bifunctional substituted ligand (or bifunctional substituted metal complex) with another bifunctional comonomer, or (b) to employ a bifunctional ligand or precursor of a ligand with a multivalent metal ion. The structure of the polymeric ligand can be evaluated and the metal ion complexation is carried out. In special cases it is useful to react a ligand/chelate precursor in the presence of a metal ion as template directing the synthesis in the direction of the metal complex formation. If the number of ligand units in mmol/g is known, the degree of metalation can be determined by metal analysis.

2.3.1.1 Noncyclic Organic Ligands

A typical example of noncyclic organic ligands are Schiff-base chelates as part of a polymer chain of the general formula (Scheme 2.41). They have been reviewed in some previous papers [223, 224].

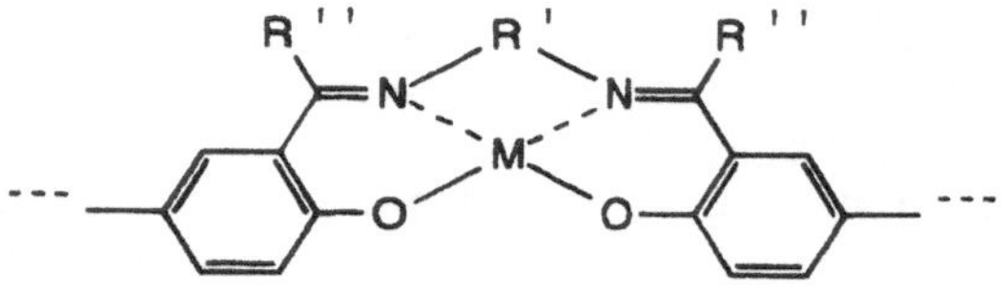

Scheme 2.41

Two possibilities exist for the preparation of these yellow, brown or black polymers: Reaction of a bifunctional Schiff-base ligand or chelate with another bifunctional monomer (route 1) or conversion of a bifunctional ligand precursor in a one-step procedure directly to the polymer ligand or chelate (route 2). Route 1 exhibits the advantage that side reactions, which may occur during ligand formation, are not possible because the ligand moiety is employed for the polyreaction. An example is the polycondensation reaction shown in (Scheme 2.42) from dihydroxy-substituted metal Schiff-bases with dicarboxylic acid dichlorides [225, 226].

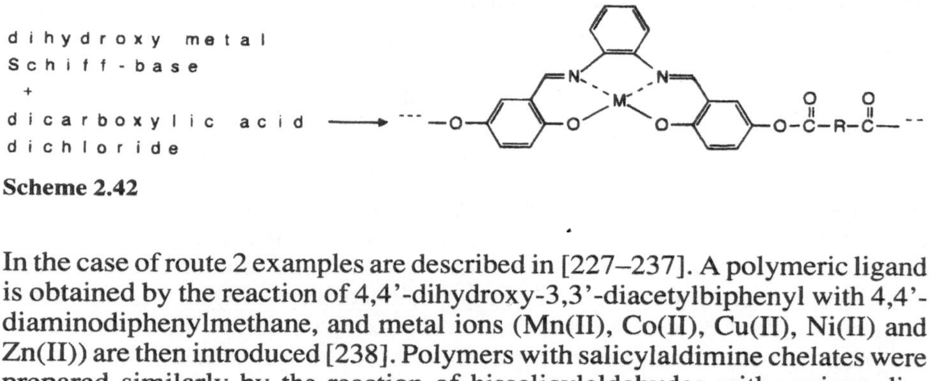

Scheme 2.42

In the case of route 2 examples are described in [227–237]. A polymeric ligand is obtained by the reaction of 4,4'-dihydroxy-3,3'-diacetylbiphenyl with 4,4'-diaminodiphenylmethane, and metal ions (Mn(II), Co(II), Cu(II), Ni(II) and Zn(II)) are then introduced [238]. Polymers with salicylaldimine chelates were prepared similarly by the reaction of bissalicylaldehydes with various diamines and then metallization (Scheme 2.43) [229–231, 239]. Some polymeric ligands are soluble in polar organic solvents, which is a good presupposition for the metallation [236]. In addition, in situ procedures use the reaction of a ligand precursor of the diamine and a metal ion or an o-hydroxyaldehyde and a metal ion, respectively, and to add then an o-hydroxyaldehyde and a diamine, respectively [234, 235, 240–242]. For cross-linked Co(salenes) and Mn(salenes) see [243, 244].

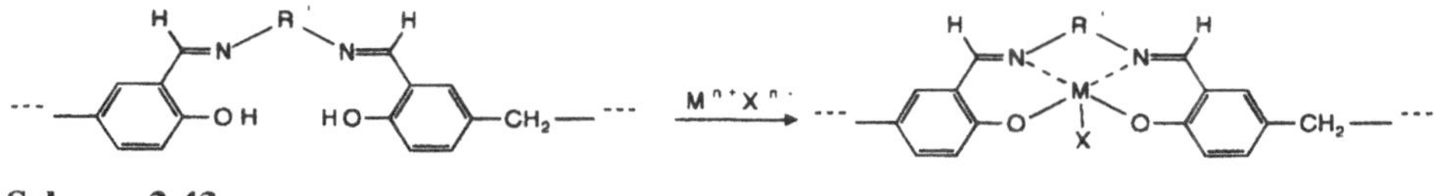

Scheme 2.43

Thermotropic liquid crystalline polymers containing β-diketonato groups - (-R-CO-CH$_2$-CO-R) either in the main of a polyester or in the ester group of poly(acrylate esters) or in the ester group of poly(acrylate esters) were prepared [245]. In polyesters the interchain complexation of the diketonato groups with Cu^{2+} and Ni^{2+} resulted after X-ray in better monoaxial orientation and after ESR in the orientation of square planar Cu complexes parallel to the fibre axis.

Examples of other polymeric ligands for the binding of metal ions are poly(thiosemicarbazides) [246, 247] and poly(terephthaloyloxalic-bisamidrazones) [248]. Selective binding of Cu(II) has been described [249]. Polymeric metal complexes were obtained by the reaction of 1,2,4,5-benzenetetracarboxylic acid dianhydride and bipyridylbipyridylamin in the presence of ZnCl$_2$ [250]. Also Fe(III), Cr(III), Co(II), Ni(II) and Cu(II) complexes were synthesized. The thermal stability, dielectric constants and electrical conductivity ($\sigma \sim 10^{-7} - 10^{-10}$ S $\cdot$ cm^{-1}) were investigated. Water-soluble carrier with bound platinum and iron compounds for biomedical applications, e. g. chemotherapy of cancer, have been developed [251]. In relation of *cis*-diaminedichloroplatinum(II), a polymeric Pt(II) (Scheme 2.44) compound were prepared by the reaction of poly(ethyleneimine) with PtCl$_4^{2-}$.

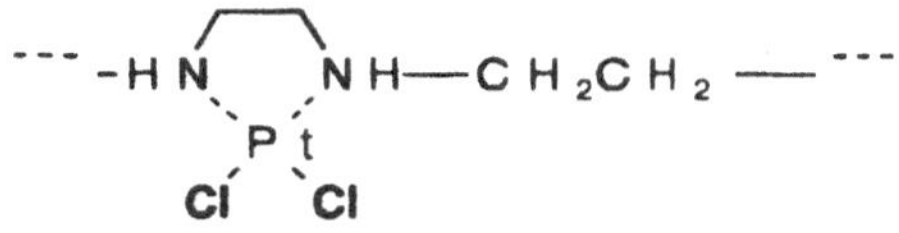

Scheme 2.44

2.3.1.2 Cyclic Organic Ligands

Porphyrins, phthalocyanines, hemiporphyrazines and tetraazaannulenes have been intensively investigated as cyclic ligands for polymeric metal chelates [252–254]. Theoretically the polymers exhibit a two-dimensional plane structure, but examples for a chain structure have also been reported. The polymeric chelates are generally obtained as insoluble brown-to-black powders. In some cases for device construction film formation during the preparation process is described.

According to route 1 (see Chap. 2.3.1.1) low molecular weight higher functional substituted macrocyclic metal complexes (M=Co, Ni, Cu, Zn) were converted with other bifunctional compounds to polymers. By the reaction of tetraaminophthalocyanine in the presence of another diamine with benzenetetracarboxylic acid dianhydride, at first in dimethylsulfoxide (DMSO) soluble amide-carboxylic acid copolymers were obtained, and after film casting and

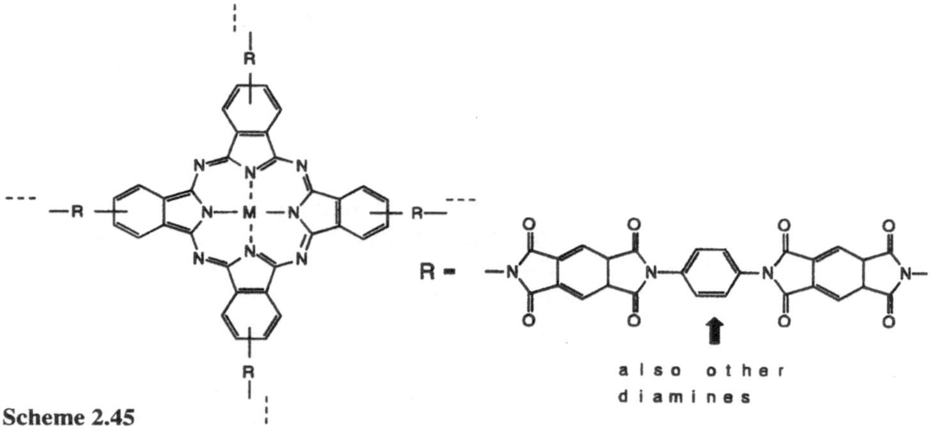

Scheme 2.45

heating to ~ 325 °C converted into films of insoluble poly(metal phthalo-cyanine) imide copolymers (Scheme 2.45) [255]. A high thermal stability of these coloured polyimides were found.

The interfacial polycondensation technique in which reactive comonomers are dissolved in separate immiscible solutions, and thereby constrained to react only at the interface between two solutions, was applied for the synthesis of chemi-cally asymmetrical polymeric porphyring (M=Zn, Cu, Ni, Pd, 2H) films containing structural elements as shown exemplarily in Scheme 2.46 [256–258]. Tetrakis(4-aminophenyl)-, tetrakis(4-hydroxyphenyl)porphyrins or aliphatic diamines in one solvent were reacted with tetrakis(4-chlorocarboxy-phenyl)porphyrin or aliphatic diacyl chlorides, respectively, in the other solvent. Typical film thicknesses are in the range of 0.01–10 μm. The unique chemical asymmetry is shown by distinctive differences in the concentration and type of functional groups present. Photoactivities of the polymeric porphyrin films were measured in dry sandwich cells (see. Chap. 5.3). It would be interesting to apply this preparative method also to other substituted metal complexes.

The electrochemical polymerization of π-electron rich aromatics, such as aniline, pyrrole and thiophene, to obtain electrically conducting polymers is well known. Tetrakis(aminophenyl)-, tetrakis(dimethylaminophenyl)-, tetrakis (hydroxyphenyl)-, tetrakis(N-pyrrolylphenyl)-, tetrakis(3-methoxy-4-hydro-xylphenyl)porphyrins, paracyclophenylporphyrins and tetraaminophthalo-cyanine, either metal-free or metal-containing (Co(II), Ni(II)), were anionical-ly polymerized on Pt, SnO$_2$ or glassy carbon electrodes from electrolyte solu-tion to cross-linked systems [259–263]. Depending on the reaction conditions, the film thicknesses varied between few to several hundred monolayer thick-ness. The electrochemical and electronic properties are similar to those of the monomers in solution. The polymer films remain electroactive at the elec-tropolymerization potential so that the oxidation and reduction current enve-lope grows with each successive potential cycle. Figure 2.7 shows exemplar-ily the gradual increase in the Faradaic current which is indicative of the oxi-dative polymerization of Co(II)-tetraaminophthalocyanine [261]. Two quasi-

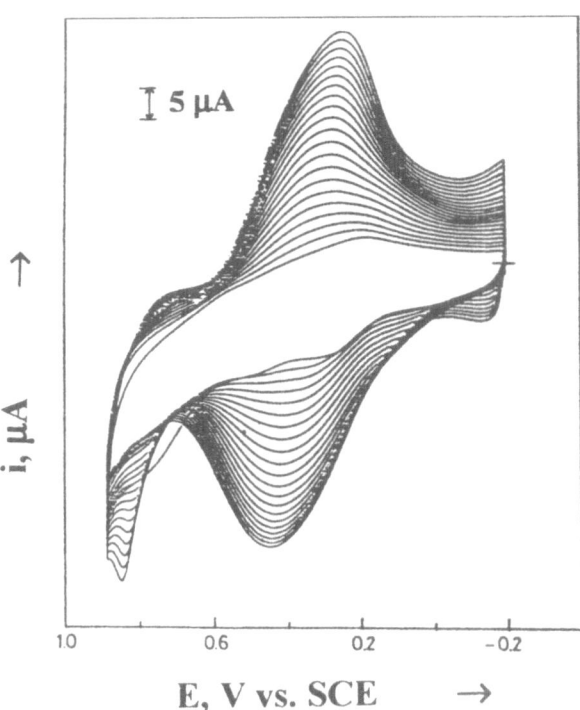

porphyrin polyester membran

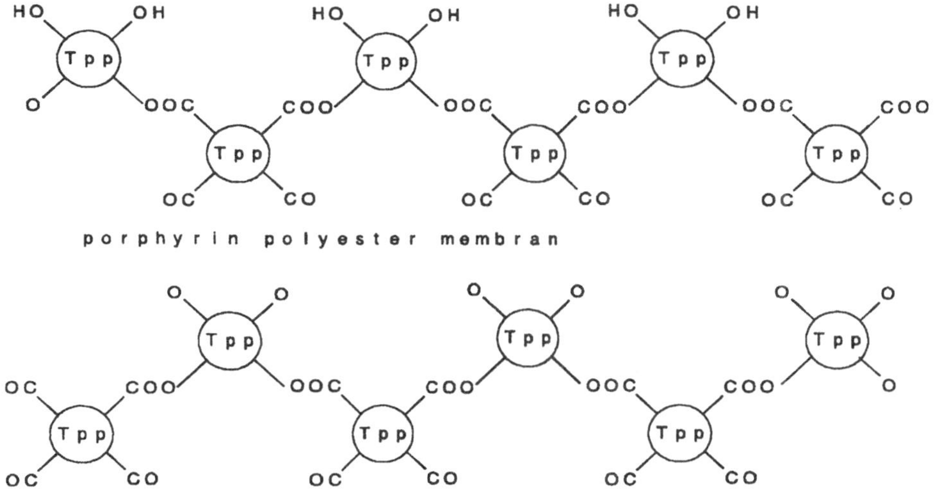

Scheme 2.46

Fig. 2.7. Repetive cyclic voltammograms (−0,2 to +0,9 V vs SCE) of 10^{-3} M cobalt(II) tetraaminophthalocyanine in DMSO (0,1 M tetraethylammonium perchlorate as conducting salt). Scan rate 200 mV/s; every fifth scan recorded. (From [261])

5 µA

i, µA

E, V vs. SCE   →

1.0   0.6   0.2   − 0.2

reversible redox couples at ~0.5 and ~0.7 V vs SCE occur. The polymerization process for a pyrrolylphenyl-substituted porphyrin is shown in (Scheme 2.47) [259]. In the case of hydroxyphenyl-substituted porphyrins the mechanism is more complex and includes a quinoid-type system [263].

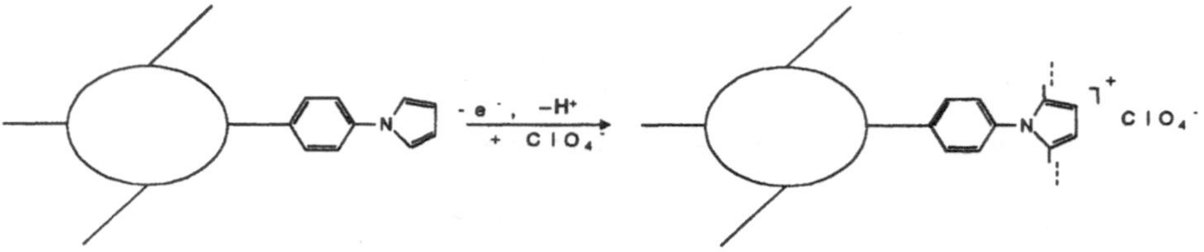

Scheme 2.47

For films of Co(II)-tetrakis(2-aminophenyl)porphyrin activity for electro-catalytic O_2 reduction is described [259]. This polymer exhibits a two-mono-layer film and excellent molecular sieving properties with permeabilities of low molecular weight redox active monomers that fall in a regular order of permanent molecular volume [260]. For other polymeric porphyrin derivatives electrocatalytic oxidation of water, methanol and hydrazine is mentioned [262, 263].

Thin films of cross-linked chelates are obtained also by homopolymerization of tetravinyl-substituted metal porphyrins and phthalocyanines ([264] and references cited therein). If copolymerizations with, for example, styrene, are carried out in solution to low yield, linear non-cross-linked copolymers are isolated. Chain structured homo- and copolymers are also synthesized by polymerization reactions of monovinylporphyrins ([223, 264] and references cited therein). Water-soluble neutral, positively charged and negatively charged polymeric Zn(II)porphyrins, phthalocyanines and naphthalocyanines have been prepared also by various polymer-analogous reactions [265]. These polymer chelates were intensively investigated for electron transfer and photon-induced electron transfer reactions (see Chap. 5.1). Other possibilities to synthesize polymeric-network metal phthalocyanines according to route 1 (Chap. 2.3.1.1) are thermal elimination reactions of tetracarboxyphthalo-cyanines [266] and plasma polymerization reactions of evaporated low molecular weight metal phthalocyanine [267, 268]. The plasma polymerization leads to thin films of polymers containing moieties of phthalocyanine units.

Only very few examples have been reported on the synthesis of ladder-type structured polymer metal complexes with macrocyclic ligands. This is due to the fact that asymmetrically substituted macrocyclic metal complexes are difficult to prepare an purify (by column chromatography). For a phthalocyanine ladder polymer (oligomer) and asymmetrically metal complex is treated in a Diels-Alder reaction with 2,3,5,6-tetramethylene-7-oxanorborane [269] to an oligomer shown in (Scheme 2.48) which had to be converted into a system containing conjugated π-electrons. Also, in the case of hemiporphyrazines it was possible to prepare by Diels-Alder reactions between dienophiolic and enophilic monomers the corresponding oligomers and by repetitive reactions analogous polymers [270].

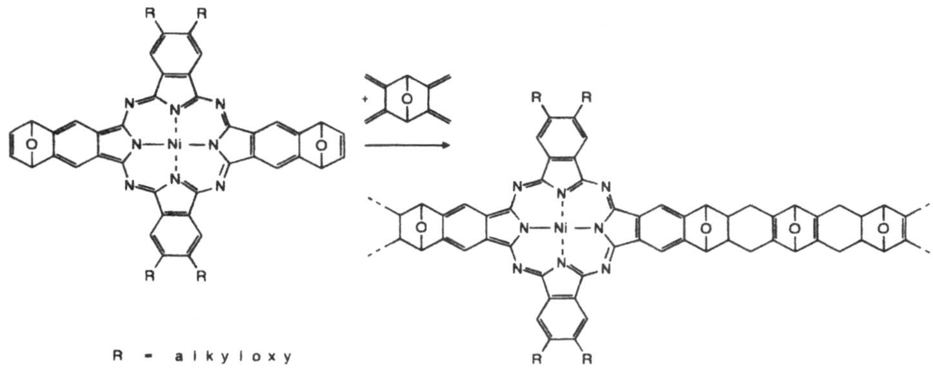

R = a l k y l o x y

Scheme 2.48

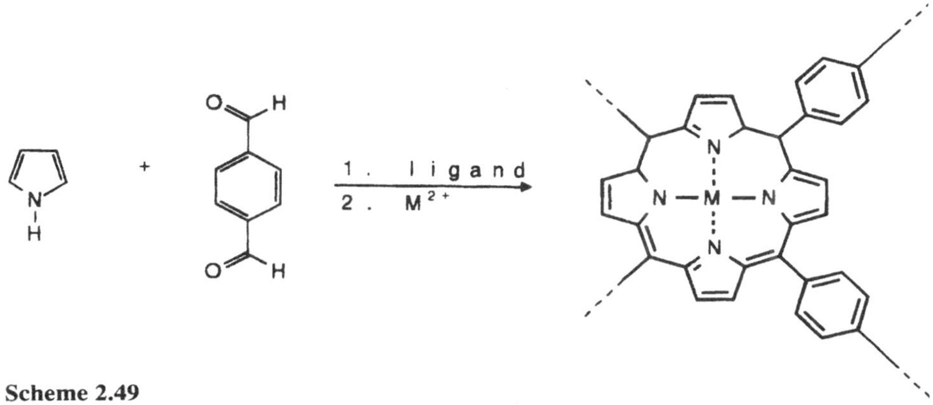

Scheme 2.49

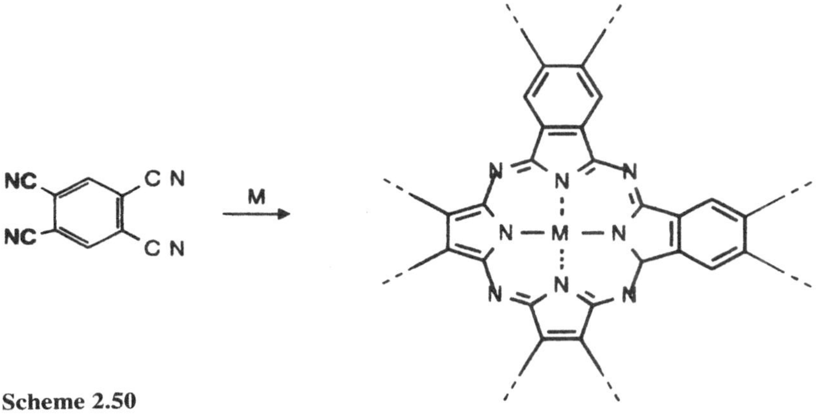

Scheme 2.50

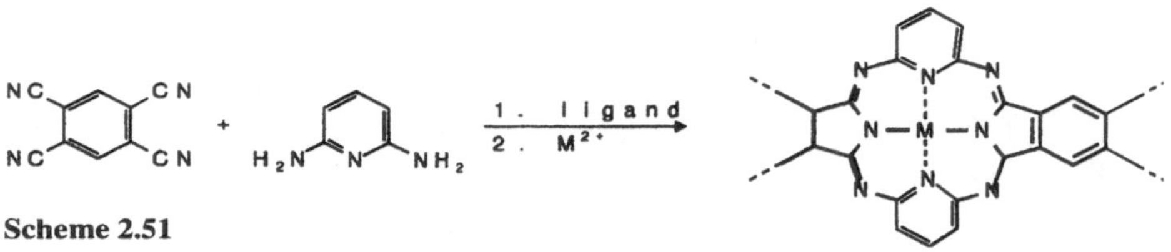

Scheme 2.51

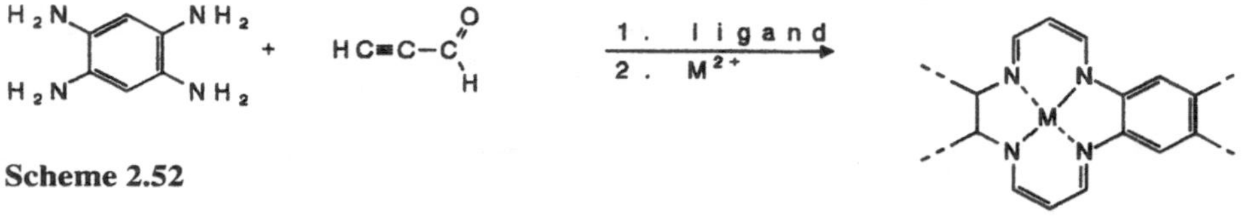

Scheme 2.52

Bifunctional ligand or chelate precursors must be employed following route 2 (Chap. 2.3.1.1). Examples for polymeric macrocyclic metal complexes according to this route are listed in Schemes 2.49–2.52.

The synthesis of poly(tetra-5,10,15,20-phenylporphyrins) (Scheme 2.49), analogous to the classical Rothemund reaction for the conversion of aromatic aldehydes with pyrroles to low molecular weight porphyrins, cannot be applied employing bifunctional aromatic aldehydes successfully. The reaction with monofunctional aldehydes results in only approximately a 15% yield of the porphyrin and is accompanied by a lot of side products. If terephthalaldehyde and pyrroles are employed an insoluble black polymer with few porphyrin units that cannot be purified is obtained in high yield. Therefore, it is a basic requirement for the synthesis of polymeric macrocyclic metallochelates from bifunctional ligand/chelate precursors that the comparable reaction to the analogous low molecular model reaction runs in high yield to pure products.

The preparation of polymeric hemiporphyrazines (polyhexazocyclanes) from various tetranitriles with diamines (Scheme 2.51) [223, 271, 272] and polymeric tetraaza-(14)annulenes (Scheme 2.53) [273, 274] were described several years ago. Polymers with ligand/chelate structure could be assumed, but the analytical purity has to be analyzed again with newer instrumental techniques.

More detailed, synthesis, structure and properties of polymeric phthalocyanines were investigated and reviewed (Scheme 2.50) [223, 252–254]. These polymers are obtained as insoluble powders, but are also available as thin-structured films, which is more interesting for a device construction. The investigation of properties concentrates on thermal stability [275–278], electrical conductivity and redox behaviour of thin films [279–283], catalytic activity [284], electrocatalytic activity for the O_2 reduction [223, 279] and photochemical properties [281].

Various tetracarbonitriles, such as 1,2,4,5-tetracyanobenzene [276, 285, 286] and tetracarboxylic acid derivatives, such as 1,2,4,5-benzenetetracarboxylic acid dianhydride [277, 278, 287] were employed in the reaction with different transition metals or metal salts such as Cu and $CuCl_2$ [223, 252–254]. Scheme 2.53 shows the reactions of tetracyanobenzene with Cu and $CuCl_2$. A detailed analysis has to consider structural uniformity (side products, poly-nitriles and polytriazines), uniform end groups (nitrile or carboxylic acid derivatives), degree of metallization (metal-free part) and molecular weight. Structural uniform polymers with cyano end groups were synthesized by bulk reaction with $CuCl_2$ at higher temperatures [276, 278, 285, 286]. Increasing flexibility of the bridge between the two reacting 1,2-dicyano groups results in molecular weights up to (inlinite plane). The high molecular weight phthalocyanines possess molecular sieve properties. Because of the insolubility of the powdered samples, end-group analysis (IR spectroscopy, titration) and elemental analysis have been used to determine the molecular weight. A model describing the structural features, such as degree of polymerization, size and shape, is described and discussed in detail [76, 77] (see also Chap. 2.1.7).

$$2n\ C_{10}H_2N_4 + nCu^\circ \rightarrow [(C_{20}H_4N_8)^{2-}\ Cu^{2+}]_n$$

$$2n\ C_{10}H_2N_4 + nCuCl_2 \rightarrow](C_{20}H_4N_8)^{2-}\ Cu^{2+}]_n + nCl_2$$

Scheme 2.53

Very thin films of polymeric phthalocyanines were obtained by in situ synthesis of 1,2,4,5-tetracyanobenzene with TiO_2 modified Ti electrodes [280, 281]. The electrodes exhibit high Faradaic activity (comparable with the Pt electrode), anodic photocurrents upon irradiation with visible light. Films with greater thicknesses (0.05–1.3 µm) of poly(Cu-phthalocyanines) on various electrodes are synthesized by the reaction of tetracarbonitriles with thin Cu films [283]. The electrochromic electrochemical reduction/reoxidation was investigated. SiO_2, Al_2O_3 and active charcoal loaded with Co^{2+} were reacted with various tetranitriles to get poly(Co-phthalocyanines) on these carriers (range of loading 0.3–10 wt.%) [284]. These composite materials are highly active as catalysts in the mercaptan oxidation (Merox process). Also, water-soluble polyphthalocyanines (with Co(II), Fe(II) and Ni(II)) has been synthesized and investigated for the mercaptan oxidation [287].

2.3.2 Ligand and Metal as Part of a Polymer Chain or Network (Metal Coordination Polymers)

A bivalent ligand with donating groups **L** and a metal ion **M** are constructing a polymer backbone (Scheme 2.54).
In coordination polymers a bifunctional (or higher functional ligand) is surrounding the metal ion in order to construct a polymer chain (or network). Depending on the atoms of the ligand in the repeating unit of the polymer it is possible to divide into inorganic coordination polymers or organic coordination polymers.

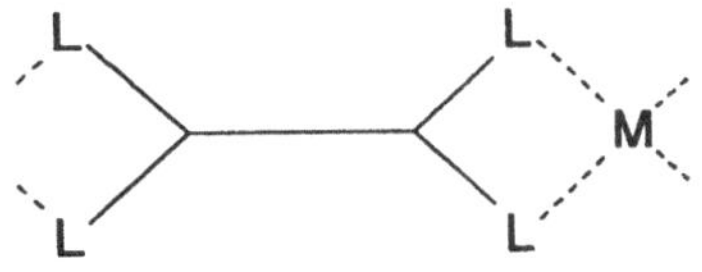

Scheme 2.54

2.3.2.1 Inorganic Coordination Polymers

Inorganic polymers are reviewed in [288–292] of transition metals with very simple formulae (e.g. AuJ, $PdCl_2$, MoJ_3, AuCN) are, in fact, as solid linear or cross-linked owing to coordination of halide, or pseudohalide bridges [291, 292]. The structure of cyano complexes exemplarily are shown (Scheme 2.55): chain structure of AuCN [293], plane structure of $Ni(CN)_2 \cdot NH_3 \cdot \frac{1}{4} H_2O$ (metal ion coordinated by four cyano groups), NH_3 coordinating perpendicular to the polymer plane, H_2O located between the layer) [294, 295], Prussian Blue $Fe_4[Fe(CN)_6]_3 \cdot H_2O$ with three-dimensional network (Chap. 5) [296]. Another material with a two dimensional ordered structure consisting of $Ni(CN)_4^{2-} - Cu^{2+}$ aggregates was recently described [297].

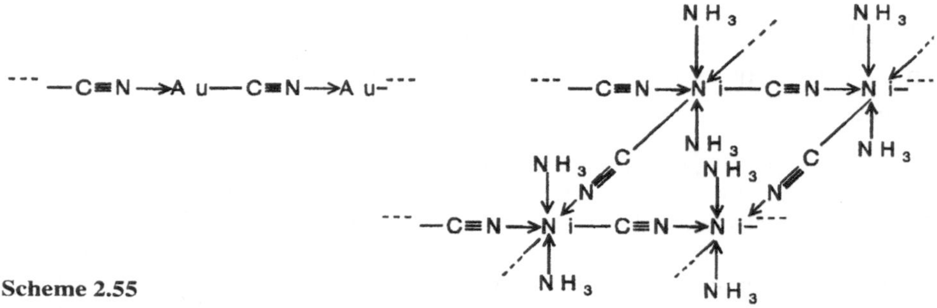

Scheme 2.55

Prussian Blue is obtained by mixing dilute equimolar solutions of $K_4Fe(II)(CN)_6$ and $FeCl_3$ as colloid with an average diameter of 23 nm and M of approximately $7 \cdot 10^6$ [296]. This compound can be coated on electrodes by cathodic electrodeposition. Properties as photosensing device, rechargeable battery material, memory device and electrochromic display has been described [298–300]. A polymer with (Fe–OH) units and $\bar{M}$ of approximately $1.4 \cdot 10^5$ is prepared by hydrolysis of ferric citate with bicarbonate [301–303]. Thermally stable polymeric metal phosphinates prepared from an appropriate metal compound and phosphinic acid derivatives (RR'(P(OH)) with $\bar{M} \sim 150000$ are interesting as viscosity stabilizers for oil, grease thickness, antistatic agent and corrosion resistant coating for metal surfaces ([304] p. 185) [305, 306].

2.3.2.2 Organic Coordination Polymers

A bifunctional ligand containing two separate ionic groups (carboxylate, alcoholate, thiolate and others for ionic/electrostatic interactions) or two coordinating donor groups (= O, =S, =NR, –NR$_2$ and others for Lewis base– Lewis acid interactions) or both kinds of groups in one ligand system are the presupposition for constructing a polymeric metal complex. The synthesis of several of these complexes have been described [223, 279, 304, 307–311, 353]. Approximately 20 years ago the main interest depended on the expectation of high thermal stability of these coordination polymers. But in many cases thermal and also moisture and oxidative stabilities fell below expectation. Other properties such as catalytic activity, conductivity and photoconductivity electron and photon-electron transfer of modified electrodes, selective ion-binding capabilities of ligands and visible-light-energy conversion, have been reported. All these properties depend on the kind of ligand, metal ion, solubility or film-processing possibilities.

Aliphatic and aromatic dicarboxylic acids are employed as bifunctional ligands, e.g. the reaction of $SnCl_2$ with sodium salts of dicarboxylic acids [312]. Depending on the pH and temperature of the reaction mixture these polymers are spherolithic, can be spined to fibres or used as adhesives. Polymeric bis(carboxylato)dioxouranium(VI) (Scheme 2.56) prepared according to two different methods (homogeneous or interfacial reaction conditions) with $\overline{M} \sim 1.8 \cdot 10^4$ are interesting to elucidate the effect of heavy metals as radiation chemistry of macromolecules and to remove uranyl ion from water ([308] p. 15; [313]).

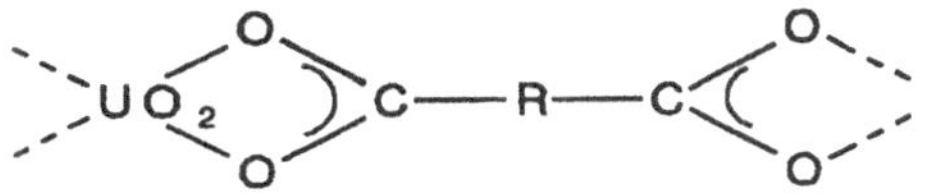

Scheme 2.56

Some other polymeric chelates consisting only of oxygen-containing functional groups (for a review see [314]) are briefly mentioned: Polymers from aromatic bis(o-hydroxy acids) and transition metal ions [315]; polymers of tetraacetylethane with Cu^{2+}, Ni^{2+}, Co^{2+}, Zn^{2+}, Cd^{2+} and Mo^{2+} ([316] p. 213); polymers from oxalic acid, squaric acid, 2,5-dihydroxy-p-benzoquinone [317, 318], 5,8-dihydroxy-1,4-naphthochinon [319], Al(III) complexes from 1,4-dihydroxyanthraquinone [320], 2,4-dihydroxybenzaldehyde-formaldehyde [321] and 2,4-dihydroxybenzophenone–urea–formaldehyde condensates with Cu^{2+}, Ni^{2+}, Co^{2+}, Mn^{2+}, Zn^{2+}, VO^{2+} and UO_2^{2+} [322]. Most polymers are insoluble and infusible. Potential applications are not known. If these coordination polymers can be coated on surfaces they may be interesting as catalysts and electrocatalysts.

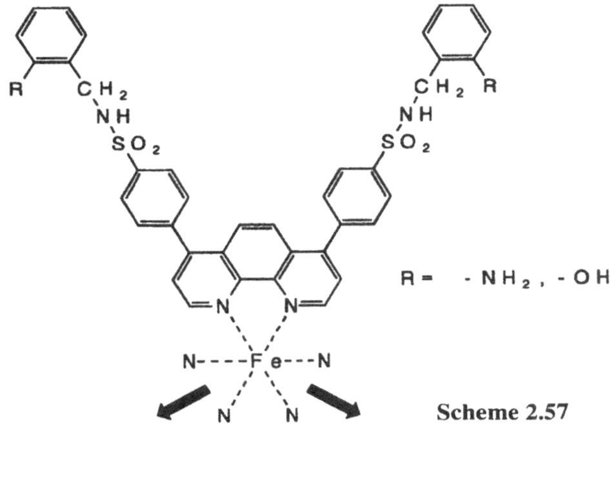

R = -NH₂, -OH

Scheme 2.57

Fe–phenanthroline complexes [Fe(Phen)$_3$]$^{2+}$ (Scheme 2.57) bearing aminophenyl or hydroxyphenyl groups connected via a spacer group were elec-
trochemially polymerized to thin films on ITO by potential sweep electrolysis [323]. The choice of such suitable low molecular weight metal complexes
and the film-forming techniques is a good progress. The films exhibit electrochromic properties. The properties were investigated by several electrochemical techniques.

Coordination polymers of tetrathiolates with $M = Fe^{2+}$, Co^{2+}, Cu^{2+}, Pd^{2+} and
Pt^{2+} have been prepared from tetrathiooxalate (Scheme 2.58) [324], tetrathio-
squarate [325], tetrathiofulvalene tetrathiolate [326; 327], benzene-1,2,4,5-
terathiolate [328] and naphthalene tetrathiolate [329]. Disadvantageous is that
the black powders are insoluble and infusible. The polymer metal complexes
exhibit conductivities up to 30 S/cm. The electronic structures were investigated theoretically [330]. Solubility could be achieved if the tetrathiolate
group are separated by flexible spacer groups such as alkylene carbonate or
urethane groups. The Ni complexes exhibit absorption at 830–940 nm and
three different oxidation states.

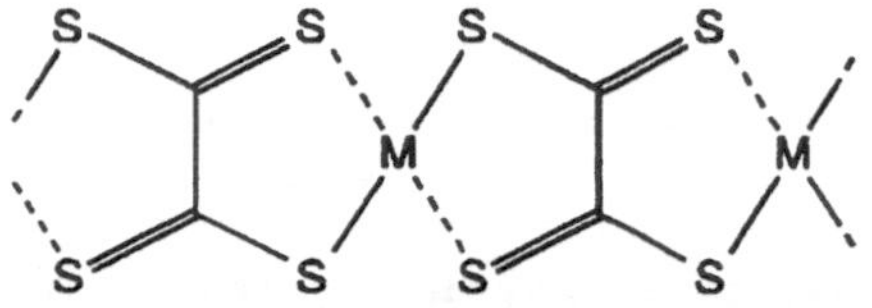

Scheme 2.58

Several coordination polymers from ligands with two different donor
atoms such as O and N, both in one bifunctional ligand, and M(II) or M(III)
salts, are known. Bifunctional ligands are: 5,8-dihydroxyquinoxalines
[331–335], bis(8-hydroxyquinolines) [336–341], bis(anthranilic acid) [342],
bis- or tetra(hydroxyazomethines) [343–346]. Most of these polymeric chelates are insoluble. Soluble polyesters prepared by polycondensation of bis[N-
alkyl-4-hydroxysalicylaldiminato-Cu(II)] with 1,10-bis[(chloroformyl)phe-
noxy]decane show different phase properties [347]. Polymeric metal complexes with bifunctional N- and S-ligand atoms are prepared from rubeanic
acid (Scheme 2.59) [348, 349], 2,5-diamino-1,4-benzenethiol [350] and poly-
(butanethiooxamide) [351]. It is interesting that electrochemical anodization
of Cu electrodes in aqueous solution of rubeanic acid leads to adherent films
of Cu-chelate polymer [352].

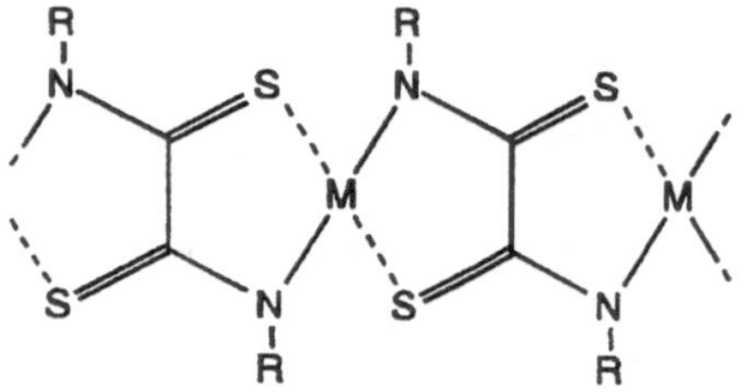

Scheme 2.59

2.3.3 Homochain Polymers with Covalent Bonds Between Metals (Homometallic Polymers)

Polymers with a homometallic M Chain are considered:

$$\cdots\cdots—M—M—M—\cdots\cdots$$

Polymers containing only a metal and, to mention in addition, semimetals in the main chain, are common with 12 elements of the groups IIIB, IVB and VB, for example, B, Si, Ge, Sn, As, Sb, Bi, Th and Po [289, 290, 353]. The bond in the main chain has covalent character, but due to weak binding energies E_b in comparison with the C–C bond ($E_b = 335$ kJ/mol), the chains often exhibit no great chemical and photocatalytical stability (Si–Si $E_b = 188$ kJ/mol; Sb–Sb $E_b = 178$ kJ/mol). Because semimetals are mainly involved in these homoatomic polymers, only a brief summary is given.

A general procedure for the preparation is the polycondensation of bifunctional monomers such as dihalogen compounds (Scheme 2.60). As known for carbon-heterochain polycondensates (e.g. polyesters and polyamides) cyclic compounds may exist in an equilibrium with linear polymers

n hal-M(R_x)-hal + 2 nA → (M(R_x))$_n$ + 2n Ahal

with R = alkyl, aryl (x = 1,2) and A = alkali metals

Scheme 2.60

Polydiorganosilicones (polysilanes) have been most intensively investigated: new thermal precursors to β-silicon carbide, new class of photo- and charge-conducting materials, radiation-sensitive materials for microlithography, photo-initiators for free radical vinyl polymerization and materials with interesting nonlinear optical properties. Preparations and properties have been reviewed [290, 353–356] and are not discussed here. The synthesis is based on the coupling reaction of Wurtz-type between an excess of alkali metals (mainly Na) and different dichloro-substituted silanes with two organic-carbon residues.

Poly(diorganogermylenes) are obtained for example from diphenyldichlorogermane with sodium under formation of a cyclic tetramer, which is then termally treated [357–360]. A mixture of cyclic and linear poly(dimethylstannylenes) have been prepared by the reaction of dimethyltin dichloride with sodium or [(CH$_3$)$_3$Si]$_2$Hg [361–363]. The synthesis of poly(methylarsinidine) and poly(methylstibylene) was realized by condensation reactions employing CH$_3$AsH$_2$ or CH$_3$SbH$_2$. For these two polymers a ladder structure in the solid by intermolecular polymer interaction is assumed [357, 364–369]. In Scheme 2.61 the structure unit of these polymers are shown.

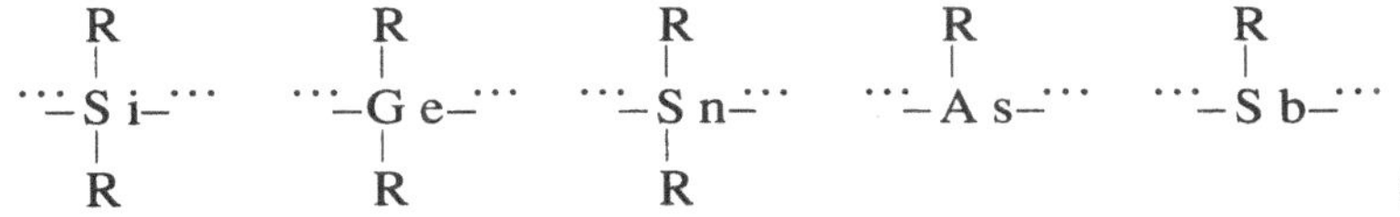

$$\begin{array}{ccccc}
\text{R} & \text{R} & \text{R} & \text{R} & \text{R} \\
| & | & | & | & | \\
\cdots-\text{S i}-\cdots & \cdots-\text{G e}-\cdots & \cdots-\text{S n}-\cdots & \cdots-\text{A s}-\cdots & \cdots-\text{S b}-\cdots \\
| & | & | & & | \\
\text{R} & \text{R} & \text{R} & &
\end{array}$$

Scheme 2.61

2.3.4 Heterochain Polymers with Covalent Bonds Between Metals and Another Element (Heterometallic Polymers)

In this case a metal **M** and another Element or group **X** are sharing the backbone of a polymer chain under formation of a polar covalent bond:

$$\cdots\cdots—M—X—M—X—M—X—\cdots\cdots$$

Polymers that posses a covalent bond between a metal, and to mention shortly in addition, semimetals are considered. Because of different electronegativities of the involved elements in the chain the bonds have a polar character. In most cases a polycondensation reaction is used for the synthesis of these polymers. One possibility for the preparation of linear polymers is the reaction of a bifunctional metal halide (mainly Cl, but also Br and J) with a bifunctional Lewis base. Diols, diamines, dihydrazines, dihydrazides, dithiohydrazides, dioximes, diamidoximes, dithiols and diacetylenes are empolyed as Lewis bases. Besides halides also hydroxyl and amino can be the leaving group in the organometallic compound. The mechanism involves nucleophilic attack of the Lewis base at the metal compound. In Scheme 2.62 a survey for the reactions is given [353].

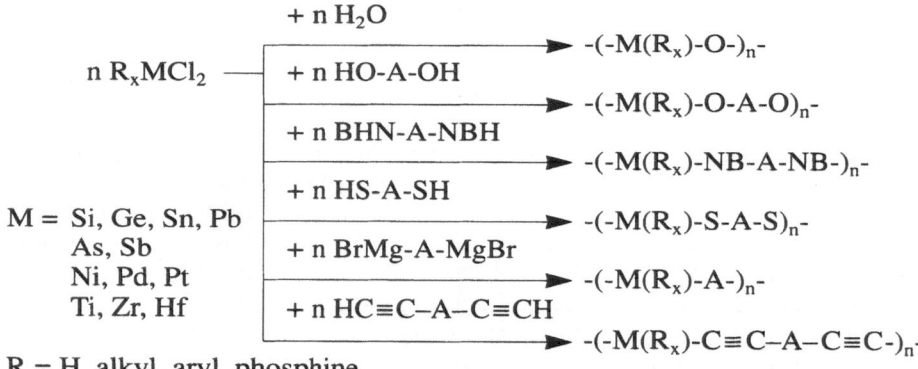

R = H, alkyl, aryl, phosphine
B = H, alkyl, aryl,
A = alkylene, arylene or other -M(R$_x$)-

Scheme 2.62

Polyaddition reactions of bifunctional metal hydrides with bifunctional alkenes can also lead to metal containing heterochain polymers (Scheme 2.63) [353].

$$nH-\underset{\underset{R}{|}}{\overset{\overset{R}{|}}{M}}-H + nH_2C=CH\text{-}R'\text{-}CH=CH_2 \rightarrow \cdots-\underset{\underset{R}{|}}{\overset{\overset{R}{|}}{M}}-H_2C-CH_2\text{-}R'\text{-}CH_2-CH_2-\cdots$$

Scheme 2.63

Another possibility uses polycondensation or polyaddition reactions of dicarboxylic acids, dialcohols or diamines containing a metal/semimetal as part of this bifunctional compound with another bifunctional compound [353]. The model reactions of monofunctional compounds for condensation or addition reactions allow study of the structure of products by instrumental analysis and to use the experienes for the analysis of the polymers. Some reviews summarize results in the field of heterometallic chain polymers ([307] pp 79–86, 101–106; [370, 371], [309] p. 208).

Polymers with boron [316, 353, 372] and poly(organosiloxanes) (silicones) [353, 372–382], polysilylalkylenes/arylenes) (polycarbosilanes) ([290, 372] p. 1783; [353]; [377] pp 478, 487; [292] p. 332; [383]) and poly(oxyorgano-germylenes) (and others) ([359]; [292] p. 353; [384–389] are not discussed here.

For the preparation of polymeric organotin compounds the general methods mentioned previously can be used: reaction of dichlorodiorganostannylenes and diols ([292] p. 368; [390]), polyaddition of diorganotin dihydrides with dienes or diynes ([292] p. 377; [391–393]), poly(alkynylstannylenes) by polycondensation [353]. The molecular weights M of the polymers (with bulky groups soluble) are on the order of 3000–7000. These polymers are interesting as stabilizers of plastics, antibacterial and antifungal properties, additives in clothings or paints to inhibit rot and mildew, and – in the case of the alkynyl derivative – for microlithographic applications [391].

An increasing interest finds polymers with transition metals in the main chain mainly σ-bonded to carbon atoms [305, 353, 394]. The stability of σ-bond depends largely on the kind of central metal, alkyl groups and other ligands. CO, PR_3 and bulky organic rests as ligands at the metal stabilize the bond [395, 396]. Also, interaction between the d-orbital and the π-orbital of the ethynyl carbon of $[M(C \equiv C)_n]$ results in an increased stability of the metal carbon bond.

Three examples for transition metal poly-yne polymers, for example those shown in Scheme 2.64, also without an additional ligand at the metal are mentioned: dehydrohalogenation between α, ω-bisethynyl compounds (Cu(I) – halide catalyzed) [305, 397–399], oxidative coupling of bisethynyl compounds containing a transition metal (CuCl catalyzed) [400], nickel-containing polymers by alkynyl-ligand exchange reaction (Cu – halide catalyzed) [401]. The polymers, if the organic residue Y contains no bulky groups, are often insoluble and decompose in time at room temperature or by heating.

$$[-M-C \equiv C-Y-C \equiv C]_n-$$

Scheme 2.64

By the reaction of *cis-* and *trans-*L_2MCl_2 (L = various phosphines, M = Pt, Pd) with bisacetylides (also containing better leaving groups than hydrogen) soluble metal-poly-ynes are obtained (Scheme 2.65) [305, 402, 403]; ([372] p. 1314 and references cited therein). By measurements of electronic spectra and luminescense it is shown that electron conjugation is expanded over the whole polymer chain [404]. Soluble polymers related to the structure in Scheme 2.64

with $M=Pt(PC_4H_9)_2$ and $Y=1,4$-phenylene (Scheme 2.65) have lyotropic nematic *meso*-phases [405]. Results of viscosity and sedimentation velocity led to persistence length of 13 ± 3 nm and suggest worm-like chains in solution. In the solid state polymers with $M=Pt(AsC_4H_9)_3$ and $Y=1,4$-phenylene form crystallites with a diameter of ~50 nm [406]. Unit cell parameters were determined. Nonlinear optical properties (third harmonic) are greater in comparison with analogous poly(diacetylenes) [407].

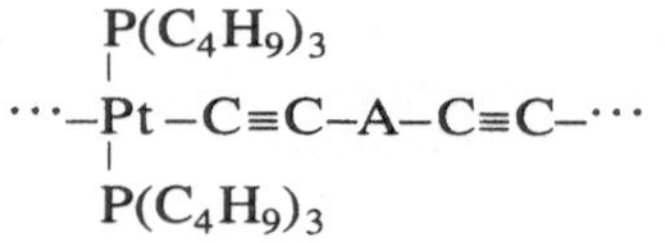

$$P(C_4H_9)_3$$
$$\cdots-Pt-C\equiv C-A-C\equiv C-\cdots$$
$$P(C_4H_9)_3$$

Scheme 2.65

Poly-ynes with iron and two metals or semimetals (Pt, Si) have been described [408, 409]. Poly(arylene cobaltcyclopentadienylenes) as π-conjucated polymers were obtained by the reaction of various diacetylenes with $CpCo(PPh_3)_2$ (Scheme 2.66) [410]. The polymers possess small band gaps and low ionization potentials.

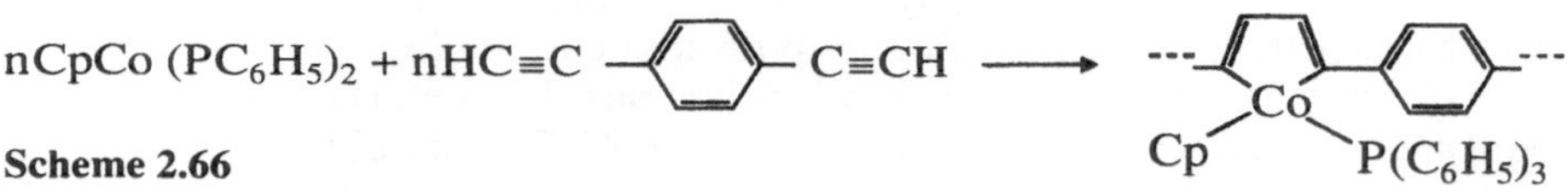

$$nCpCo\,(PC_6H_5)_2 + nHC\equiv C\text{—}\langle\bigcirc\rangle\text{—}C\equiv CH \longrightarrow$$

Scheme 2.66

2.3.5 Cofacial Stacked Polymeric Metal Complexes

A metal ion **M** surrounded by a multiligand are parts of a polymer chain forming stacked arrangements by different bonds to an element or residue R:

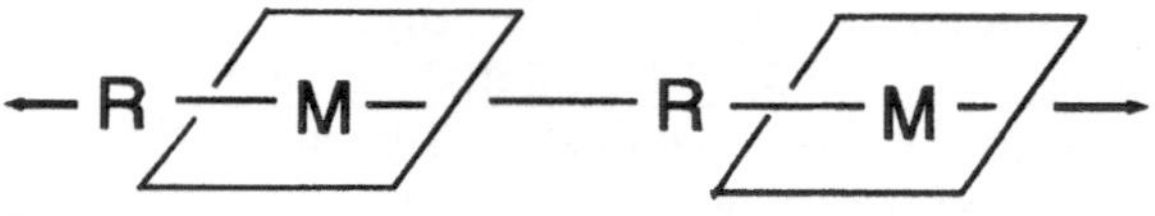

$$-R\diagdown M-\diagup-R\diagdown M-\diagup-$$

Scheme 2.67

Several papers have been published concerning a polymeric arrangement of metal complexes in a stacked or face-to-face orientation. Preferable are metal complexes where the ligand is surrounding the central metal ion in a planar arrangement. The driving force for such an arrangement is the interaction between the metal ions in most cases separated by a bridging group L. Simply, a classification can be given as follows:

1. Covalent/covalent bonds between central metal ions

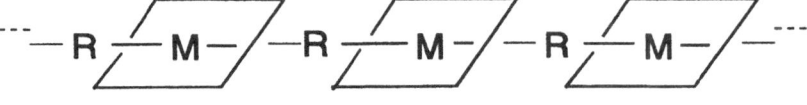

M = Si(IV), Ge(IV), Sn(IV)
R = -O-, -S-, -O-R$_1$-O-, -C≡C-, -N=C=N-

Scheme 2.68

2. Covalent/coordinative bond between central metal ions

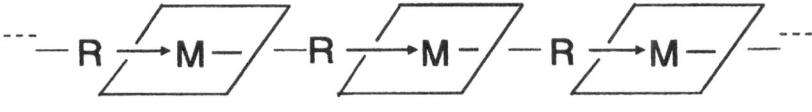

M = Al(III), Ga(III), Cr(III)   R = -F$^-$
M = Fe(III), Co(III), Rh(III), Mn (III), Cr(III)   R = C≡N$^-$, S-C≡N$^-$, N$_3^-$

Scheme 2.69

3. Coordinative/coordinative bonds between central metal ions

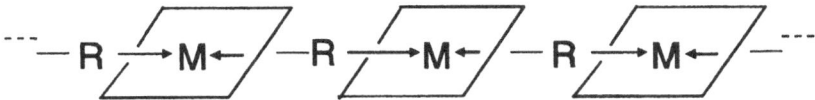

M = Fe(II), Co(II), Ru(II); Os (II)   R = pyrazine, tetrazine, triazine,
p-diisocyanobenzene, p-phenylenediamine, fumaronitrile

Scheme 2.70

4. Self-organization, discotic crystalline liquids

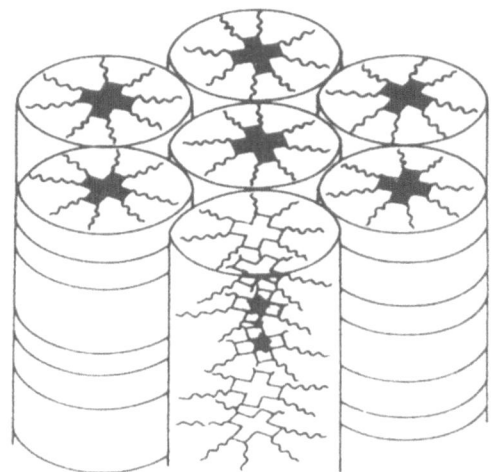

Scheme 2.71

Some of the employed metal complexes for these polymeric compounds are exemplarily phthalocyanines, 2,3- and 1,2-naphthalocyanines, 5,10,15,20-tetraarylporphyrines, hemiporphyrazines, tetraaza(14)annulenes, etioporphyrin, etc. The interest in the properties, which are discussed in Chapter 5.3, concentrates on conductivity, photoconductivity, nonlinear optical properties, photovoltaic cells, electroluminescent dioles, gas detectors.

Some reviews summarize preparations and properties of these cofacially stacked materials [253, 328, 353, 372, 411, 412]. The first polymers of this type described several years ago consist of unsubstituted macrocycles. Because of the insolubility, a full characterization was difficult. Meanwhile, substituted macrocycles were employed to enhance the solubility of the corresponding polymers. Now structural information on the intensively coloured polymers can be derived from IR, UV/VIS, solid-state NMR, X-ray, Mössbauer method (for Fe-containing polymers), etc. The stability of these polymeric metal complexes is given by the bond in the stacking chain and increases therefore in the order (Scheme 2.71) < (Scheme 2.70) < (Scheme 2.69) < (Scheme 2.68).

2.3.5.1 Covalent/Covalent Bonds Between Central Metal Ions

Simple μ-oxo-bridged IVB polymers (Scheme 2.68) are prepared by polycondensation of the dihydroxides M(OH)macrocycle (M = Si(IV), Ge(IV) and Sn(IV); macrocycle = phthalocyanine, hemiporphyrazine, tetrabenzporphyrin, tetraphenylporphyrin, etioporphyrin) through heating in bulk at $325 - 440\,°C$ or high boiling solvents (1-chloronaphthalene, quinoline) [413 – 428]. One example is catena-poly[phthalocyaninatosilicium(IV)-μ-oxo]. A topotactic polymerization mechanism has been proposed (Chap. 2.1.3) [429]. The degree of polymerization determined by IR or tritium labelling increases with longer reaction time: 1 h $\bar{P}_n = 66$ and 120 h $\bar{P}_n = 139$ shown for $[Si(O)Pc]_n$. Working in solvents, dehydrating additives or catalysts, such as benzoylchloride, $ZnCl_2$, P_2O_5 or $FeCl_3$, ban be added [417, 419].

The presence of bulky or longer-chain substituents in the macrocycles, such as phthalocyanine, enhances the solubility of the polymers: For example, the μ-oxo-phthalocyaninatosilicon compound $[Si(O)PcR_8]_n$ contains R like longer-chain oxyalkyl, methyleneoxyalkyl, crown ether and ester groups [425, 430–439]. Instead of $Si(OH)_2PcR_8$ trifluoracetylated monomers $Si(OCOCF_3)_2PcR_8$ were employed [433]. The degree of polymerization of these soluble polymers can be higher than 150. $[Si(O)PcR_8]_n$ with ester groups R exhibit as determined by GPC $\bar{M}_n$ 140000 and $\bar{M}_w$ 360000 [438]. The rotational dynamic behaviour of $[Si(O)PcR_8]_n$ was invetigated in the solid state by two-dimensional NMR spectroscopy [440]. It was shown that the SiPc moieties can rotate round their covalent Si−O bond with a single macromolecule, rather than the molecules as a whole around their columnar axis.

Polymers with longer-chain groups R were investigated for their behaviour of Langmuir-Blodgett film formation [432, 437, 438] and liquid crystalline properties [433, 436, 438]. Alkoxy substituted poly[phthalocyaninatosilicium-

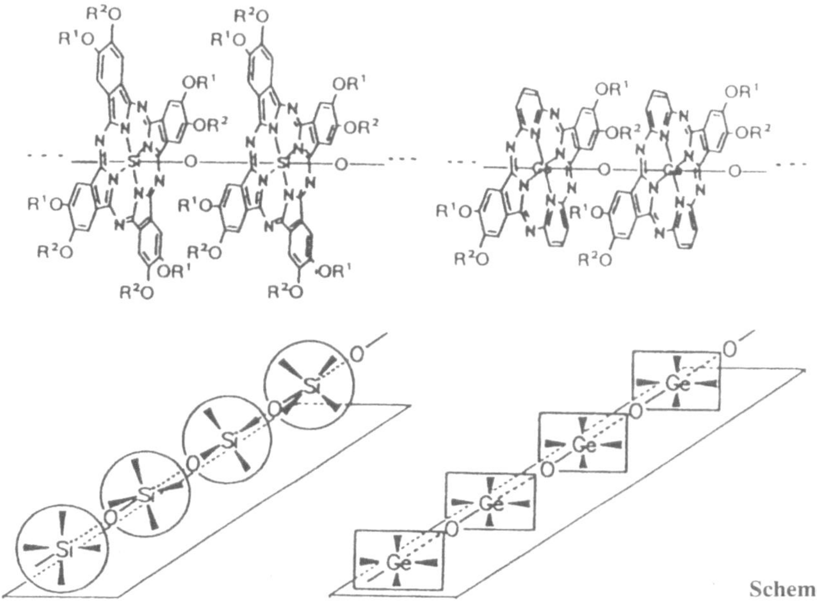

Scheme 2.72

IV)-μ-oxo] and poly[hemiporphyrazinatogermanium(IV)-μ-oxo] were prepared as thin films by the LB technique and were investigated by UV/VIS (polarized light), dichroism in the IR and X-ray under different angles. Arrangements as shown in (Scheme 2.72) were found resulting in supramolecular structures on Si wafers [434].

The following polymers with analogous stacked structure were described: $[Si(S)Pc]_n$ [428, 441, 442], $[Si(O(CH_2)_xO)Pc]_n$ [417, 418, 427], $[Si(Oary\text{-}leneO)Pc]_n$ [417, 418, 427], $[Si(C{\equiv}C)Pc]_n$ [424, 444, 445] and incorporation into commercial polycondensates [443]. Some metal-oxo-bridged polymers are exceptionally stable materials. They are unaffected by aqueous HF at 100 °C, aqueous 2 M NaOH at reflux, and concentrated H_2SO_4 at room temperature.

2.3.5.2 Covalent/Coordinative Bonds Between Central Metal Ions

Other stacked phthalocyanines are the bridged poly(fluorophthalocyaninato) metal complexes $[M(F)Pc]_n$ with M = Al(III), Ga(III) and Cr(III) prepared by thermal treatment at higher temperatures (Scheme 2.69) [446–448].

Another type of polymer contains three-valent transition metal ions such as Co(III), Fe(III), Mn(III), Cr(III) and Rh(III) with -C≡N as bridging group (described in [353, 372, 411, 412] and in detail in [449–454]) (Scheme 2.69). Different routes have been described for the preparation of these cyano-linked cofacial stacked polymers [353]. A more general route is the splitting off of alkali-metal cyanide from alkali $[M(CN)_2Pc]^-$. The polymeric structure is destroyed when treating the materials with a competing ligand (pyridine, butylamine, etc.) under formation of $M(L, CN)_2Pc$. A cyano-bridged polymer with eight methyleneoxyalkyl groups at the phthalocyanine exhibit liquid crystalline properties [449]. Thiocyanato-bridged polymers are obtained analogously [454].

2.3.5.3 Coordinative/Coordinative Bonds Between Central Metal Ions

Polymers of this type contain macrocycles with transition metal ions in the oxidation state +2 capable of hexa-coordination and neutral organic donors containing two groups or heteroatoms for coordination (Scheme 2.70). Examples of unsubstituted and substituted macrocycles are summarized in [353, 372, 411, 412] and in detail described in [455–471]. The preparation of the poly[phthalocyaninato)-μ-pyrazine-iron(II)] is exemplarily mentioned. One possibility is the heating of Fe(II)Pc with pyrazine in a closed bomb vessel and afterwards removing the excess of pyrazine in vacuo [456, 457]. Another way uses heating of Fe(pyz)$_2$Pc in an organic solvent. At the phthalocyanine with bulky groups substituted polymers are soluble in most common organic solvents. The degree of polymerization determined, for example, by NMR, depends on the preparation method between 20 and 50. By an excess of a strong donor these polymers are destroyed by formation of hexa-coordinated bis-adducts.

2.3.5.4 Self-Organization, Discotic Crystalline Liquids

Metallo-*meso*-genes are metal complexes forming liquid crystalline *meso*-phases, and several examples are reviewed [472]. Phthalocyanines substituted in the peripheric positions by long-chain substituent form, depending on the structure, liquid discotic crystalline phases or columnar phases with metal – metal contact (Scheme 2.71) [412, 473–477]. Liquid crystalline behaviour is seen for side chains longer than C$_4$ or C$_6$ depending also on the metal ion and substituent. The arrangement in detail depends on the nature of the side chains [476, 478]. Chains connected via O or S to the aromatic macrocycle exhibit an arrangement with the molecular planes to the columnar axis. Above ~ 80 °C these molecules form the columnar *meso*-phases. This transition from the solid to liquid crystalline phases corresponds to the melting of the flexible chains and the macrocycles retain positional and orientational order. The transition from the *meso*-phase to the isotropic liquid corresponds to the destruction of the columns. Extension and branching of the aliphatic chains depresses the transition temperature, branching in the middle of the chains depresses the melting point and branching of the chains near the aromatic core depresses the clearing point [412, 479, 480]. Polymerization of the liquid-crystalline phthalocyanines in their *meso*-phase can lead to supramolecular structures in which the columnar orientation is fixed [474, 481, 482].

Stacking is also realized by the help of crown ether rings, which are attached to the phthalocyanines molecule [474, 483]. Addition of metal ions with diameters exceeding those of the crown ethers induces the formation of linear stacks of metallophthalocyanines. Other examples are water-soluble amphilic porphyrine amides, amines and carboxylates forming in aqueous media fibrous assembles [484]. The porphyrin fibres are approximately 4–6 nm thick and up to several μm long. Metal chelates can be anisotropically aligned in bilayer assemblies [485].

2.3.6 Metallocenes as Part of a Polymer Chain (Polymetallocenes)

Bifunctional π-electron-rich charged aromatics are connected by π-bonds in a polymer chain:

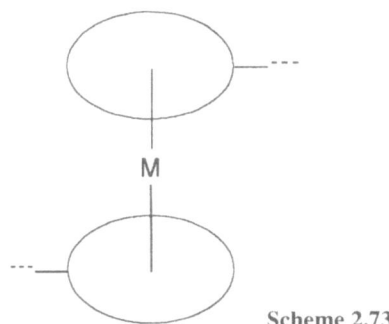

Scheme 2.73

After the discovery of ferrocene 1950 and then similar metallocenes, such as ruthenocene, osmocene, titanocene and cobalticenium, the technological interest in areas such as thermal stability, radiation protection, combustion catalysis, rubber vulcanization and redox reactions stimulated scientists to synthesize various kinds of polymers containing metallocenes (for reviews see [309, 353, 486–491]. Some examples are now mentioned.

Metallocenylene polymers containing only metallocenes in the positions 1,1' or 1,2 (Scheme 2.74) range from orange – tan to dark brown (Fe compounds) or white (Ru compounds) in colour, soluble in polar solvents and high melting. The synthesis was critically reviewed [488]. Bifunctional metallocenes were applied as starting materials in a step-growth polycondensation. Low-temperature coupling of 1,1'-dilithioferrocene with 1,1'-diiodoferrocene results in poly-1,1'-ferrocenylenes with molecular weight of ~4000 (10000 upon subfractionation) [488, 492]. Also, poly-1,1'-ruthenocenylenes with $\overline{M}_n$ <2000 were described [488, 492, 493]. Poly(1,2–ferrocenylenes) were obtained in low yield [494, 495].

A number of polymeric metallocenes (Scheme 2.75) in which the repeating metallocene units are held face to face (polydecker sandwich complexes) by a naphthalene spacer are prepared by a monomer dianion through its reaction with a transition metal ion [489, 496]. Also, polymeric cobalt complexes with chiral heptahelicene ring systems containing cyclopentadienyl rings as bridging elements were reported [489]. Air-sensitive black polydecker sandwich

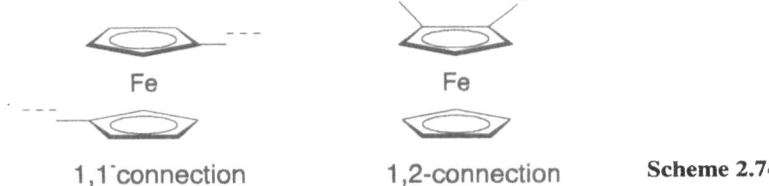

1,1'connection   1,2-connection   **Scheme 2.74**

complexes containing 2,3-dihydro-1,3-diborole ligands and Rh^{2+} are described [491]. Poly(metallocenylenemethylenes) are obtained, for example, by polycondensation of -CH_2-X-substituted metallocenes (e.g. R-CH_2OH, R-$CH_2^+BF_4^-$) [487, 497−501]. A new promising concept for soluble high molecular weight polyferrocenylenearylenes uses the Pd catalyzed polycondensation of 1,1'-bis(4-bromophenyl)ferrocene with 2,5-dialkylbenzene-1,4-diboronic acid [645].

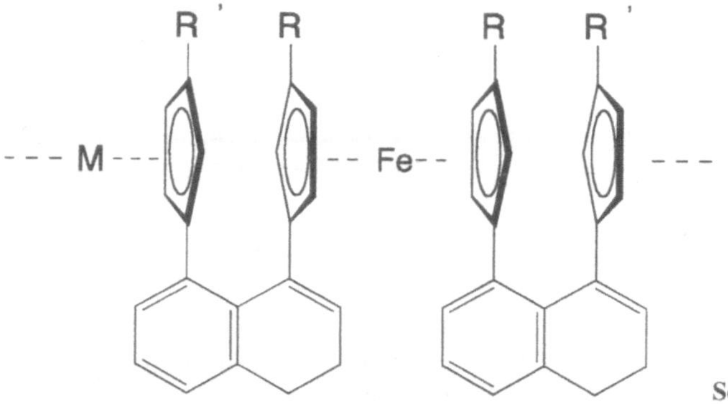

Scheme 2.75

Poly(ferrocenylethylene) and polymers with ferrocene units bridged by SiR_2, GeR_2 instead of $(CH_2)_n$ of high molecular weight can be readily prepared by stoichiometric ring-opening polymerization reactions of the corresponding ferrocenophanes [490, 502]. Electrical conductivity and redox behaviour measured by cyclic voltammetry are investigated and indicate cooperative interactions between the iron centres.

Numerous polymers containing beside metallocenylenes, functional groups are described (for reviews see [353, 487, 503]). Polyesters, polyamides, polyurethanes, polymers with soft segments based on silicon are prepared by normal polycondensation or polyaddition reactions of 1,1'-disubstituted metallocenylenes and the corresponding bifunctional other monomers. These polymers combine the advantages of the hard segment metallocene (for good stability, resistance to UV and γradiation) and advantages of the organic bridging groups (for flexibility, solubility and thermal transitions). X-CpMo(CO)$_3$-Mo(CO)$_3$-CpX (Cp-cyclopentadienyl; X = -CH_2CH_2OH, -$COCH_2OH$) were reacted with diisocyanates to polyurethanes [504]. The metal−metal bond can be split by irradiation in solution.

2.4 Type III: Metal Complexes, Zero-Valent Metals and Metal Clusters Physically Connected with Macromolecular Compounds

D. Wöhrle

This chapter consider metal complexes and metal clusters not covalently bonded at a or as part of a molecular chain/network. The "physical" incorporation into or anchoring on a macromolecular system produces interactions between the host–guest materials that consist mainly of various secondary binding forces: physical adsorption by van der Waals interactions, coulombic or electrostatic interactions, charge-transfer or hydrophobic interactions, coordination or hydrogen bonding [505]. These multiple and dynamic interactions influence the properties of the composite material and are difficult to describe. Of great interest is the incorporation into organic polymers as host materials. In addition, few examples of the interaction of metal complexes with high molecular inorganic hosts, such as the surfaces of inorganic oxides, the framework of molecular sieves and clays are discussed. Besides the guest-compound metal complexes, metal clusters stabilized by the organic or inorganic high molecular environment is also a point of discussion.

2.4.1 Combination with Organic Polymers

2.4.1.1 Metal Complexes

In principle every metal complex (or metal salt) can be incorporated into an organic polymer monomolecular or aggregated with the result of a solid composite material. Because of several hundred papers that have been published in this field, the following survey considers only few examples.

Film preparation from a solution containing the polymer and metal complex homogeneously dissolved or the dissolved polymer and dispersed metal complex on a suitable carrier [Pt, Au, carbon (in the form of graphite or glassy carbon) and inorganic semiconductors (ITO, SnO_2, Si, GaAs)] were carried out. The layer thicknesses varied a great deal, extending from approximately, 50 nm to a few μm. The number of active centers of a metal complex in the films coated on a carrier is on the order of 10^{-10}–10^{-6} mol/cm^2. The apparent concentration of a metal complex in a polymer is often as high as 01.–5 M. The different methods as reviewed in [506] are:

1. Casting from solution. The solution is spread onto a carrier followed by careful evaporation of the solvent.
2. Spin coating. A small amount of solution is dropped onto a spinning carrier. The thickness of a film depends on the rotation speed, the evaporation rate of the solvent and the initial viscosity of the solution.
3. Dip coating. A carrier is dipped into a solution and then dried. Higher concentration and longer soaking time yield thicker coatings.
4. Coating and adsorption process. First, a layer of the polymer is coated onto a carrier from solution. Then, in a second step, this coated material is placed in contact with an aqueous solution containing the metal complex for a few minutes or several hours. In Fig. 2.8 the time dependence of the coordinative attachments of Ru(III)-EDTA from solution to a poly(vinylpyridine) coating is shown [507]. After 2000 s 40% of the pyridine groups exhibit coordination of the metal complex.

In order to obtain a homogeneous distribution of the metal complex and a smooth film formation, the experimental conditions must be optimized carefully. Low

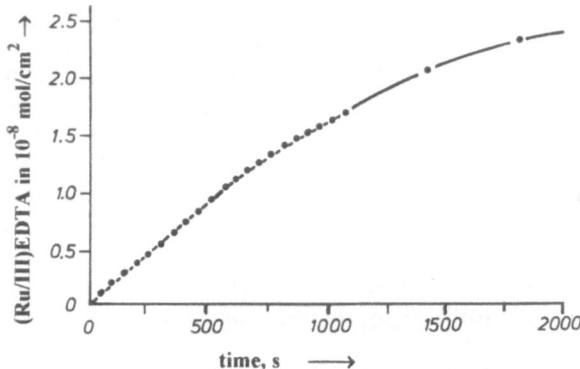

Fig. 2.8. Time dependence of the attachment of Ru(III)-EDTA to a graphite electrode coated with $6 \cdot 10^{-8}$ mol/cm^2 pryridine groups in poly(vinylpyridine). $2 \cdot 10^{-4}$ M Ru(III)-EDTA dissolved in an aqueous solution containing 0.2 M CF$_3$COONa (After [507])

concentrations of a metal complex ($<1-10$ wt.%) in a polymer can result in mononuclear distribution. The polymer appears in this case to function as a solvent that minimizes complex–complex interaction by separation. Depending on the solubility of the metal complex in the solvent used for casting higher concentrations ($>1-10$ wt.%) can result in aggregation or microcrystal formation.

Few examples for the combination of a metal complex with a polymer have been given. Tris(2,2'-bipyridine)ruthenium(II) (Ru(bpy)$_3^{2+}$) is dispersed in a Nafion membrane coated on an ITO electrode (see Chap. 5.2). By cyclic voltammetry oxidation and re-reduction of Ru(byp)$_3^{2+}$/Ru(byp)$_3^{3+}$ was studied. A critical distance of the Ru-complexes for efficient charge hopping between redox centres of approximately 1.3 nm is found and calculated. The maximum of a Poisson distribution of distances is ≥ 0.1 M of Ru(byp)$_3^{2+}$. Ru-, Re- and Co-complexes in Nafion membranes were investigated as electrocatalysts for water oxidation and carbon dioxide reduction (see Chap. 5.2.3). Intermolecular interactions of metal complexes can be seen by photophysical and photochemical measurements. Some metal complexes, such as Ru(bpy)$_3^{2+}$ or Ru(II)tris(4,7-diphenyl-1,10-phenanthroline)$^{2+}$, exhibit a high quantum yield and long life time of excited states, which is quenched by acceptors such as oxygen (energy transfer under formation of singlet oxygen) or methylviologen (1,1'-dimethyl-4,4'-bipyridinium) (electron transfer and formation of methylviologen cation radical) [508–511]. Inclusion in polymers can enhance the lifetimes because of the microenvironmental effect of the surrounding polymer and reduced bimolecular annihilation processes [509, 510].

Photo-induced electron transfer from polymer embedded Ru(bpy)$_3^{2+}$ to methylviologen (eigher coadsorbed or in solution) can be very efficient. Ru(bpy)$_3^{2+}$ in a poly(vinylalcohol) matrix exhibits excellent luminescence [508]. The polymer reduces or eliminates O$_2$-quenching. As judged by the nearly exponential decays, the local sites seen by the complex are nearly homogeneous. At very high concentration aggregation, microcrystal formation gave rise to a second exponential decay. Different binding sites lead to another behaviour. For ionically bound metal complexes the luminescence decays are somewhat nonexponential. Different Ru-complexes were adsorbed on fumed silica (surface area 200 m²/g) [508]. Only approximately 1 % of the surface area is suitable for binding interactions with metal complexes, and little apparent variation in the adsorption spectra of the complexes indicates monomolecular distribution. The quenching of the luminescence by O_2 of Ru-complexes in different polymer environments is interesting as O_2-sensor [508, 510].

The excitation and emission spectra of lanthanide metal ions, such as Tb^{3+} or Eu^{3+} in polycarboxylates (ionic interaction), were investigated in detail [513, 514]. The interest is to study the structure of ionomers in solution and as solid. Because of a linear increase of the luminescene intensity with increasing concentrations of lanthanides in a polymer film, a mononuclear distribution at lower concentrations is shown. In a copolymer of styrene and acrylic acid a decrease of the luminescence intensity at >4–6 wt.%, and in poly(acrylic acid) or poly(styrene-maleic acid) at >15 wt.%, shows formation of ionic aggregates [512].

Luminescence behaviour and electron transfer of various transition metal complexes with 4,4'-bipyridyl, EDTA or 1,2-diaminoethane in the interaction with polyelectrolytes were studied in solution in detail [515, 516, 643, 644]. The systems are prepared by mixing solutions of the polymers (in excess) and the metal complexes. The measurements show monomolecular distribution of the metal complexes, and allow study depending on the pH-value-dependent conformational transition of the polymer chains as shown in Fig. 2.9.

Co(II) and Fe(II) Schiff bases and porphyrin complexes were included by casting in various copolymers containing vinylpyridine or *N*-vinylimidazole units. Details on the preparation of thin films and their use in O_2-transporting membranes are given in Chap. 4.1 (see Chap. 4.2, 4.3).

Metal complexes such as sulfonated Co-phthalocyanines [517–522] Mn(III)-porphyrins [520] and Co–Schiff-base [522] supported on colloidal

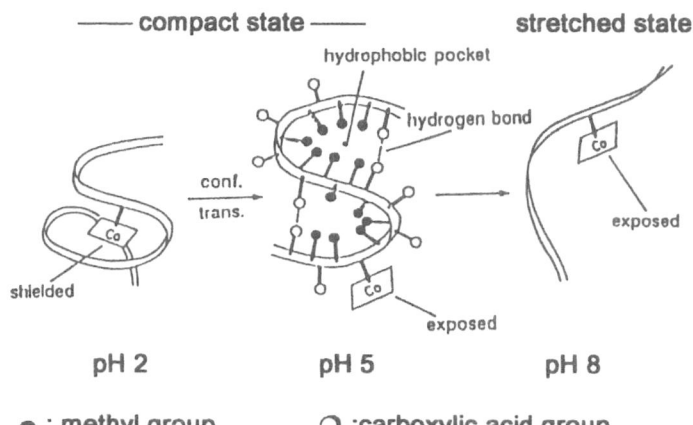

polymers with particle size diameter less than 1 μm were investigated as catalysts for oxidation reactions. The composite systems were prepared by mixing of a colloidal anion exchange resin (obtained by emulsion copolymerization of Cl-methylstyrene with 5 % divinylbenzene followed by quarternization with trimethylamine [521, 522] with an aqueous solution of the oppositely charged metal complex followed by ultrafiltration.

Co(II)-phthalocyanines are active in the electrocatalytic reduction of CO_2 to CO. Electrodes with the metal complex incorporated into a polymer exhibit higher activity and selectivity in comparison with electrodes covered only with the phthalocyanine [523, 524]. Basal plane pyrolytic graphite were coated with the phthalocyanine (10^{-10} mol/cm^2) from DMF solution in the absence or presence of poly(4-vinylpyridine-co-styrene) (ratio CoPc: pyridine residues = 1:5000). The phthalocyanine is mononuclear-distributed in the polymer, due to the coordination to the pyridine group. The CO_2 reduction studied in aqueous solution shows for the polymer modified electrode a 30-times higher CO formation in comparison with the electrode without a polymer. The high local concentration of CO_2 in the polymer and the coordination to the mononuclear CoPc favours the CO_2 reduction.

Anionic metal complexes, such as tetrasulfophthalocyanine or disulfoferrocene, are incorporated from solution during the oxidative electrochemical polymerization of pyrrole [525–528]. After re-reduction of the polypyrrole film the phthalocyanine derivative remains in the film, and the observed electrochemical conductivity is explained by ion transport of small electrolyte cations [527].

Casting of a polymer solution containing a dissolved or dispersed metal complex in a high amount leads to films containing aggregates or microcrystallites of the employed metal complex. One example is given. 10 μL of a dimethylacetamide solution containing $0.5 \cdot 10^{-4}$ mol/L zinc(II)-phthalocyanine and 0.04 g/L of a polymer (poly(vinylidine fluoride) and others) were dropped on a cleaned ITO plate [529]. The solvent was removed at 10^{-3} mbar and 70 °C. A transparent film of ~250 nm thickness and ~40 nm roughness was obtained containing 50 wt.% of metal complex. Depending on the polymer used and pretreatment of the film, different polymorphic structures, are possible [529–531]. In comparison with films of the metal complex without polymer (prepared also by casting or vapor deposition), the combination with polymers offer the advantage of higher mechanical stability. In addition, polar polymers can improve conductivity, photoconductivity [532–534] and activity in photoelectrochemical [529, 535–537] and photovoltaic cells [538] (see Chap. 5). For ≤10 wt.% metal-free octacyanophthalocyanine in polyimide, a decrease of the electrochemical activity (reduction/reoxidation of the film) is observed, which is due to the loss of interparticle contact [539].

Transition metal salts are stabilized by interaction with part of a polymer chain. Examples are poly(styrene) AlCuCl$_4$–Co complexes [540, 541], CuCl complexes at poly(styrene)-modified with amino groups [540], PdCl$_2$ and RhCl complexes at poly(styrene-co-butadiene] [542, 543], PdCl$_2$ at poly(acrylonitrile), [544], RhH(CO)(PPh$_3$)$_3$, PtCl$_2$-SnCl$_2$, RuCl$_3$-CoCl$_2$, Rh$_2$(CO)$_4$Cl$_2$ at different polymer phosphine ligands [545], PdCl$_2$ at different organic polymers (poly(benzimidazole), cyanomethylated cross-linked poly(styrene), cross-linked poly(acrylonitrile)) [547], carboxylated Co(II) and Fe(III)-phthalocyanines as rayon fibres [546]. These complexes are investigated mainly as catalysts in different reactions. Polymer (e.g. poly(ethylene oxide))/inorganic salt (e.g. alkali salts) complexes have been actively investigated for solid-state ionic conductivity to develop materials for commercial applications (battery, electrochromic devices, moisture or gas sensors) (see Chap. 4.1) [548–551]. Films are prepared easily by casting a solution of the polymer and metal salt followed by drying.

2.4.1.2 Metal Clusters

In recent years considerable reseach has been directed towards studying dispersions of metal clusters in homogeneous solution in the presence of a protecting polymer or heterogenized in a high molecular host matrix. A metal cluster is composed of several to several hundred metal atoms. Methods for the preparation of polymer-immobilized noble metal clusters were recently summarized [552]. These nanoscopic materials [553] exhibit chemical and physical properties very different from those of bulk metals and metal atoms. Besides many papers for research on catalysts with specific activity and selectivity metal clusters (and also semiconductor clusters) are interesting materials for the quantum size effect, which may be applied for nonlinear optics and quantum-size electronic devices [554–557].

For the preparation of metal clusters in homogeneous solution a general method uses the reduction of metal salts in the presence of polymers to prevent aggregation of the metals. Colloidal Pt was prepared by the reduction of K_2PtCl_6 with sodium citrate in aqueous solution [558] followed by mixing with an ion exchange resin to remove citrate ions. The liquid contains small Pt particles (average siez 3 nm) that can be stabilized by poly(vinyl alcohol) (determination of particle size by photon-correlation spectroscopy or SEM)

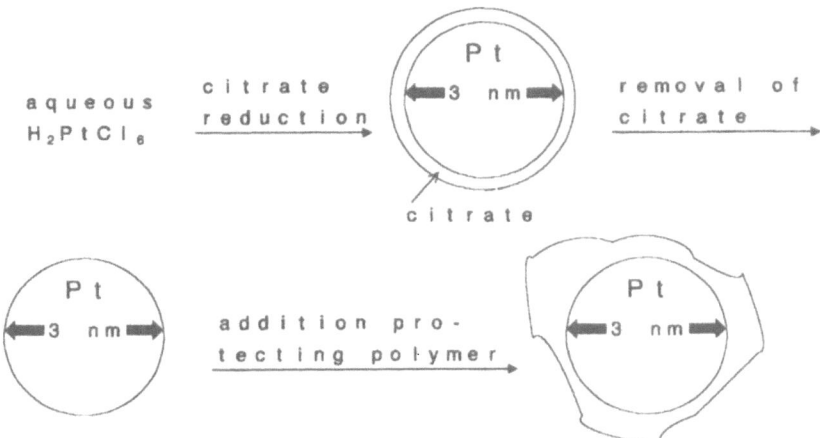

Fig. 2.10. Procedure for the preparation of ultrafine Pt clusters (After [559])

[559, 560]. The procedure is shown in Fig. 2.10. Also, colloidal particles of inorganic semiconductors are stabilized by polymers [553, 561], e. g. yellow soll of CdS is prepared from solutions of $Cd(NO_3)_2$ and CH_3CSNH_2 at pH 0.75 in the presence of poly(vinylalcohol) [556, 562]. Colloidal particles of Prussian Blue ($KFe^{III}[Fe^{II}(CN)_6]$) with an average diameter of 23 nm, which are in aqueous solution in the presence of $Ru(bpy)_3^{2+}$ active in the simultaneously evolution of O_2 and H_2 under irradiation, are mentioned (see Chap. 5) [509].

Small metal clusters are also prepared by alcohol reduction (Scheme 2.76) [555, 563] or photoreduction (Scheme 2.77) [564, 565] of metal ions in solution in the presence of a water-soluble polymer such as poly(N-vinylpyrrolidone). For the alcohol reduction $PdCl_2$, H_2PtCl_6 or $RhCl_3$ in water/ethanol (1/1 v/v) in the presence of poly(N-vinylpyrrolidone) were heated under reflux for 1–2 h under inert gas, which results in homogeneous solution of ultrafine particles of average diameter of 1–5 nm with narrow size distributions. The clusters are stable for months at room temperature. Figure 2.11 shows the preparation by the alcohol-reduction method. Reduction of $PdCl_2$ adsorbed on polyheteroarylenes by $NaBH_4$ results in Pd clusters of 1–3 nm size [565].

$$PdCl_2 + C_2H_5OH \rightarrow Pd + 2\,HCl + CH_3CHO \qquad \textbf{Scheme 2.76}$$

$$PdCl_2 \rightarrow Pd + Cl_2 \qquad\qquad\qquad\qquad \textbf{Scheme 2.77}$$

Colloidal dispersions of bimetallic clusters (Au/Pd, Pd/Rh, CuPd, Pt/Rh, Pd/Pt, etc.) are prepared by simultaneous reductions of alcoholic solutions containing two noble metal salts in the presence of poly(N-vinylpyrrolidone) [563, 565, 566]. These clusters are very stable and have a size distribution of 2–7 nm. The structure of the polymer-surrounded bimetallic clusters were investigated by EXAFS, electronic spectra, micrographs, X-ray and STM. After EXAFS measurements core structures were found. Pt/Rh bimetallic clusters can be described by an assembly model composed of a microstructure having a Pt core.

The catalytic properties (unusually high activities and selectivities) of monometallic and bimetallic clusters are described in detail [509, 555, 556, 559, 560, 563–566] and are not discussed here. The influence of particle size of metal clusters [509, 566–568], their optimal concentration [569] and the kind of protecting polymer [509, 570, 571] on the catalytic activity was determined. CO-complexes of Ir, Os, Rh and Ru clusters were immobilizes at posphory-

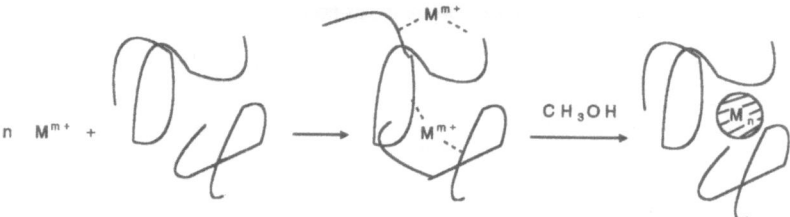

Fig. 2.11. Preparation of colloidal dispersions of metal clusters in the presence of polymers by reduction of metal salts with alcohols (After [566])

lated cross-linked polystyrene ([552] and referenes cited therein). A recently described method of broad application used the polymerization of cluster-containing monomers [552].

Inclusion of Ag and Cu into films of an oligomer of poly(thiophene) is realized by immersion of the film into dilute solutions of $AgCF_3SO_3$ or $Cu(CF_3SO_3)_2 2$ in CH_3CN in the presence of $LiClO_4$ as conducting salt followed by an electrochemical pulse-potential method (between $+0.6$ V and -0.2 V vs SEC with 200 ms) [572]. By this method formation of small aggregates of Ag or Cu in the films are obtained. Electrochemical characterizations showed improved carrier mobility by a factor of ~700, which is important for enhanced charge-carrier transport in polymer films with low mobility of charge carriers.

A general method for the preparation of metal particles (Pd, Ag, Au, Zn, Cd, Ga, In, Ge, Sn, Sb, Bi) in various polymers are described by vacuum co-deposition of the metals and vinyl monomers (vinyl acetate, styrene, etc.) at liquid nitrogen temperature followed by normal radical polymerization [552, 573]. Metal carbon composites, which contain ultrafine metal particles (Fe, Co, Ni, Pd, Pt, Rh, Cu, Au, etc.) uniformly dispersed in a polymer/carbon matrix, are produced by pyrolysis of a variety of macromolecular metal complexes [552] at approximately $300-1400\,°C$ under inert gas or hydrogen. This single preparation method allows large-scale production of each composite in various forms (thin films, fibres, powder, pellets, etc.). Possible broad applications in catalysis, adsorbents, antibacterial agents and electronic devices are envisaged [574, 575]. Different precursors for these composites are described in [574]. Pyrolysis were conducted under Ar or N_2 up to $1000\,°C$ at a rate of $0.3-0.6°/min$ and maintained at $1000\,°C$ for 2 h [574]. The TEM, SEM, X-ray, EPMA, ^{57}Fe Mössbauer methods and others were used for analysis. Particle sizes vary to a great extent depending on the precursor being between $160-2$ nm. The results of the preparation of Pt, Pd, Ru, Mn and Ti aggregates on various organic and inorganic fibres, and the activity in catalysis, is reviewed in [575]. The catalytic activity is also the aim of Pt clusters with an average diameter of 1.6 nm obtained by pyrolysis ($600\,°C$) of $(Ph_3P)_2Pt$ (C_2H_4) and $Co_2(CO)_8$ complexes at poly(phenylendiacetylene) [576, 577]. High-quality thin films of glassy carbon with high conductivity are obtained by the mild pyrolysis of the soluble polyacetylene derivative. Crystalline Pt clusters (average diameter 3.2 nm) are obtained as shown in Fig. 2.12. Au clusters of average diameter 2 nm were deposited in polydiacetylene by vapor deposition of Au [578].

Plasma polymers for metal clusters relies on the simultaneous plasma polymerization of a monomer and deposition of a metal by sputtering [579, 580]. Plasma-polymer Au clusters are obtained by using a metal cluster source (Au) separated from the plasma discharge region (monomer vinyltrimethylsilane) [580]. Depending on the evaporation rate of Au and the pressure of the inert gas Ar, gold clusters of 1.4 nm size in the polymer film are prepared. Plasma-polymer silver composite films were obtained by simultaneously or alternating plasma polymerization of benzene and Ag evaporation [581]. The optical properties of the composites were investigated in detail under consideration of theoretical simulations.

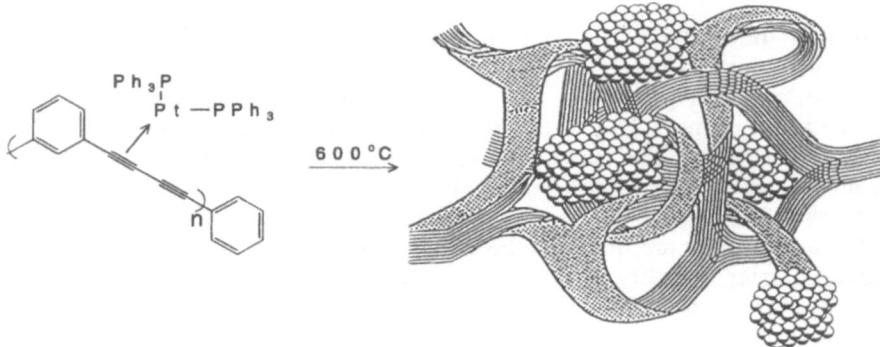

Fig. 2.12. Preparation of Pt clusters on glassy carbon by pyrolysis of a Pt componend on poly(phenylenediacetylene) (After [572])

2.4.2 Combination with Inorganic High Molecular Systems

2.4.2.1 Metal Complexes

The interaction with surfaces of inorganic carriers (oxides such as silica, alumina and titandioxide), the intercalation into layers of clays (e.g. layer silicates) and encapsulation into the framework of molecular sieves (e.g. zeolites, aluminium phosphates) exhibit advantages for metal complexes or metal clusters. On one side, the inorganic systems are thermally, mechanically and often chemically more stable in comparison with organic polymers. On the other side, specific carrier/metal complex (metal cluster) interactions provide improved properties. Only few examples for the combinations part of this chapter are mentioned.

Metal complexes or salts can interact with surfaces of modified inorganic semiconductors such as silica (see Chap. 2.1.1, Scheme 2.8; Chap. 2.2.4, 2.2.5). Additional examples are $PdCl_2$ at chitosan grafted on SiO_2 [582], $CuCl_2$ at copolymers of styrene and N-vinylimidazole grafted on silica [583], Ti, Zr and Ni complexes on silica and alumina [584], silica-supported polyalumazane-Rh [585], Fe(II) and Cu(II) on modified silica [586] and carboxylated Ru-bipyridyl complexes on TiO_2 [587–589]. The most convenient procedure uses simple uptake of metal salts from solution. Low molecular weight and polymeric phthalocyanines on SiO_2, Al_2O_3 and carbon in loadings of $0.2–12$ wt.% are prepared by the reaction of Co(II)-loaded carriers with 1,2-dicyanobenzene or 1,2,4,5-tetracyanobenzene from the gas phase [590], whereas coordinative binding is achieved by the reaction of the porphyrin derivatives with imidazolyl-modified silica [591] (for catalysis see Chap. 4). Inorganic molecular sieves of three-dimensional arrangement and layered materials of two-dimensional structure find increasing usage as host systems with highly organized structure. Unlike silica and other amorphous minerals of high surface areas, layered materials and molecular sieves possess a well-defined internal structure in either arranged simple galleries or uniform cages, cavities and channels. Cationic metal-free and Fe(II)-porphyrins such as tetrakis(N-methyl-4-pyridinium)porphyrin were incorporated in layer silicates [592]. Aqueous gels consisting of a low amount of the porphyrin and LiF, $Mg(OH)_2$ and SiO_2 were heated under reflux, which leads to crystallization of porphyrin-containing hectorite clays within 2 days. SANS and XRD indicates that the porphyrins are aggregated in the unheated gel and than dispersed up on thermal treatment giving rise to an intercalation parallel between the layers. Adsorption and intercalation of porphyrins and $Ru(bpy)_3^{2+}$ is reviewed in [593]. Careful choice of the silicate [594] allows adsorption on the exterior surface or within the clay layers. Photophysics of adsorbed and intercalated molecules have been investigated [593, 594]. Figure 2.13 illustrates after X-ray the position of the porphyrin nucleus within the gallery region of a montmorillonite [593, 595]. Polyoxometalates $[Co_9(OH)_3(H_2O)_6(HPO_4)_2 (PW_6O_{34})_3]^{16-}$ containing a Co_9-cluster were obtained by heating aqueous solution of Na_2WO_4, Na_2HPO_4 and Co(II)-acetate [596].

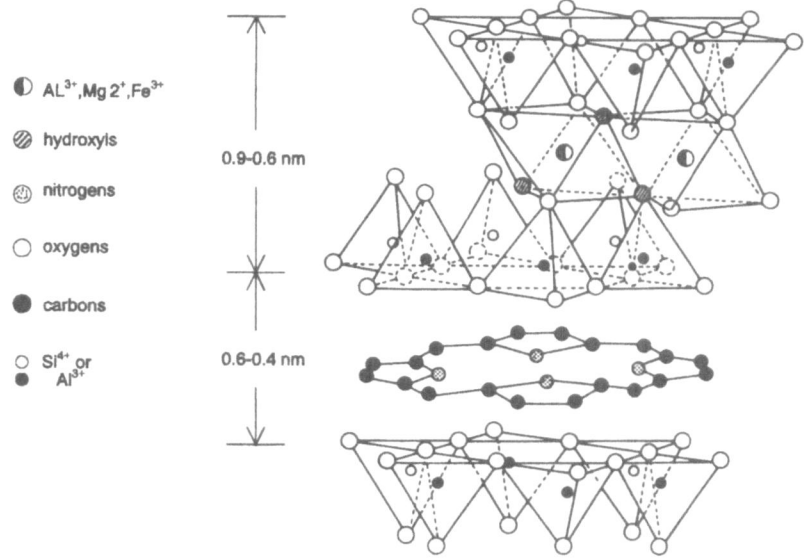

Fig. 2.13. Illustration of the position of a porphyrin within the gallery region of montmorillonite (After [593, 597])

Numerous molecular sieves with different cages/cavities/channels can be prepared by variation of the kind of tetrahedra atoms, counter ions, templates, solvents and reaction conditions [557, 594, 597–600]. Examples are anionic zeolite faujasite X, Y (aluminosilicate with SiO_4^{4-} and AlO_4^{5-} tetrahedra, pore-window size 0.78 nm, three-dimensional channels with supercages of 1.3 nm diameter) and neutral aluminium phosphate – AlPO – (with PO_4^{3-} and AlO_4^{5-} tetrahedra, pore-single channel size of 0.73 nm). A wide range of encapsulated guests have been studied, e.g. metal complexes [601–603], metal clusters [604–607], semiconductor nanoclusters [557, 608–609] and small organic molecules [594, 610, 611]. The constrained dimensions of the molecular sieves change properties and reactivities of encapsulated guest molecules and clusters. With these composites, besides more "classical" applications, such as gas separation, selective catalysis, removal of pollutants and ion exchange, new "unusual" potential applications are evaluated, e.g. data storage, quantum electronics, nonlinear optics, chemoelective devices, nanoreaction chamber and energy conversion systems.

For the incorporation of a homogeneous metal chelate into a molecular sieve different routes were realized:

1. Ion exchange. A macromolecular molecular sieve with big enough openings of channels or cages enables the introduction of a cationic metal complex (or other cationic organic molecules) [601, 603]. By this way [Cu(ethylenediamine)$_2$]$^{2+}$ was introduced into faujasite zeolite [612].
2. Adsorption and anchoring of complex in dehydrated zeolites. Metal carbonyl complexes can be adsorbed from the gas phase [613]. Also, heterobinuclear organometallic were employed for incorporation [614].
3. Synthesis of metal complexes inside the molecular sieve. In this case it is essential first to immobilize one guest precursor in the molecular sieve by diffusion and ionic or electrostatic interactions with the host system. Then, in a second step, another precursor can diffuse inside the framework to perform dye or metal chelate synthesis ("ship-in-bottle synthesis"). Most intensively the synthesis of phthalocyanines have been investigated. First the metal is introduced, e.g. by ion exchange or vapour deposition of a metal complex (carbonyl or metallocene complexes), and then, in the second step, the metal-loaded zeolite is reacted with 1,2-dicyanobenzene (Fig. 2.14) [597, 602, 615]. Co(smdpt) in zeolite Y and EMT was obtained by heating mixtures of the Co^{2+}-exchanged zeolite with the ligand bis[(3-salicylideneamino)propyl]methylamine (smdpt) [601, 616]. The authors pointed out that the previously reported synthesis of Co(salen) in zeolites [617] was not reproducible. The reason for the successful preparation of Co(smdpt) in the host materials is explained with the more flexible ligand smdpt in comparison with the ligand salen. Other examples also show the success of metal chelate formation in molecular sieves: bis(dimethylglyoximato)-Co^{2+} -Co(dmgH)$_{2-}$ (from Co^{2+}-exchanged zeolite X with dimethylglyoxim by gas phase diffusion) [618], aliphatic polyamine-Co^{2+} or -Cu^{2+} complexes, and 2,2'-bipyridyl-Ru^{2+}, -Fe^{2+} and -Co^{2+} complexes [601]. The complexation of NO, CO, cyanide NH$_3$ and amines (alkylamines, arylamines, pyridine, 2,2'-bipyridine, ethylendiamine, tetraethylerepentamine), phosphine from the gas phase or solution to metal-exchanged molecular sieves (mostly zeolites) are reviewed in [601].
4. Crystallization inclusion. In the crystallization inclusion metal chelates or other organic molecules are added to batches for hydrothermal crystallization of molecular sieves. In some cases this directs the synthesis, even resulting in new phases [619]. It is difficult to predict whether an incorpora-

Na-X $\xrightarrow[\text{Na}^+ \text{ against Co}^{2+}]{\text{ion exchange}}$ Co-X zeolite 3,3 wt.-% Co^{2+}, every supercage one Co $\xrightarrow[\text{phthalodinitrile}]{\text{reaction with}}$ Co-phthalocyanine in zeolite 0,7−3 wt.-%, every 10th supercage one Co-phthalocyanine

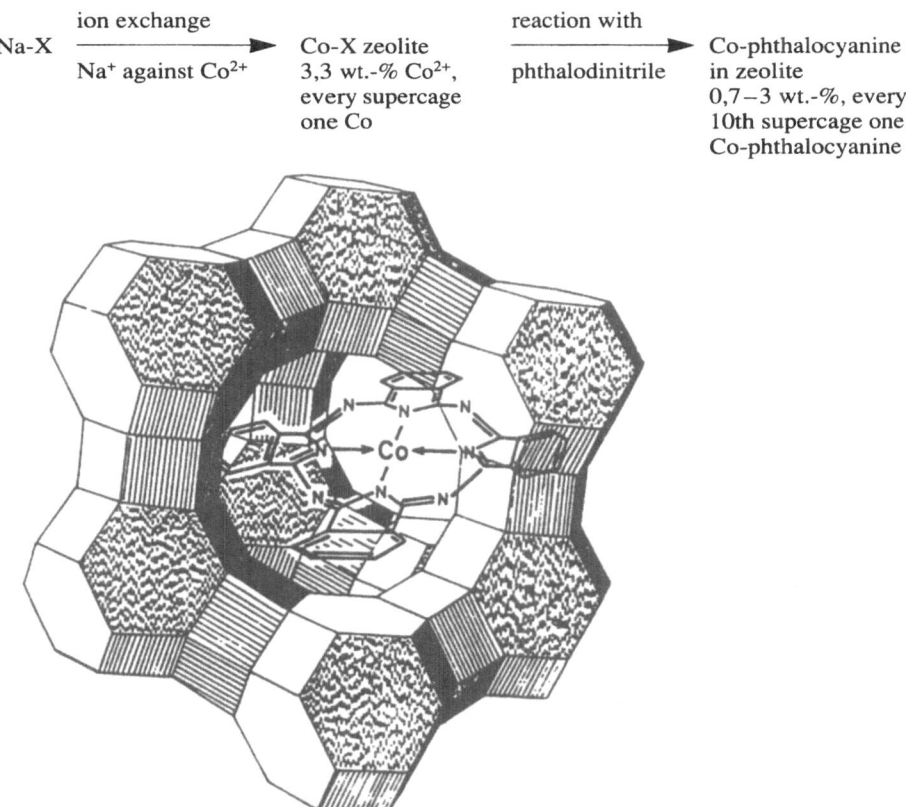

Fig. 2.14. Synthesis of phthalocyanines in the zeolite supercage

tion during hydrothermal synthesis will occur. But some prerequisites for a successful incorporation are as follows:

(a) The metal complex has to be soluble, possibly as a monomer in the synthesis mixture, to achieve a monomeric inclusion.

(b) A major problem is that the guest molecules have to be stable during all phases of the molecular sieve synthesis process. Especially high pH values and high temperature over a long period of time may result in decomposition of the organic additive. Examples are demethylation of methylene blue [620] and decomposition of phthalocyanines [621].

(c) As in the case of loading by ion exchange, only positively charged molecules can be incorporated monomolecularly in a considerable amount during the crystallization process of anionically charged or neutral molecular sieves. Positively charged porphyrins and phthalocyanines, such as ZnTaP and ZnTpyP (also their metal-free derivatives), decompose during zeolite synthesis, but can be successfully included in AlPO in an amount of 10^{-6} mol/g (ca. 0.015 molecules per unit cell) [621, 624]. Inclusion of unsubstituted phthalocyanines during zeolite and AlPO crystallization have been described [625]. Other examples are the incorporation of dyes such as methylene blue in the AlPO family, with approximately 0.1–1 dye molecules per unit cell (ca. $6 \cdot 10^{-5}$ mol/g) [622, 623].

A detailed analysis is necessary to decide (a) on the preferential dispension of the dye, i. e. either the interior of the molecular sieve or on the surface of the crystals, (b) on the influence of the incorporation into the structures of the dye and the molecular sieve, and (c) on the precise location and host guest interaction. Experimental techniques for these investigations are XPS, UV-VIS, FT-IR, XRD, SEM, XANES, EXAFS or Rietveld refinement. XPS provides important data as to whether the dye or metal chelate is deposited onto the external surface or in core/channel domains of molecular sieve crystals. For phthalocyanines synthesized in zeolite Y from precursors the location in the supercages shown in comparison with N/Si ratios and chemical analysis is clearly revealed [626]. In addition, the changed N1s signal structure indicates the interaction with the zeolite matrix. For a dye molecule binding energies of N1s electrons and N/Si atomic ratios estabished the encapsulation inside cages and channels of zeolite Y and AlPO [624]. In addition, distribution of aggregated and mononuclear dye molecules can be evaluated depending on the preparation technique. Strong host/guest interactions explains unusual inhomogeneous line broadening in the VIS spectrum [627]. The ESR method is very essential to show the electronic interaction with O_2. For Co(smdpt) in zeolite it was estimated that the distribution of the chelate over the cages prevents dimerization and irreversible oxygenation processes [616]. Depending on the treatment of the zeolite/Co(dmgH)$_2$ system, square planar, distorted, hydrated and oxygenated chelates could be identified by detailed EPR characterization [618]. One method, which can be used to evaluate the detailed location of the guests, is the analysis of X-ray powder diffractograms by means of Rietveld refinement in combination with electron density maps [620, 628]. Molecular sieve-encaged metal chelates exhibit a high reactivity (a) in the binding of O_2 under formation of mononuclear superoxo complexes and (b) in various catalytic reactions. This can be related to a high dispersion of the active guest molecules in the host lattices [601, 602].

2.4.2.2 Metal Clusters

A large potential of research on metal clusters in molecular sieves, such as zeolites, is based on the possibility of prearing clusters that are uniform in size and arranged in superlattice relationship, i. e. in crystal-lattice positions of the host. Beyond the objective to get a better insight into the structure property correlation under the viewpoint of catalysis (stereoselectivity, shape selectivity), the study of quantum-size effects in metal clusters for advanced confirmed electronic, optical and opto-electronic devices is an actual task.

A large number of parameters reviewed in [629] influence the target to prepare tailor-made clusters in molecular sieves. The formation of metal clusters starting from metal ion-exchanged zeolites is characterized by consecutive processes of metal reduction, migration, agglomeration and zeolite framework reconstruction. Transition metal ions after ion exchange can populate well-defined in situ zeolite matrices depending on parameters such as degree of ion exchange, T, type of ligand or degree of ligand (H_2O) removal, e.g. dehydration processes as confirmed by X-ray, ESR and diffuse reflectance spectroscopy [629–631]. The reduction of transition metal ions by hydrogen occurs with an increase of the proton activity in the zeolite [632]. Noble metals that have been ion-exchanged into zeolites as amine complexes can be reduced without addition of H_2 by decomposition at elevated temperatures (Schemes 2.78 and 2.79) [633, 634]. The influence of calcination conditions on the reduction process and the final dispersion of a Pt metal phase is a complicated process (Fig. 2.15) [629, 635, 636].

$$M^{n+} + n/2\,H_2 \rightarrow M^0 + nH^+ \qquad\qquad \textbf{Scheme 2.78}$$

$$[Pt(NH_3)_4]^{2+} \rightarrow Pt^0 + 2H^+\;\tfrac{1}{2}\,N_2 + \tfrac{1}{2}\,H_2 + NH_3 \qquad \textbf{Scheme 2.79}$$

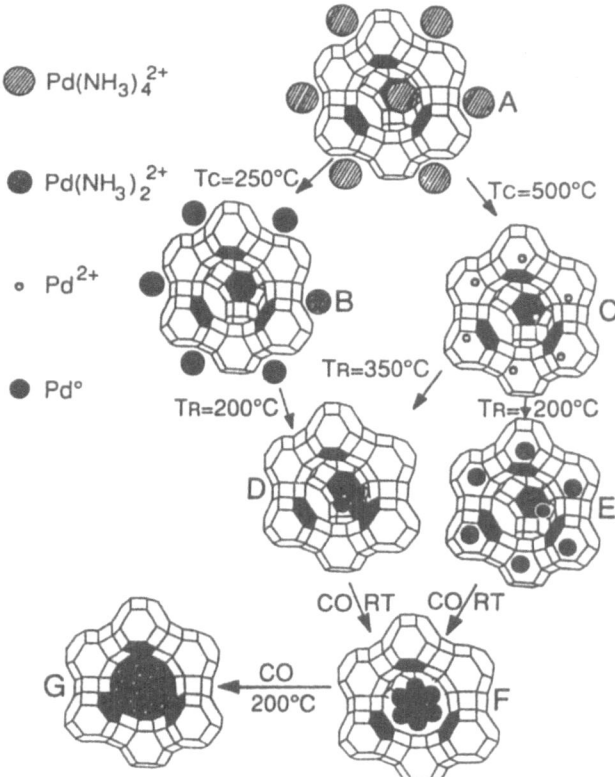

Fig. 2.15. Illustration of location and sizes of Pd particles after different calcination, reduction and CO exposure conditions (After [629, 635, 636])

The addition of CO to faujasites containing $[Pt(NH_3)_4]^{2+}$ results in the formation of small Pt-carbonyl dianions [637, 638]. Aggregation of the trinuclear Pt-complex may be via reoxidation by charge-compensating protons (Schemes 2.80 and 281) [638].

$$3 [Pt(NH_3)_4]^{2+} + 10 CO + 4H_2O \rightarrow$$
$$[Pt_3(CO)_6]^{2-} + 4CO_2 + 8NH_4^+ + 4NH_3 \qquad \text{Scheme 2.80}$$

$$2 [Pt_3(CO)_6]^{2-} + 2H^+ \rightarrow [Pt_3(CO)_6]_2^{2-} + H_2 \qquad \text{Scheme 2.81}$$

Also, the aggregation of metal atoms produced in reduction processes and the resulting metal dispersion depend on many parameters [639]. Pt clusters of 2 – 6 atoms have been detected under mild conditions [639, 640], but also metal phase, e. g. of Pt exceeding the dimension of a supercage under formation of *meso*-pores, have been observed [641]. Clusters of different metal ions, such as Ag, Pt and alkali metals, have been reviewed [629].

After reduction of H_2PtCl_6 the colloidal Pt can be stabilized by deposition onto TiO_2 and others [556, 560]. Additional doping of TiO_2 – Pt with RuO_2 was achieved by suspending the powder in an aqueous solution of $RuCl_3 \cdot 3$ H_2O and heating for several hours at 95 °C [509, 556, 560]. Heating of a Ti(IV)-modified silica with H_2PtCl_6 ad transition metal salts ($MnCl_2$, $FeCl_3$, $SnCl_2$) results in bimetallic catalysts $SiO_2/Ti-Pt-Me$. Similar bimetallics were also prepared on chitosan-modified silica [642].

References

1. Hodge P, Sherrington DC (eds) (1980) Polymer supported reactions in organic synthesis. Wiley, Chichester
2. Ise N, Tabushi I (ed) (1980) Speciality polymers. Iwanami Shoten, Tokyo
3. Merrifield RB (1965) Science 150:178
4. Guyot A, Bartholin M (1982) Progr Polym Sci 8:227
5. Pepper KW, Paisley HM, Yong MA (1953) J Chem Soc, 4097
6. Pomogailo AD (1988) Polymer immobilized metal complex catalysts. Nauka, Moscow (in Russian)
7. Nicolaus V, Wöhrle D (1992) Angew Makromol Chem 198:179
8. Geckeler K, Lange G, Eberhartt H, Bayer E (1980) Pure Appl Chem 52:1883
9. Wöhrle D, Bohlen H, Meyer G (1984) Polym Bull 11:143, 151
10. Kritskaya DA, Pomogailo AD, Ponomarev AN, Dyachkovskii FS (1980) J Appl Polym Sci 25:349
11. Garnett JM, Levot R, Long MA (1981) J Polym Sci Polym Lett Ed 19:23
12. Hartley FR, Murray SG, Nicholson PN (1982) J Mol Catal 16:363
13. Allum KG, Hancock RD, Howell IV et al. (1975) J Organomet Chem 87:203
14. Pomogailo AD (1986) In: Complex organometallic catalysts for olefines polymerization. Institute of Chemical Physics Russian Academy of Sciences, Chernogolovka, 10:63 (in Russian)
15. Wegscheider W, Knapp G (1981) CRS Crit Rev Anal Chem 11:79
16. Muzzarelli RAA (1977) Chitin. Pergamon Press, Oxford
17. Tsuchida E, Nishide H (1977) Adv Polym Sci 24:1
18. Wöhrle D (1983) Adv Polym Sci 50:45
19. Konstantinos GB, Stephen HC (1989) J Polym Sci C Polym Lett 27:355

20. Gregor HP, Luttinger LB, Loebl EM (1955) J Phys Chem 59:34
21. Marinsky JA (1976) Coord Chem Rev 19:125
22. Scatchard G (1949) An NY Acad Sci 51:660
23. Barrow GM (1974) Physical chemistry for life sciences. McGraw-Hill, New York
24. Katchalsky A, Kunzle O, Kuhn W (1950) J Polym Sci 5:283
25. Brenerman ML, Barabanov VP, Usachev AA, Goldman VV (1988) Vysokomol Soedin 30A:1575, 1582 (in Russian)
26. Scatchard G, Yap WT (1964) J Am Chem Soc 84:3434
27. Rossotti FJC, Rossotti H (1960) The determination of stability constants and other equilibrium constants in solution. McGraw-Hill, New York
28. Bjerrum J (1957) Metal amine formation in aqueous solution. Theory of reversible step reactions. P. Haase and Son, Copenhagen
29. Szymanski R (1991) Makromol Chem 192:757
30. Wong K, Smid J (1977) J Am Chem Soc 99:5637
31. Arai K, Ogiwara Y (1986) J Polym Sci 24:2027
32. Filippov AP (1983) Theor Eksp Khim 19:463
33. Vainshtein EF (1981) Izuchenie osobennostei kompleksoobrazovaniya tsepnykh molekul v razbavlennykh rastvorakh (doctoral thesis). Institute of Chemial Physics USSR Academy of Science, Moscow
34. Morawetz H (1965) Macromolecules in solution. N.Y. Interscience Publishers, New York
35. Arkhipovitch GI, Dubrovskii SA, Kazanskii KS, Shupik AN (1981) Vysokomolek Soedin 23:1653 (in Russian)
36. Polinskii AS, Pshezhetskii VS, Kabanov VA (1981) Dokl Acad Nauk SSSR 256:129; (1982) Vysokomol Soedin A25:72; (1985) A27:2295
37. Pomogailo AD, Sokol'skii DV, Baishiganov E (1974) Dokl Acad Nauk SSSR 218:1111
38. Schwarzenbach G (1952) Helv Chim Acta 35:2344
39. Yatsimirskii KB (1980) Theor Eksp Khim 16:34
40. Pomogailo AD, Uflyand IE (1991) Makromolekulyarnye metallokhelaty. Khimiya, Moscow (in Russian)
41. Uflyand IE, Vainshtein EF, Pomgailo AD (1991) Zh Obshch Khim 61:1790
42. Barbucci R, Champbell MJM, Casolaro M (1986) J Chem Soc Dalton Trans 2325
43. Takenchi A, Yamada S (1974) Inorg Chim Acta 8:225
44. Kalalova E, Populova O, Stokrova S, Stopka P (1983) Collect Czech Chem Commun 49:2021
45. Rosthauser JW, Winston A (1981) Macromolecules 14:538
46. Graessley WW (1974) Adv Poly Sci 16:1
47. Warshawsky A, Kalir K (1979) J Appl Poly Sci 24:1125
48. Todokoro H, Chatani Y, Yoshihara T et al. (1964) Macromol Chem 73:109
49. Iwamoto R, Saito Y, Ishikara H, Iadikoro H (1968) J Poly Sci A-2, 6:1509
50. Parker JM, Wright PV, Lee CC (1981) Polymer 22:1305
51. Wetton RE, Jamess DB, Whitting W (1976) J Poly Sci Polym Lett Ed 14:577
52. Higuchi N, Hiraoki T, Hikichi K (1979) Polym J 11:139
53. Barbucci R, Furruti P (1979) Polymer 20:1061
54. Nishide H, Tsuchida E (1976) Makromol Chem 177:2295
55. Khamraeva AL, Abdurashidov TR, Namazov MB, Yusupbekov AK (1972) Uzbekskii Khim Zh:47
56. Popov AA, Vainstein EF, Entelis SG (1977) J Macromol Sci 11:859
57. Pomogailo AD, Kuzaev AI, Dyachkovskii FS, Enikolopyan NS (1981) Dokl Akad Nauk SSSR 256:132
58. Pomogailo AD, Sokol'skii DV, Mambetov UA (1972) Dokl Akad Nauk SSSR 207:882
59. Kuzaev AI, Pomogailo AD, Mambetov UA (1981) Vysokomol Soedin 23A:213; (1982) 24A:1199
60. Roman E, Valenzuella G, Gargallo L, Radic D (1983) J Polym Sci Polym Chem Ed 21:2057

61. Joppien GR (1978) Angew Macromol Chem 70:189, 199
62. Ikada Y, Nishizaki Y et al. (1976) J Poly Sci Poly Chem Ed 14:2251
63. Echmaev SB, Iveleva IN, Raevskii AV, Pomogailo AD (1983) Kinet Katal 24:1428
64. Ungurenasu C, Cotzur C (1982) Polym Bull 6:299
65. Baishiganov E, Pomogailo AD, Sokol'skii DV (1972) Kompleksnaya pererabotka mangyshlakskoi nefti. Alma-Ata:Nauka. 4:127
66. Biedermann HG, Graf W (1974) Z Naturforsch B 29:65
67. Kawata N, Mizoroki T, Ozaki A, Ohkawara M (1973) Chem Lett 1165
68. Ozin GA, Francis CG (1980) J Mol Struct 59:55
69. Deratani A, Sebille B, Hommel H, Legrand AP (1983) React Polym 1:261
70. Windler SCH (1984) Nachr Chem Techn Lab 32:392
71. Grubbs RH, Gibbons C, Kroll LC et al. (1973) J Am Chem Soc 95:2373
72. Regen SL, Lee DP (1977) Macromolecules 10:1418
73. Grinberg OY, Nikitaev AT, Zamaraev KI, Lebedev YS (1969) Zh Strukt Khim 10:230
74. Kokorin AI (1992) Stroenie koordinatsionnykh soedinenii s makromolekulyarnymi ligandami (doctoral thesis) Institute of Chemical Physics Russian Academy of Sciences, Moscow
75. Pomogailo AD, Nikitaev AT, Dyachkovskii FS (1984) Kinet Katal 25:166
76. Knothe G, Wöhrle D (1989) Makromol Chem 190:1573. Knothe G (1992) Makromol Chem, Theory Simil 1:187
77. Knothe G (1993) Macromol Chem, Theory Simil 2:917; (1994) J Inorg Organomet Polym 4:325
78. Tsuchida E, Nishide H (1977) Polymer–metal complexes and their catalytic activity. Advances in polymer science, vol 24. Springer, Berlin Heidelberg New York
79. Tsuchida E et al. (eds) (1978) Macromolecule metal complexes. Kagaku-Dojin Kyoto (in Japanese)
80. Tsuchida E, Nishide H (1980) Selective adsorption of metal ions to polymeric amines immobilized by template reaction. In: Goethals EJ (ed) Polymeric amines. Pergamon Press, New York; chap. 8
81. Tsuchida E (ed) (1991) Macromolecular complexes: dynamic interactions and electronic processes. VCH Publishers, New York
82. Tsuchida E, Takeoka S (1994) Interpolymer complexes and their ion conduction. In: Dubin P et al. (eds) Macromolecular complexes in chemistry and biology. Springer, Berlin Heidelberg New York, chap. 5
83. Biedermann HG, Wichtmann K, Griessl E (1973) Z Naturforsch B 28:182; Makromol Chem 172:49; (1974) Z Naturforsch B 29:132, 267; Chem Ztg 98:109
84. Nishide H, Tsuchida E (1976) Makromol Chem 177:2295
85. Agnew NH (1976) J Polym Sci Polym Chem Ed 14:2819
86. Polinskii AS, Pshezhetskii VS, Kabanov VA (1981) Dokl Akad Nauk SSSR 256:129, (1983) Vysokomol Soedin 25A:72
87. Polinskii AS, Izumrudov VA, Kabanov VA (1985) Vysokomol Soedin 27A:1014
88. Kirsh YuE, Kovner VYa, Kokorin AI, Zamaraev KI, Chernyak VYa, Kabanov VA (1974) Eur Polym J 10:671
89. Okubo T, Enokido A (1983) J Chem Soc Faraday Trans (part 1) 79:1639
90. Clear JM, Kelly JM, Vos JG (1983) Makromol Chem 184:613
91. Shimidzu T, Izaki K, Akai Y, Iyoda T (1981) Polym J 13:889
92. Kabanov BA, Lukovkin GM, Starodubtsev SG, Golina LV (1977) Vysokomol Soedin Ser B 19:95
93. Pomogailo AD (1988) Polimernye immobilizovannye metallokompleksnye katalizatory. Nauka Moscow
94. Ivleva IN, Echmaev SB, Pomogailo AD et al. (1977) Dokl Akad Nauk SSSR 233:903; (1983) Kinet Katal 24:1428
95. Pomogailo AD, Klyuev MV (1985) Izv Akad Nauk SSSR Ser Khim 1716
96. Chanda M, O'Driscoll KF, Rempel GL (1981) J Mol Catal 12:197
97. Kopylova VD, Saldadze KM (1980) Plast Massy 8:19

98. Geckeler K, Lange G, Eberhardt H et al. (1980) Pure Appl Chem 52:1883
99. Bauer E, Geckeler K, Weingartner K (1980) Makromol Chem 181:585
100. Patsevich IV, Ogorodnikov IA, Zhuk DS (1979) Vysokomol Soedin 21B:803
101. Perchenko VN, Mirekova IS, Nametkin NS (1980) Dokl Akad Nauk SSSR 251:1437; Neftekhimiya 20:518
102. Antonelli ML, Bucci R, Carunchio V, Cernia E (1980) J Polym Sci Polym Chem Ed 18:179
103. Shupik AN, Kalashnikova IS, Perchenko VN (1984) Zh Fiz Khim 58:1313
104. Parmon VA (1984) Novosibirsk. Institut kataliza Sibirskogo Otdeleniya Akad Nauk SSSR (doctoral thesis)
105. Pearson PG (1968) J Chem Educ 45:581
106. Medved ZN, Zhegalova NN (1978) Vysokomol Soedin 20B:524
107. Arkhipovich GN, Shupik AN (1985) Zh Fiz Khim 59:1725
108. Pomogailo AD, Mambetov UA, Sokol'skii DV et al. (1978) Zh Obshch Khim 48:2101; (1972) Dokl Akad Nauk SSSR 207:882
109. Kuzaev AN, Pomogailo AD, Mambetov UA (1981) Vysokomol Soedin 23A:213; (1982) 24:1199
110. Roshchupkina OS, Lisitskaya AP, Pomogailo AD et al. (1978) Dokl Akad Nauk SSSR 243:1223
111. Biedermann HG, Grat W (1974) Z Naturforsch 28b:65; (1975) 30b:226
112. Becturov EA, Kudaibergenov SE, Zhaimina GM (1984) Eur Polym J 20:1113
113. Reed J, Eisenberger P, Teo B-K, Kincaid BM (1978) J Am Chem Soc 100:2375
114. Bruner H, Bailar JC Jr (1973) Inorg Chem 12:1465
115. Hartley FR, Murray SG, Nicholson PN (1982) J Mol Catal 16:363; J Polym Sci Polym Chem Ed 20:2395
116. Garnett JM, Levot R, Long MA (1981) J Polym Sci Polym Chem Ed 19:23
117. Barker H, Garnett JL, Levot R, Long MA (1978) J Macromol Sci A 12:261
118. Stille JK (1984) J Macromol Sci A 21:1689
119. Dumont W, Poulin J-C, Domg T-P, Kogan HB (1973) J Am Chem Soc 95:8295; CR Acad Sci C 277:41
120. Kabanov VA, Efendiev AA, Orudzev DD (1979) Vyskomol Soedin 21A:589
121. Bauer E, Grathwohl PA, Geckeler K (1983) Angew Macromol Chem 113:141
122. Jones MM, Coble HD, Pratt TH, Harbison RD (1975) J Inorg Nucl Chem 37:2409
123. Davies JA, Sood A (1983) Makromol Chem Rapid Commun 4:777
124. Mitchell PCH, Taylor MG (1982) Polyhedron 1:225
125. Allum KG, Hancock RD, Howel IV et al. (1975) J Organomet Chem 87:203; 107:393; (1976) J Catal 43:331
126. Bartholin M, Canan J, Guyot A (1977) J Mol Catal 2:307; 3:17; (1976) 1:375
127. Verlaan JPJ, Bootsmann JPC, Challa G (1982) J Mol Catal 14:211
128. Yanzhu Z, Yongjun Li, Lingzhi W, Yingyan J (1982) Fundamental reseach organometallic chemistry. Proc China–Japan U.S. Trilateral Semin, Peking, June 1980 N4; Beijing etc. p. 77
129. Royer GP, Wen-Sheiung Chow, Hatton KS (1985) J Mol Catal 31:1
130. Freidlin LKh, Nazarova NM, Litvin EF, Annamuradova MA (1978) Izv Akad Nauk SSSR, Ser Khim 2465
131. Uriarte RJ, Meek W (1980) Inorg Chim Acta 44:283
132. Arai H (1978) J Catal 51:135
133. Bradley DC, Hillyer MJ (1965) Preparative inorganic reactions, vol 2. (Jolly ML ed) Interscience Publishers, New York
134. Hojo N, Shirai H, Hajashi S (1974) J Polym Sci Polym Symp (47):299; (1972) J Chem Soc Jpn Chem Chem Ind 1954
135. Kakinoki H, Chong Su Cho, Higashi F, Sumita O (1977) J Polym Sci Polym Chem Ed 15:2303
136. Suzuki T, Shirai H, Shimizu F, Hojo N (1983) Polym J 15:409
137. Pomogailo AD, Lisitskaya AP, Kritskaya DA (1983) Kompleksnye metalloorganicheskie katalizatory polimerizatsii olefinov. Institut Khimicheskoi Fiziki Acad Chernogolovka, Nauk SSSR 8:78

138. Nicolaides CP, Covielle NJ (1984) J Mol Catal 24:375
139. Kolawole EG, Mathieson SM (1977) J Polym Sci Polym Chem Ed 15:2291; (1979) J Polym Sci Polym Lett Ed 17:573
140. Marinsky JA, Ansapach WM (1975) J Phys Chem 79:439
141. Nicolaides CP, Coville NJ (1984) J Mol Catal 24:375; (1981) J Organometal Chem 222:285
142. Sbrana G, Braca G et al. (1976) J Chem Soc Dalton Trans 1847; (1977) J Mol Catal 3:111
143. Zolin VF, Koreneva LG (1980) Redkozemel'nyi zond v khimii i biologii. Nauka, Moscow
144. Okamoto Y, Ueba Y, Dzhanibekov NF, Banks E (1981) Macromolecules 14:17
145. Carraher CE Jr, Schroeder JA (1975) Polym Prepr 16:659; J Polym Sci Polym Lett Ed 13:215
146. Kolawole EC, Bello MA (1980) Eur Polym J 16:325
147. David C, Depauw A, Geuskens G (1968) J Polym Sci C-1, 319
148. Morcellet M (1985) J Polym Sci Polym Lett Ed 23:99
149. Pomogailo AD, Golubeva ND (1985) Kinet Katal 26:947
150. Bravaya NM, Pomogailo AD, Vainshtein EF (1984) Kinet Katal 25:1140
151. Travers C, Marinsky JA (1974) J Polym Sci Polym Symp 285
152. Fung Ti Shi, Astanina AN, Bystrov GV et al. (1984) Zh Fiz Khim 58:1818
153. Gustafson RL, Sirio IA (1968) J Phys Chem 72:1502
154. Kopylova VD, Saldadze KM (1972) Zh Fiz Khim 46:990; (1985) Koord Khim 11:41
155. Noskova MP, Radionov BK, Vasenina NS, Lipunov IN, Kazantsev EI (1983) Zh Fiz Khim 57:707
156. Koide M, Tsuchida E, Kurimura Y (1985) Makromol Chem 182:259
157. Imai N, Marinsky JA (1980) Macromolecules 13:275
158. Mathenitis C, Morcellet-Sauvage J, Morcellet M, Loucheux C (1983) Proc Intern Symp IUPAC Macro 1983. Bucharest Sect 4, 711; Macromolecules 16:1564; (1984) Polym Bull 12:133, 141
159. Lekchiri A, Morcellet J, Morcellet M (1987) Macromolecules 20:49
160. Nakai M, Yonoyama M, Hatano M (1971) Bull Chem Soc Jpn 44:874
161. Bernadyuk SZ, Kudryavtsev GV, Lisichkin GV (1979) Koord Khim 5:1834
162. Rollman LD (1972) Inorg Chim Acta 6:137
163. Xiang-Yao Guo, Hui-Juan Zong, Yong-Jun Li, Ying-Yan Jiang (1984) Makromol Chem Rapid Commun 5:507
164. Pomogailo AD, Uflyand IE (1991) Makromolekulyarnye metallokhelaty. Khimiya Moscow
165. Pomogailo AD, Uflyand IE (1988) Koord Khim 14:147; (1990) Adv Polym Sci 97:61; Platinum Metals Rev 34:185
166. Kaneko M, Yamada A (1984) Adv Polym Sci 55:2
167. Myasoedova GV, Savvin SB (1986) CRC Crit Rev Anal Chem 17:1
168. Sahni SK, Reedjik J (1984) Coord Chem Rev 59:1
169. Davydova SL, Plate NA (1975) Coord Chem Rev 16:195
170. Warshawsky A (1988) Synthesis and separations using functional polymers. In: Wiley J, Sherrington DC, Hodge P (eds) Chichester, p. 325
171. Kaneko M, Nemoto S, Yamada A, Kurimura Y (1980) Inorg Chim Acta 44:L289
172. Furue M, Sumi K, Nazakura S (1982) J Polym Sci Polym Lett Ed 20:291
173. Potts KT, Usifer DA, Guodalupe A, Abruna HD (1987) J Am Chem Soc 109:3961
174. Barbucci R, Casolaro M, Barone V et al. (1983) Macromolecules 16:1159
175. Pittmann CU Jr (1973) J Org Chem 43:4928; (1978) 48:640
176. Teyssie MT, Teyssie P (1961) J Polym Sci 50:253; (1963) Makromol Chem 66:133
177. Patel MN, Patel JR, Patel SH (1988) J Macromol Sci A 25:211
178. Desaraju P, Winston A (1985) J Polym Sci Polym Lett Ed 23:73
179. Patil DR, Smith DJ (1990) J Polym Sci (part A) Polym Chem Ed 28:949
180. Deratani A, Sebille B (1981) Makromol Chem 182:1875; (1983) React Polym 1:261
181. Tsuchida E (1979) J Macromol Sci A 13:545
182. Wöhrle D, Gitzel J, Krawczyk G, Tsuchida E et al. (1988) J Macromol Sci A 25:1227

183. Kabanov VA, Kozhevnikova VA, Kokorin AI et al. (1979) Vysokomol Soedin 21A:209, 1891; (1977) 19:118
184. Zezin AB, Kabanov VA (1982) Usp Khim 51:1447
185. Subramanian R, Natarajan P (1984) J Polym Sci Polym Chem Ed 22:437
186. Saegusa T, Kobayashi S, Jamada A (1977) J Appl Polym Sci 21:2481
187. Mustafaev MI, Kabanov VA (1981) Vysokomol Soedin Ser A 23:271
188. Korshak VV, Shtil'man MI (1984) Polimery v protsessakh immobilizatsii i modifikatsii prirodnykh soedinenii. Nauka, Moscow
189. Sakura B J, Nakai S (1980) J Food Sci 45:582
190. Kham K, Deratani A, Sebille B (1987) Now J Chem 11:709
191. Lisichkin GV, Kudryavtsev GV, Ivanov VM (1979) Zh Vsesoyusnogo Khim Obshestva Mendeleeva 24:294
192. Skopenko VV, Lishko TP, Sukhan TA, Trofimchuk AK (1980) Ukr Khim Zh 46:1029
193. Kendall DC, Leuder DE, Burggraf LW, Pern FJ (1982) Appl Spectroscop 36:436
194. Tashkova K, Andreev A (1985) Koord Kim 11:650
195. Miki S, Maruyama T, Ohno T et al. (1988) Chem Lett 861
196. Bradshaw JS, Bruening RL, Krakowiak KE et al. (1988) J Chem Soc Chem Commun 812
197. Tatsumi T, Nakamura M, Tominaga H (1989) Chem Lett 419
198. Wöhrle D, Buck T, Schneider G, Schulz-Ekloff G, Fischer H (1991) J Inorg Organometal Polym 1:115
199. Krauss H-L, Nockl J (1965) Z Naturforsch B 20:630
200. Vol'pin ME, Novikov YuN, Lapkina ND et al. (1975) J Am Chem Soc 97:3366
201. Chandrasekaran ES, Grubbs RH, Brubaker CH Jr (1976) J Organometal Chem 120:49; (1975) J Am Chem Soc 97:2128; (1977) 99:4517
202. Smith TW, Luca DJ (1982) Proc Symp Modif Polym Las Vegas (Nev.) Abstr N 4 1983:85
203. Baukoa EYu, Bronshtein LM, Valetskii PM et al. (1992) Metalloorg Khim 5:1386
204. Filippov AP (1983) Teor Eksp Khim 19:463; (1982) Zh Neorg Khim 27:353
205. Malovikova A, Kohn R (1983) Collect Czechosl Chem Commun 48:3154
206. Shevchenko LI, Dubina AM, Tolmachev VN (1980) Vysokomol Soedin 27A:1993
207. Miroshnik LV, Dubina AM, Tolmachev VN (1980) Koord Khim 6:870; Zh Anal Khim 37:1897
208. Okamoto Y (1987) J Macromol Sci A 24:455
209. Akkaya EU, Crarnik AW (1988)) J Am Chem Soc 110:8553
210. Naoshima Y, Carraher CE Jr, Gehrke TJ et al. (1986) J Macromol Sci A23:861
211. Inoue K, Baba Y, Yoshizuka K et al. (1988) Chem Lett 1281
212. Millish F, Hellmuth EW, Huang SY (1975) J Polym Sci Polym Chem Ed 13:2143
213. Lopez-de-Alba PL, Urbina B, Alvarado GC et al. (1987) J Radioanal Nucl Chem Lett 118:99
214. Tanaka K (1975) J Natl Chem Lab Ind 70:203
215. Kostromina NA, Davydova SL, Shaposhnikova AD, Tikhonov VP (1975) Teor Eksp Khim 15:297
216. Zhorobekova Sh Zh (1987) Makroligandyne svoistva guminovykh kislot Frunze. Ilim
217. Gamble DS, Underdown AW, Langford CH (1980) Anal Chem 52:1901
218. Pomogailo AD, Savost'yanov VS (1983) Usp Khim 52:1688; (1985) J Macromol Sci Rev C 25:375
219. Pomogailo AD, Savost'yanov VS (1988) Metallosoderzhashchie monomery i polimery na ikh osnove. Khimiya, Moscow
220. Pomogailo AD, Savost'yanov VS (1994) Synthesis and polymerization of metal-containing monomers. CRC Press, Boca Raton
221. Nomenclature for regular single-strand and quasi-single strand inorganic and coordination polymers (1985) Pure Appl Chem 57:151
222. Wilkinson G, Gillard RD, McCleverty JA (eds) (1987) Comprehensive coordination chemistry, vols 1–7. Pergamon Press, New York
223. Wöhrle D (1983) Adv Polym Sci 50:45

224. Dey AK (1986) J Indian Chem Soc 63:357
225. Spiratos M, Rusu GI, Airinei A, Ciobanu A (1982) Angw Makromol Chem 107:33
226. Macru M, Lazarescu S, Grigoriu GE (1986) Polymer Bull 16:103
227. Patel RD (1986) Macromol Chem 187:1871
228. Sawodny W, Grünes R, Reitzle H (1982) Angew Chem Int Engl 21:775
229. Sawodny W, Riederer M (1977) Angew Chem 89:897
230. Riederer M, Urban E, Sawodny W (1977) Angew Chem 89:898
231. Riederer M, Sawodny W (1978) Angew Chem 90:642
232. Bottino FA, Finocchiaro P, Libertini E, Mamo A, Recca A (1983) Polymer Commun 63
233. Marvel CS, Tarkoy N (1957) J Am Chem Soc 79:6000
234. Manecke G, Wille R (1970) Makromol Chem 133:61
235. Manecke G, Wille R (1972) Makromol Chem 160:111
236. Grünes R, Sawodny W (1983) Inorg Chim Acta 70:247
237. Patel MN, Patel SH, Setty MS (1981) Angew Macromol Chem 97:69
238. Patel MN, Jani BN (1985) J Macromol Sci Chem A 22:1517; J Indian Chem Soc 63:278
239. Sawodny W, Riederer M, Urban E (1978) Inorg Chim Acta 29:63
240. Rheingold AL, Choudhury P (1977) J Organomet Chem 128:155
241. Goodwin HA, Bailar JC (1961) J Am Chem Soc 83:2467
242. Patel MN, Patel SH (1983) Synth React Inorg Met Org Chem 13:133
243. Wöhrle D, Bohlen H, Meyer G (1984) Polymer Bull 11: 143, 151; Wöhrle D, Bohlen H,
     Blum K (1986) Makromol Chem 187:2081, Wöhrle D, Aringer C, Pohl D, Bohlen H
     (1984) Makromol Chem 185:669
244. Campestrini S, Meunier B (1992) Inorg Chem 31:1999. Fujii Y, Matsutanik K,
     Kikutik K (1985) J Chem Soc Chem Commun 415
245. Hanabusa K, Tanimura Y, Suzuki T, Koyama T, Shirai H (1991) Makromol Chem
     192:233; (1992) Makromol Chem 193:2149
246. Donaruma LG, Kitoh S, Walsworth G, Depinto JV, Edzwald JK (1979) Macromolecules
     12:435
247. Tomic EA, Campbell TW, Foldi VS (1962) J Polym Sci 62:379, 387
248. Berkel van FCAA, Grotjahn H (1973) Appl Polym Symp 21:67
249. Donaruma LG (1981) Polymer Preprints 22(1)
250. Majumdar A, Biswas M (1991) Polym Bull 26:145, 151
251. Neuse EW (1994) Macromol Symp 80:111
252. Wöhrle D (1972) Adv Polym Sci 10:35
253. Wöhrle D (1989) Phthalocyanines: Properties and applications. In: Leznoff CC, Lever
     ABP (eds) VCH Publishers, New York
254. Berlin AA, Sherle AJ (1971) Inorg Macromol Rev 1:235
255. Achar BN, Fohlen GM, Parker JA (1985) J Polym Sci Polym Chem Ed 23:801
256. Wamser CC, Bard RR, Senthilathipan V (1989) J Am Chem Soc 111:8485
257. Wamser CC (1991) Mol Cryst Liq Crist 194:65
258. Wamser CC, Senthilathipan V, Li W (1991) SPIE 1436:114
259. Bettelheim A, White BA, Raybuck SA, Murray RW (1987) Inorg Chem 26:1009
260. Pressprich KA, Maybury SG, Thomas RE, Linton RW, Irene EA, Murray RW (1989) J
     Phys Chem 93:5568
261. Li H, Guarr TF (1989) J Chem Soc Chem Commun 833
262. Bennett JE, Wheeler DE, Czuchajewski L, Malinski T (1989) J Chem Soc Chem Com-
     mun 723
263. Malinski T, Ciszewski A, Bennett JE, Fish JR (1991) J Electrochem Soc 138:2009
264. Eichhorn H, Sturm M, Wöhrle D (1995) Makromol Chem 196:115
265. Wöhrle D, Krawczyk G (1986) Makromol Chem 187:2535. Wöhrle D, Gitzel J (1988)
     Makromol Chem Rapid Commun 9:229. Wöhrle D, Gitzel J, Krawczyk G, Tsuchida E,
     Okura I, Nishisaka T (1988) J Macromol Sci Chem A25:1227. Wöhrle D, Krawczyk G,
     Paliuras M (1989) Makromol Chem 189:1001, 1013. Wöhrle D, Paliuras M, Okura I
     (1991) Makromol Chem 192:819
266. Achar BN, Fohlen GM, Parker JA (1983) J Polym Sci Polym Chem Ed 21:589

267. Osada Y, Mizumoto A, Tsurutu H (1987) J Macromol Sci Chem A24:403
268. Miyata S, Park YH, Soeda Y, Itoh R, Tasaka S (1987) Jpn J Appl Phys 26:L1632
269. Feucht C, Linssen T, Hanack M (1994) Chem Ber 127:113
270. Rack M, Hanack M (1994) Angew Chem 106:1712
271. Korshak VV, Vinogradova SV (1975) Faserf Textiltechnik 26:318
272. Kossmehl G, Rohde M (1976) Makromol Chem 180:345
273. Wöhrle D, Müller R (1976) Makromol Chem 177:2241
274. Müller R, Wöhrle D (1978) Makromol Chem 179:2161
275. Wöhrle D, Schulte B (1985) Makromol Chem 186:2229
276. Wöhrle D, Marose U, Knoop R (1985) Makromol Chem 186:2209
277. Wöhrle D, Preussner E (1985) Makromol Chem 186:2189
278. Wöhrle D, Hündorf U (1985) Makromol Chem 186:2177
279. Kaneko M, Wöhrle D (1988) Adv Polym Sci 84:141
280. Wöhrle D, Bannehr R, Jaeger N, Schumann B (1983) Angew Makromol Chem 117:103
281. Wöhrle D, Bannehr R, Jaeger N, Schumann B (1983) J Mol Catal 21:255
282. Wöhrle D, Schumann B, Schmidt V, Jaeger N (1987) Makromol Chem Macromol Symp 8:195
283. Wöhrle D, Schmidt V, Schumann B, Yamada A, Shigehara K (1987) Ber Bunsenges Phys Chem 97:975
284. Wöhrle D, Buck T, Hündorf, Schulz-Ekloff G, Andreev A (1989) Makromol Chem 190:961
285. Wöhrle D, Schulte B (1988) Makromol Chem 189:1167, 1229
286. Djurado D, Tadlaoui S, Hamwi A, Cousseins JC (1991) Synth Met 41–43:2595. Liao MS, Kuo KT (1993) Polymer J 25:947
287. Achar BN, Fohlen GM, Parker JA (1982) J Polym Sci Polym Chem Ed 20:1785
288. Buck T, Preussner E, Wöhrle D, Schulz-Ekloff G (1989) J Mol Catal 53:L17
289. Rheingold A (1987) Inorganic polymers. In: Kroschwitz J et al. (eds) Encyclopedia of polymer socience and engineering vol 8, Wiley-Interscience, New York, p. 138
290. Carraher C, Pittmann CU (1989) In: Elvers B et al. (eds) Ullmanns Encyclopedia of industrial chemistry vol A 14, VCH Publishers, New York, Weinheim, p. 241
291. Ray NH (1978) Inorganic Polymers. Academic Press, New York
292. Stone FGA, Graham WAG (eds) (1962) Inorganic Polymers. Academic Press, New York
293. Sharpe AG (1976) The chemistry of the cyano complexes of the transition metals. Academic Press, New York p. 275
294. Rayner JH, Powell HM (1958) J Chem Soc 3412
295. Kuroda R, Sasaki Y (1974) Acta Crystallogr Sect B 30:687
296. Buser HJ, Schwarzenbach D, Petter W, Ludi A (1977) Inorg Chem 16:2704. Kaneko M, Hou X-H, Yamada A (1984) Bull Chem Soc Jpn 57:156
297. Kimizuka N, Handa T, Ichinose I, Kunitake T (1994) Angew Chem 106:2576
298. Sonogashira K, Fujikura Y, Yatake T, Toyoshima N, Takahashi S, Hagihara N (1978) J Organomet Chem 145:101
299. Kaneko M (1986) J Polym Sci Polym Lett 14:435
300. Kaneko M (1987) J Macromol Sci Chem A 24:357
301. Allerton SE et al. (1966) J Am Chem Soc 88:3148
302. Spiro TG et al. (1966) J Am Chem Soc 88:2721
303. Spiro TG et al. (1967) J Am Chem Soc 89:5559
304. Foxman BM, Gersten SW (1986) Coordination polymers. In: Kroschwitz J et al. (eds) Encyclopedia of polymer science and engineering vol 4 Wiley-Interscience, New York, pp. 175–191
305. Hagihara N, Sonogashira K, Takahashi S (1981) Adv Polym Sci 41:149
306. Block BP (1970) Inorg Macromol Rev 1:115
307. Carraher CE, Sheats JE, Pittmann CU (eds) (1978) Organometallic polymers. Academic Press, New York
308. Sheats JE, Carraher CE, Pittmann CU (eds) (1985) Metal-containing polymeric systems. Plenum Press, New York

309. Sheats JE (1981) Metal containing polymers. In: Grayson M et al. (eds) Encyclopedia chemical technology vol 15, Wiley-Interscience, New York, p. 184
310. Tsuchida E, Nishide H (1977) Adv Polym Sci 24:1
311. Kaneko M, Yamada A (1984) Adv Polym Sci 55:1
312. Ibidapo TA (1990) Adv Polym Sci 30:1151
313. Carraher CE, Schroeder JA (1975) J Polym Sci Polym Lett Ed 13:215
314. Dey AK (1986) J Indian Chem Soc 63:357
315. Srivastava PC, Pandeya KB, Nigam HI (1973) J Inorg Nucl Chem 35:3613
316. Charles R (1963) J Polymer Sci A-1:267
317. Wrobleski JT, Brown DB (1979) Inorg Chem 18:498, 2738
318. Theocharis CR (1987) J Chem Soc Chem Commun 80
319. Rao TR, Rao PR, Lingaiah P, Deshmukh LS (1991) Angew Makromol Chem 191:177
320. Allen NS, Richards AM (1990) Eur Polym J 26:1229
321. Patel BK, Patel MM (1990) J Indian Chem Soc 67:186
322. Patel GC, Pancholi HB, Patel MM (1991) J Polym Mater 8:127, 339
323. Oyama N, Kitagawa M, Inaba H, Kawase K (1994) Makromol Symp 80:337
324. Reynolds JR, Lillya CP, Chien JCW (1987) Macromolecules 20:1184
325. Schumater RR, Engler EM (1977) J Am Chem Soc 99:5521
326. Rivera NM, Engler EM, Schumater RR (1979) J Chem Soc Chem Commun 184
327. Ribas J, Cassoux PC (1981) R Seances Acad Sci 293:665
328. Dirk CW, Mintz EA, Schoch KF, Marks TJ (1986) In: Carraher CE, Sheats JE, Pittmann CU (eds) Advances in organometallic and inorganic polymer science. Marcel Dekker, New York, p. 275
329. Tec BK, Wudl F, Hauser JJ, Krüger A (1977) J Am Chem Soc 99:4862
330. Böhm MC (1984) Phys Stat Sol (B) 121:255
331. Kawai S, Hamaguchi H, Tatsumoto M (1956) Bunseki Kagaku 5:165
332. Kidani Y (1958) Chem Pharm Bull Jpn 6:563
333. Berg FW, Alam A (1963) Anal Chim Acta 28:128
334. Adachi J (1955) Nippon Kagaku Zasshi 76:311
335. Patel RD (1986) Makromol Chem 187:1871
336. Korshak VV, Vinogradova SV, Bachmistu TM (1960) Vysokomol Soedin 2:498
337. Berg EM, Alam A (1962) Anal Chim Acta 27:454
338. Poddar SN, Saha N (1970) Indian J Appl Chem 33:244
339. Banerjie V et al. (1979) Polymer Bull 1:685
340. Banerjie V et al. (1980) Makromol Chem Rapid Commun 1:41
341. Kumari VD, Khave M, Munishi KN (1985) Indian J Chem 24A:72
342. Patel MN, Patel JB (1979) Angew Makromol Chem 80:171
343. Hassan MK (1990) J Macromol Sci A27:1501
344. Archer RD, Illingworth ML, Rau DN, Hardiman CJ (1985) Macromolecules 18:1371
345. Rana AK et al. (1981) Makromol Chem 182:3387
346. Maurya PL, Agarwala BU, Dey AK (1982) Makromol Chem 183:511
347. Marcos M, Oriol L, Serrano JL, Alonso PJ, Puertolas J (1990) Macromolecules 23:5187
348. Hurd NN et al. (1960) J Am Chem Soc 82:4455
349. Hurd NN et al. (1961) J Org Chem 26:3980
350. Kim O-K, Tsai T, Yoon TH, Choi LS (1989) J Chem Soc Chem Common 740
351. Khandwe M, Bajpai A, Bajpai UDN (1991) Macromolecules 24:5203
352. Zotti G, Mengoli G, Decker F (1985) Mol Cryst Liq Cryst 121:337
353. Wöhrle D (1991) In: Kricheldorf HR (ed) Handbook of polymer synthesis (Part B) Marcel Dekker, New York, p. 1113
354. Miller RD (1989) Angew Chem Adv Mater 101:1773
355. West R (1983) In: Wilkenson G, Stone FGA, Abel EW (ds) Organopolysilanes, comprehensive organometallic chemistry, vol 9. Pergamon Press Oxford, pp. 365–397
356. West R (1986) J Organomet Chem 300:327
357. Rheingold AL (1977) Homoatomic rings, chains and macromolecules of main-group elements. Elsevier Amsterdam

358. Ingham RK, Gilman H (1962) In: Stone FGA, Graham WAG (eds) Inorganic polymers. Academic Press, New York, p. 321
359. Satge J, Massol M, Rieviere R (1973) J Organomet Chem 56:1
360. Shu H-C, Wunderlich B (1980) J Polym Sci Poly Phys Ed 18:443
361. Brown TL, Morgan GL (1963) Inorg Chem 2:736
362. Mitchell TN (1975) J Organomet Chem 92:311
363. Grugel C, Neumann WP, Leifert P (1977) Tetrahedron Lett 2205
364. Rheingold AL, Lewis JF, Bellama JM (1973) Inorg Chem 12:2845
365. Daly JJ, Sanz F (1970) Helv Chim Acta 53:1879
366. Choudhury P, El-Shazly MF, Spring C, Rheingold AL (1979) Inorg Chem 18:543
367. Wieber M (ed) (1981) Organo-antimon compounds. In: Gmelin handbook of inorganic chemistry, Springer, Berlin, 8th ed, part 2, p. 150
368. Rheingold AL, Choudhury P (1977) J Organomet Chem 128:155
369. Breunig HJ, Kanig W (1980) J Organomet Chem 186:C5−C8
370. Carraher CE (1981) J Chem Ed 58:921
371. Carraher CE, Scott WJ, Schroeder JA (1981) J Macromol Sci A 15:625
372. Methoden der Organischen Chemie (Houben-Weyl) (1987) Thieme, Stuttgart, vol E20; Wulff G, pp. 1772−1793, Kochs P, pp. 2219−2236, Zipplies T, Hanack M, pp. 2237−2249
373. Moretto E et al. (1993) Silicones. In: Elvers B et al. (eds) Ullmanns Encyclopedia of Industrial Chemistry vol A 24, VCH Publishers, New York, Weinheim, p. 57. Hardman B, Torkelson A (1989) Silicones. In: Kroschwitz J et al. (eds) Encyclopedia of polymer science and engineering vol 15, Wiley-Interscience, New York, p. 204
374. Rochow EG (1987) Silicon and silicones. Springer, Berlin Heidelberg New York
375. Wick M, Kreis G, Kreuzer F-H (1982) In: Ullmanns Encyclopädie der technischen Chemie, Verlag Chemie, vol 21, pp. 511−543
376. Wright PV (1984) In: Ivin K, Saegusa T (eds) Ring opening polymerization. Elsevier, Amsterdam p. 1055
377. Lichtenwalner HK, Sprung MN (1970) In: Mark F et al. (eds) Encyclopedia of polymer science and technology. Wiley-Interscience, New York, vol 12, pp. 464−577
378. Semylen JA (ed) (1986) Cyclic polymer. Elsevier, London
379. Voronkov MG, Mileshkevich VP, Yzhelevskii YA (1976) The siloxane bond. Consultants Bureau, New York
380. Lynch W (1978) Handbook of silicon rubber fabrication. Van Nostrand Rheinhold, New York
381. Noll W (1968) Chemie und Technik der Silikone, Verlag Chemie, Weinheim
382. Ranney MW (1977) Silicones. Noyes Data Corp, Park Ridge, NY, vols 1, 2
383. Dumnavant WR (9171) Inorg Macromol Rev 1:165
384. Brown MP, Rochow EG (1960) J Am Chem Soc 82:4166
385. Ross L, Dräger M (1984) Z Naturforsch 39b:868
386. Trefonas P, West R, Miller RD, Hofer D (1983) J Polym Sci Polym Lett Ed 21:823
387. Riviere P, Castel AL, Satge J (1980) J Am Chem Soc 102:5413
388. Baceiredo A, Bertrand G, Mazerolles P (1981) Tetrahedron Lett 22:2553
389. Kobayashi S, Iwata S, Yajimak K, Yagi K, Shoda S (1992) J Am Chem Soc 114:4929
390. Carraher CE, Schroeder JA (1975) J Polym Sci Polym Lett Ed 13:215
391. Labadie JW, MacDonald SA, Willson CG (1986) Polym Bull 16:427
392. Pearce AA (1959) J Polym Sci 40:273
393. Adrova NA, Koton MM, Klages VA (1960) Vysokomol Soedin 3:1041
394. Green MLH (1968) In: Coat GE, Green MLH (eds) Organomettalic compounds Methuen, London, 3rd ed vol 2, p. 203
395. Davidson PJ, Lappert MF, Pearce R (1976) Chem Rev 2:219
396. Schrock RR, Parshall GW (1976) Chem Rev 2:243
397. Hay AS (1969) J Polym Sci A-1 7:1625
398. Matsuda H, Nakanishis N, Karo M (1984) J Polym Sci Polym Lett Ed 22:107
399. Sonogashira K, Takahashi S, Hagihara N (1977) Macromolecules 10:879

400. Takahashi S (1980) J Polym Sci Polym Chem Ed 18:661
401. Sonogashira K (1980) J Organometal Chem 188:237
402. Sonogashira K, Fujikura Y, Yatake T, Toyoshima N, Takahashi S, Hagihara N (1978) J Organomet Chem 145:101. Takahashi S, Kariya M, Yatake I, Sonogashira K, Hagihara N (1978) Macromolecules 11:1063. Takahashi S, Murata E, Kariya M, Sonigashira K, Hagihara N (1979) Macromolecules 12:1016
403. Lang H (1994) Angew Chem 106:569
404. Johnson BFG (1991) J Mater Chem 1:485
405. Abe A, Tabata S, Kimur N (1991) Polym J 23:69
406. Dray AT et al. (1992) Macromolecules 25:3473
407. Blau WJ (1991) J Mater Chem 1:245
408. Johnson BFG (1991) J Organomet Chem 409:C12
409. Kotani S, Shiina K, Sonogashira K (1991) Appl Organomet Chem 5:417
410. Nishihara H, Shimura T, Ohkubo A, Matsuda N, Aramaki K (1993) Adv Mater 5:752 Tomita I, Nishio A, Igarashi T, Endo T (1993) Polym Bull 30:179
411. Schultz H, Lehmann H, Rein M, Hanack M (1991) Struct Bond 74:41
412. Hanack M, Lang M (1994) Adv Mater 6:819
413. Joyner RD, Kenney ME (1960) J Am Chem Soc 82:5790
414. Joyner RD, Kenney ME (1962) Inorg Chem 1:717
415. Owen JE, Kenney ME (1962) Inorg Chem 1:334
416. Wöhrle D, Meyer G (1974) Makromol Chem 175/3:715
417. Meyer G, Hartmann M, Wöhrle D (1975) Makromol Chem 176:1919
418. Hartmann M, Meyer G, Wöhrle D (1975) Makromol Chem 176:831
419. Berezin BD, Akopov AS (1974) Vysokomol Soyed A-16:450, 2334
420. Dirk CW, Inabe T, Schoch KF, Mark TJ (1983) J Am Chem Soc 105:1539
421. Diehl BN, Inabe T, Lyding JW, Schoch KF, Kannewurf CR, Marks TJ (1983) J Am Chem Soc 105:1551
422. Orthmann EA, Enkelmann V, Wegner G (1983) Makromol Chem Rapid Commun 4:687
423. Orthmann EA, Wegner G (1986) Makromol Chem Rapid Commun 7:243
424. Hanack M, Metz J, Pawlowski G (1982) Chem Ber 115:2836
425. Metz J, Pawlowski G, Hanack M (1983) Z Naturforsch 38b:378
426. Hanack M, Zipplies T (1985) J Am Chem Soc 107:6127
427. Meyer G, Wöhrle D (1977) Z Naturforsch 32b:723
428. Hanack M, Münz X (1985) Synth Met 10:357
429. Beltsios KG, Carr SH (1989) J Polym Sci C27:355
430. Shirlin C, Bosio L, Simon J (1988) J Chem Soc, Chem Commun 236; (1988) Mol Cryst Liq Cryst 155:231
431. Caseri W, Sauer T, Wegner G (1988) Makromol Chem Rapid Commun 9:651; (1990) Mol Cryst Liq Cryst 183:387
432. Orthmann E, Wegner G (1986) Angew Chem 98:1114
433. Sauer T, Wegner G (1989) Makromol Chem, Macromol Symp 24:303
434. Ferenz A, Ries R, Wegner G (1993) Angew Chem 105:1251
435. Sielken OE, van de Kuil LA, Drenth W, Schoonman J, Nolte RJM (1990) J Am Chem Soc 112:3086
436. Kentgens APM, Markies BA, van der Pol JF, Nolte RJM (1990) J Am Chem Soc 112:8800
437. Crockett RGM, Campbell AJ, Ahmed FR (1990) Polymer 31:602
438. Dulog L, Gittinger A, Roth S, Wagner T (1993) Makromol Chem 194:493; (1993) Mol Cryst Liq Cryst 237:235
439. Schwiegk S, Fischer H, Xu Y, Kremer F, Wegner G (1991) Makromol Chem, Macromol Symp 46:211
440. Schwiegk S, Werth M, Leisen J, Wegner G, Spiess HW (1993) Acta Polym 44:31
441. Meier H, Albrecht W, Zimmerhackl E, Hanack M, Metz J (1985) Synth Met 11:333
442. Fischer K, Hanack M (1983) Chem Ber 116:1860
443. Meyer G, Plieninger P, Wöhrle D (1978) Angew Makromol Chem 72:173

444. Hanack M, Mitulla K, Pawlowski G, Subramanian LR (1980) J Organomet Chem 204:315
445. Mitulla K, Hanack M (1980) Z Naturforsch B 35:1111
446. Linski JF, Paul TR, Nohr RS, Kenney ME (1980) Inorg Chem 19:3131
447. Nohr RS, Kuznesof PM, Wynne KJ, Kenney ME, Siebenmann PG (1981) J Am Chem Soc 103:4371
448. Klofta TJ, Rieke PC, Linkous CA, Buttner WJ, Nathakumar A, Newborn TD, Armstrong NP (1985) J Electrochem Soc 132:2134
449. Hanack M, Beck A, Lehmann H (1987) Synthesis:703
450. Metz J, Hanack M (1983) J Am Chem Soc 105:828
451. Schwartz M, Hatfield WE, Joesten MD, Hanack M, Datz A (1985) Inorg Chem 24:4188
452. Schneider O, Hanack M (1984) Z Naturforsch 39b:265
453. Ziener U, Fahmy N, Hanack M (1993) Chem Ber 126:2559
454. Hedtmann-Rein C, Hanack M, Peters K, Peters E-M, Schnering HG von (1987) Inorg Chem 26:2647
455. Metz J, Hanack M (1987) Chem Ber 120:1307
456. Schneider O, Hanack M (1980) Angew Chem Int Engl Ed 19:392
457. Schneider O, Hanack M (1983) Chem Ber 116:2088
458. Keppeler U, Hanack M (1986) Chem Ber 119:3363
459. Kobel W, Hanack M (1986) Inorg Chem 25:103
460. Diehl BN et al. (1984) J Am Chem Soc 106:3207
461. Deger S, Hanack M (1986) Isr J Chem 27:347
462. Koch JW, Hanack M (1983) Chem Ber 116:2109
463. Meier H, Albrecht W, Hanack M, Koch J (1986) Polym Bull 16:75
464. Koch JW, Hanack M (1987) Chem Ber 120:1853
465. Hanack M, Keppeler U, Schulze H-J (1987) Synth Met 20:347
466. Hanack M, Leverenz A (1987) Synth Met 22:9
467. Hanack M, Osia-Barcina J, Witke E (1992) Synthesis: 211
468. Hanack M, Gül A, Subramanian LR (1992) Inorg Chem 31:1542
469. Hanack M, Kang Y-G (1992) Synth Met 48:79
470. Hanack M, Ryu H (1992) Synth Met 46:113
471. Hanack M, Grosshans R (1992) Chem Ber 125:1243
472. Giroud-Godquin A-M, Maitlis PM (1991) Ange Chem 103:370
473. Snow AN, Barger WR (1989) In: Leznoff CC, Lever ABP (eds) Phthalocyanines, properties and applications. VCH Publishers, New York, p. 341
474. Nolte RJM, Drenth W (1992) In: Laine RM (ed) Inorganic and organometallic polymers with special properties. Kluwer Academic Publishers, The Netherlands, p. 223
475. Espinet P, Esternelas MA, Oro LA, Serrano JL, Sola E (1992) Coord Chem Rev 117:215
476. Engel MK, Bassoul P, Bosio L, Lehmann H, Hanack M (1993) Liq Cryst 15:704
477. Chandraskhar S, Ranganath GS (1990) Rep Prog Phys 53:570
478. Weber P, Guillon D (1991) Liq Cryst 9:369
479. Collard DM, Lillye CP (1991) J Am Chem Soc 113:8577
480. Schouten PG, van der Pol JF, Zwikker JW, Drenth W, Picken SJ (1991) Mol Cryst Liq Cryst 195:291
481. Van der Pol JF, Neeleman E, Nolte RJM, Zwikker JW (1989) Makromol Chem 190:2727
482. Van Nostrum CF, Nolte RJM, Devillers MAC, Oestergetel GT, Teerenstra MN, Schouten AJ (1993) Macromolecules 26:3306
483. Sielcken OE, van Lindert HCA, Drenth W, Schoonman J, Schram J, Nolte RJM (1989) Ber Bunsenges Phys Chem 93:702
484. Fuhrhop J-H, Binding U, Demoulin C, Rosengarten B (1994) Macromol Symp 80:63
485. Ishikawa Y, Kunitake T (1991) J Macromol Sci-Chem A 27:1157
486. Neuse EW (1968) In: Mark HF et al. (eds) Encyclopedia of polymer science and technology. Wiley-Interscience, New York, vol 8, p. 667
487. Neuse EW, Rosenberg H (1970) Metallocene polymers. Marcel Dekker, New York
488. Neuse EW (1981) J Macromol Sci Chem A16:3
489. Rosenblum M (1994) Adv Mater 6:159

490. Manners I (1994) Adv Mater 6:68
491. Sieber W (1991) Russ Chem Rev 60:784
492. Neuse EW, Bednarik L (1979) Macromolecules 12:187, Transition Metal Chem 4:104
493. Bednarik L, Neuse EW (1980) J Org Chem 45:2032
494. Rausch MD (1972) Pure Appl Chem 30:523
495. Roling PU, Rausch MD (1977) J Organomet Chem 141:195
496. Nugent HM, Rosenblum M, Klemarczyk P (1993) J Am Chem Soc 115:3848
497. Neuse EW, Trifan DS (1963) J Am Chem Soc 85:1952
498. Neuse EW, Quo E, Howells WG (1965) J Org Chem 30:4071
499. Neuse EW, Quo E (1965) J Polym Sci (Part A) 3:1499
500. Gal'A, Cais M, Kohn DH (1971) J Polym Sci (Part A-I) 9:1833
501. Neuse EW (1966) J Organomet Chem 6:92
502. Hmyene M, Yasser A, Escorne M, Percheron A, Garnier F (1994) Adv Mater 6:564
503. Nuyken O, Burckhardt V, Pöhlmann T, Heberhold M (1991) Makromol Chem, Macromol Symp 44:195
504. Tenhaeff SC, Tyler DR (1991) Organometallics 10:473
505. Tsuchida E (1991) Macromolecular complexes. VCH Publishers, New York
506. Kaneko M, Wöhrle D (1988) Adv Polym Sci 84:141
507. Scott NS, Oyama N, Anson FC (1980) J Electroanal Chem 110:303
508. Demas JN, DeGraff BA (1992) Makromol Chem, Macromol Symp 59:35; (1988) J Macromol Sci Chem A 25:1189
509. Kaneko M, Yamada A (1984) Adv Polym Sci 55:1; (1985) In: Sheats JE, Carraher CE, Pittman CU (eds) Metal containing polymer systems. Plenum Press, New York, p. 249
510. Kaneko M, Hayakawa S (1988) J Macromol Sci Chem A25:1255
511. Wöhrle D, Paliuras M, Okura I (1991) Makromol Chem 192:819
512. Okamoto Y, Kido J, Brittain HG, Paoletti S (1988) J Macromol Sci Chem A 25:1385
513. Okamoto Y (1992) Makromol Chem, Macromol Symp 59:83
514. Okamoto Y, Kido J (1991) In: Tsuchida E (ed) Macromolecular complexes, VCH Publishers, New York, p. 143
515. Kurimura Y, Sairenchi Y, Nakayama S (1992) Makromol Chem, Macromol Symp 59:199
516. Kurimura Y (1991) In: Tsuchda E (ed) Macromolecular complexes. VCH Publishers, New York, p. 93
517. Turk H, Ford WT (1988) J Org Chem 53:460
518. Hassanein M, Ford WT (1989) J Org Chem 54:3106
519. Van Herk AM, van Streun KH, van Welzen J, German AL (1989) Br Polym J 21:125
520. Van Streun KH, Tennebroek R, Piet R, German AL (1990) Makromol Chem 191:2181
521. Turk H, Ford WT (1991) J Org Chem 56:1253
522. Hassanein M, Aly E-SA, Abbas YA, El-Sigeny SM (1993) Makromol Chem 194:1817. Hassanein M, Salim A, El-Hamshary H (1994) Makromol Chem 195:3845
523. Yoshida T, Shirasagi T, Kaneko M (1992) Denki Kagaku 60:1108
524. Yoshida T, Kamoto K, Tsukamoto M, Iida T, Schlettwein D, Wöhrle D, Kaneko M (1995) J Electroanal Chem 385:204
525. Skotheim T, Velazquez M, Linhous CA (1985) J Chem Soc, Chem Commun:612
526. Bull RA, Fan FR, Bard AJ (1984) J Electrochem Soc 131:687
527. Choi CS, Tachikawa H (1990) NTIS Chem 90:20, 37 AD-A 217 677
528. Walton DJ, Hall CE (1991) Synth Met 45:363
529. Schlettwein D, Kaneko M, Yamada A, Wöhrle D, Jaeger N (1991) J Phys Chem 95:1748
530. Law K-Y (1985) J Phys Chem 89:2652
531. Loutfy RO (1981) Can J Chem 59:549
532. Loutfy RO (1982) J Phys Chem 86:3302
533. Fujimaki Y (1991) Proceedings of the 7th International Congress on advances in ion impact printing technology. Portland, Oregon, p. 269
534. Law K-Y (1993) Chem Rev 93:449
535. Schlettwein D, Jaeger NI, Wöhrle D (1992) Makromol Chem, Macromol Symp 59:267
536. Loutfy RO, McIntyre LF (1983) Can J Chem 61:72

537. Shimidzu T, Iyoda T (1984) Polym J 16:919
538. Minami N, Sasaki K, Tsuda K (1983) J Appl Phys 54:6764
539. Wöhrle D, Kaune H, Schumann B, Jaeger NI (1986) Makromol Chem 187:2947
540. Hirai H (1990) J Macromol Sci Chem A 27:1293
541. Toshima N, Kanaka K, Komiyama M, Hirai H (1988) J Macromol Sci Chem A 25:1349
542. Bronstein LM, Larikova IE, Valetsky PM (1987) Polym Sci USSR 29:2653
543. Mirzoeva ES, Bronstein LM, Valetsky PM (1989) Polym Sci USSR 31:2898
544. Jia C-G, Jin F-Y, Pan H-Q, Hung M-Y, Jiang Y-Y (1984) Macromol Chem Phys 195:751
545. Pan C, Zong H (1994) Macromol Symp 80:265
546. Sherrington DC, Tang H-G (1994) Macromol Symp 80:193
547. Shirai H, Hanabusa K, Koyama T, Tsuiki H, Masuda, E, Makromol Chem, Macromol Symp 59:155
548. Ono K (1985) High Polym Jpn 34:766
549. Orihara K, Yonekura H (1990) J Macromol Sci Chem A 27:1217
550. Hegenmüller P, Gool WV (eds) (1980) Solid electrolytes. Plenum Press, New York
551. Bogdanov B, Uzov C, Michaelov M (1992) Acta Polym 43:202
552. Pomogailo AD (1994) Platinum Met Rev 38:60
553. Schmid G (1992) Chem Rev 92:1709
554. Martin GA (1988) Catal Rev Sci Eng 30:519. Burch R (1982) Acc Chem Res 15:24
555. Hirai H, Toshima N (1986) In: Iwasawa Y (ed) Tailored metal catalysts. Reidel, Dordrecht, p. 87
556. Grätzel M (1989) Heterogeneous photochemical electron transfer. CRC Pess, Boca Raton
557. Ozin GA, Steele MR (1994) Macromol Symp 80:45. Ozin GA, Kuperman A, Stein A (1989) Angew Chem 101:373. Ozin GA (1994) Adv Mater 6:71
558. Bond GC (1956) Trans Faraday Soc 52:1235
559. Kiwi J, Grätzel M (1978) J Am Chem 101:7214. Turkevich J, Aika K, Ban LL, Okura I, Namba S (1976) Res Inst Catal Hokkaido Uni 24:54
560. Grätzel M (1980) Ber Bunsenges Phys Chem 84:981. Duonglong D, Borgarello E, Grätzel M (1981) J Am Chem Soc 103:4685
561. Fojitk A, Weller H, Koch U, Henglein A (1984) Ber Bunsenges Phys Chem 88:969
562. Nenadovic MT, Micic OI, Rajh T, Savic D (1983) J Photochem 21:35
563. Toshima N (1990) J Macromol Sci Chem A 27:1225. Hirai H, Nakao Y, Toshima N (1979) J Macromol Sci Chem A 13:727
564. Toshima N, Takahashi T, Hirai H (1985) Chem Lett 1245. Ohtaki M, Toshima N (1990) Chem Lett 489. Harada M, Asakura K, Toshima N (1994) J Phys Chem 98:2653. Toshima N, Wang Y (1974) Adv Mater 6:245
565. Belyi AA, Chigladze LG, Rusanov AL, Volpin ME (1989) Izv Akad Nauk SSSR Ser Khim 1961, 2678
566. Toshima N (1991) In: Tsuchida E (ed) Macromolecular complexes. VCH Publishers, New York, p. 321
567. Kiwi J, Grätzel M (1979) Nature 281:657
568. Toshima N, Kuriyama M, Yamada Y, Hirai H (1981) Chem Lett 793
569. Keller P, Moradpour A, Amouyal E, Kagan H (1980) J Mol Catal 7:539
570. Toshima N, Yamada Y, Hirai H (1981) Polym Prepr Jpn 30:416, 1500
571. Akashi M, Motomura T, Miyaushi N, O'Driscoll KF, Rempel GL (1981) Polym Prepr Jpn 30:1496
572. Xu B, Fichou D, Horrowitz G, Garnier F (1991) Adv Mater 3:150
573. Cardenas G, Munoz C (1993) Makromol Chem 194:3377
574. Miyonaga S, Yasuda H, Hiwara A, Nakumura A (1990) J Macromol Sci Chem A 27:1347
575. Cooke TF (1990) J Polym Eng 9:1
576. Neenan TX, Callstrom MR, Schneller OJA (1994) Macromol Symp 80:315
577. Pocard NL, Alsmeyer DC, McCreeny RL, Neenan TX, Callstrom MR (1991) J Am Chem Soc 114:769
578. Ohlson AW, Kafafi ZH (1991) J Am Chem Soc 113:7758
579. Perrin J, Despax B, Kay E (1985) Phys Rev B 32:719

580. Lamber R, Baalman A, Jaeger NI, Schulz-Ekloff G, Wetjen S (1994) Adv Mater 6:223
581. Heilmann A, Kampfrath G, Hopfe V (1988) J Phys D Appl Phys 21:986
582. An Y, Yuan D, Huang M-Y, Jiang Y-Y (1994) Macromol Symp 80:257. Wang X, Huang M-Y, Jiang Y-Y (1992) Makromol Chem, Macromol Symp 59:113. Tang L-M, Huang M-Y, Jiang Y-Y (1994) Macromol Rapid Common 15:527
583. Challa G, Chen W, Reedijk J (1992) Makromol Chem, Macromol Symp 59:59
584. Ciardeli F, Altomare A, Conti G, Arribas G, Mendez B, Ismayel A (1994) Macromol Symp 80:29
585. Guan S-Y, Huang M-Y, Jiang Y-Y (1992) Makromol Chem Macromol Symp 59:53
586. Kurusu Y (1992) Makromol Chem, Macromol Symp 59:313
587. Grätzel M (1991) Comments Inorg Chem 12:93. O'Reagan B, Grätzel M (1991) Nature 353:737
588. Nazeeruddin MK, Liska P, Moser J, Vlachopoulos N, Grätzel M (1990) Helv Chim Acta 73:1788
589. Amadelli R, Argazzi R, Bignozzi CA, Scandola F (1990) J Am Chem Soc 112:7099
590. Wöhrle D, Buck T, Hündorf U, Schulz-Ekloff G, Andreev A (1989) Makromol Chem 190:961. Fischer H, Schulz-Ekloff G, Buck T, Wöhrle D, Vassileva M, Andreev A (1992) Langmuir 8:2720
591. Buck T, Bohlen H, Wöhrle D, Schulz-Ekloff G, Andreev A (1993) J Mol Catal 80:253
592. Carrado KA, Thiyagarajan P, Winans RE, Botto RE (1991) Inorg Chem 30:794
593. Jones W (1991) In: Ramamurthy V (ed) Photochemistry in organized and constrained media. VCH Publishers, New York, p. 387
594. Suib SL (1993) Chem Rev 93:803
595. Cady SS, Pinnavaia TJ (1978) Inorg Chem 17:1501
596. Galan-Mascaros JR, Gomez-Garcia CJ, Baras-Almenar JJ, Coronado E (1994) Adv Mater 6:221
597. Barrer RM (1982) Hydrothermal synthesis of zeolites. Academic Press, London
598. Jacobs PA, Marten JA (1987) Synthesis of high silicon alumino silicates. In: Stud Surf Sci Catal vol 33
599. Breck DW (1974) Zeolites molecular sieves. Wiley-Interscience, Toronto
600. Ramamurthy V (1991) Photochemistry in organized and constrained media. VCH Publishers, New York
601. De Vos DE, Thibault-Starzyk F, Knops-Gerrits PP, Parton RF, Jacobs PA (1994) Macromol Symp 80:157
602. Romanovsky BV (1994) Macromol Symp 80:185
603. Wöhrle D, Schulz-Ekloff G (1994) Adv Mater 6:875
604. Gallezot P (1979) Catal Rev Sci Eng 20:121
605. Ozin GA, Gil C (1989) Chem Rev 89:174
606. Sachtler WMH, Zhang Z (1993) Adv Catal 39:129
607. Jacobs PA (1986) Stud Surf Sci Catal 29:357
608. Stucky GD, MacDougall JE (1990) Science 247:669
609. Ozin GA, Steele MR (1994) Macromol Symp 80:45
610. Ramamurthy V, Eaton DE, Caspar JV (1992) Acc Chem Res 25:299
611. Caro J, Marlow F, Wübbenhorst M (1994) Adv Mater 6:413
612. Peigneur P, Lunsford JH, De Wilde W, Schoonheydt RA (1977) J Phys Chem 81:1179
613. Bein T, Jacobs PA (1984) J Chem Soc Faraday Trans I 80:1391
614. Bornvornwattanamont A, Bein T (1992) J Phys Chem 96:6713, 9447
615. Meyer G, Wöhrle D, Mohl M, Schulz-Ekloff G (1984) Zeolites 4:30
616. DeVos DE, Thibault-Starzyk F, Jacobs PA (1994) Angew Chem 106:447
617. Herron N (1986) Inorg Chem 25:4714
618. Lubitz W, Winscom CJ, Diegruber H, Möseler R (1987) Z Naturforsch 42a:970
619. Hoppe R, Schulz-Ekloff G, Rathausky J, Stark J, Zukal A (1994) Zeolites 14:126
620. Hoppe R, Schulz-Ekloff G, Wöhrle D, Kirschhock C, Fuess H (1994) In: Stud Surf Sci Catal 84:821
621. Franke O, Sobbi A, Schulz-Ekloff G, Wöhrle D (1995) Zeolites (in press)

622. Hoppe R, Schulz-Ekloff G, Wöhrle D, Shpiro E, Tkachenko OP (1993) Zeolites 13:222
623. Wohlrab S, Hoppe R, Schulz-Ekloff G, Wöhrle D (1992) Zeolites 12:862
624. Ehrl M, Deeg FW, Bräuchle C, Franke O, Sobbi A, Schulz-Ekloff G, Wöhrle D (1994) J Phys Chem 98:47
625. Kowalak S, Balkus KJ (1992) Collect Czech Chem Commun 57:774
626. Shpiro ES, Antoshin GV, Tkachenko OP, Gudhov SV, Romanovsky BV, Minachev KM (1984) Structure and reactivity of modified zeolites. In: Stud Surf Sci Catal vol 18, p. 31
627. Hoppe R, Schulz-Ekloff G, Wöhrle D, Ehrl M, Bräuchle C (1991) Stud Surf Sci Catal 69:199
628. Hoppe R, Schulz-Ekloff G, Wöhrle D, Kirschhock C, Fuess H (1994) Langmuir 10:1517
629. Schulz-Ekloff G (1995) In: Lehn J-M (ed) Comprehensive supramolecular chemistry, vol 11, Pergamon Press, Oxford (in press)
630. Mortier WJ (1982) compilation of extra framework sites in zeolites. Butterworth, London
631. Schoonheydt RA (1993) Catal Rev Sci Eng 35:129
632. Schulz-Ekloff G, Czarnetzki L, Zukal A (1987) J Chem Soc Faraday Trans I, 83:3015
633. Exner D, Jaeger NI, Möller K, Schulz-Ekloff G (1982) J Chem Soc Faraday Trans I, 78:3537
634. Zhang Z, Sachtler WMH, Chen H (1990) Zeolites 10:784
635. Gallezot P (1979) Catal Rev Sci Eng 20:121
636. Sachtler WMH, Zhang Z (1993) Adv Catal 39:129
637. Bischoff H, Jaeger NI, Schulz-Ekloff G, Kubelkova L (1993) J Mol Catal 80:95
638. Schulz-Ekloff G, Lipski RJ, Jaeger NI, Hülstede P, Kubelkova L (1995) Catal Lett 30:67
639. Möller K, Bein T (1990) J Phys Chem 94:845
640. Xu L, Zhang Z, Sachtler WMH (1992) J Chem Soc Faraday Trans I, 88:2291
641. Schulz-Ekloff G (1991) Stud Surf Sci Catal 69:65
642. Chen C-W, Huang M-Y, Jiang Y-Y (1994) Macromol Rapid Commun 15:587. Yang XT, Tian J-X, Huang M-Y, Jiang Y-Y (1994) Macromol Rapid Commun 14:485
643. Tsuchida E, Abe K (1982) Interaction between macromolecules in solution and intermacromolecular complexes. Advances in Polymer Science, vol 45. Springer, Berlin Heidelberg New York
644. Tsuchida E, Abe K (1985) Polyelectrolyte complexes. In: Wilson P (ed) Developments in ionic polymers II. Elsevier, Amsterdam, chap. 5
645. Knapp R, Rehahn M (1993) Makromol Chem Rapid Commun 14:451

3 Polymer-Metal Complexes in Living Systems

J. Reedijk

3.1 Introduction

In all living systems metal ions, either as isolated ions or in clusters, play an essential role. This role deals with both growth and metabolism. The role of the metal ion as essential for life has been known since the 18th century, whereas the role of other elements, such as cobalt, copper, manganese and zinc, has been reported since the beginning of the 20th century.

Not only has the beneficial role of elements been known for a long time: An excess of these elements can be very dangerous. As a matter of fact, a narrow concentration window exists for most of the so-called trace elements. Although many elements of the periodic table are presently known and accepted to be essential or beneficial for life on earth, the molecular role of these elements is only beginning to be understood [1–3].

It has been known for many years that several non-essential elements of the periodic table also have an influence on the quality of life, either as a toxic pollutant or as a drug to cure certain diseases. Well-documented examples are available, although still little is understood about the details of their modes of action at the molecular level. An important challenge of modern bioinorganic chemistry is to understand the molecular basis of such interactions and to apply this knowledge in biology, environmental sciences, medicine, polymer sciences, catalysis and technology. This chapter focuses on the coordination aspects of these metals, both in their natural environments (e.g. bound to proteins and nucleic acids) and on the effect of non-essential metal compounds on these systems. Isolated natural polymers such as macrocyclic ligands, are described also and classified in Chap. 2.1.1.4 and 2.2.5 of this monograph.

As an introduction, first some general aspects of bioinorganic chemistry are presented. Table 3.1 lists a selection of important elements together with some statistical information and a few comments about their biological role. The table contents are restricted to some of the most important transition elements and to Ca and Mg. For full details the reader is referred to a comprehensive overview [3].

Bioinorganic chemistry has grown within a relatively short period into a large subdiscipline of science. Scientists from a variety of disciplines contribute to this interdisciplinary area. The important role of metals, metal ions and metal compounds in relation to living systems is a subject of common interest in this field. Furthermore, subdivision of the field has been made along different lines:

Table 3.1. Short overview of important elements in biology. Occurrence, concentration of element and biological roles

Element name	Concentrations in human species (mg)	Daily dose (in mg)[a]	Comments
Iron	4500	10	Many enzymes; respiratiory proteins
Zinc	2000	12	Hydrolytic enzymes; nucleic acid synthesis
Copper	100	2	Many enzymes; dioxygen transport
Manganese	20	3	Enzyme activation, photosynthesis
Molybdenum	5	0.2	Many redox enzymes; nitrogenase (plants)
Cobalt	1	0.3	Vitamin B-12
Vanadium	< 0.1	< 0.1	Role in bromoperoxidases
Tungsten	< 0.1	< 0.1	Role in dehydrogenases
Nickel	< 0.1	< 0.1	Role in hydrogenases
Calcium	10^6	800	Bones, teeth; muscle activation
Magnesium	$4 \cdot 10^4$	350	Photosynthesis; nucleic acid processes

[a] Generally accepted amounts for humans.

1. The details of the metal coordination environment in metalloproteins, nucleic acids, carbohydrates and membranes can be studied in great detail.
2. The process of biomineralization, i.e. the build in of metal ions into solids, is increasingly being studied.
3. The mechanism of reactions occurring at a metal centre in an enzyme is the subject of many studies.
4. Genetically modified active-site structures in metalloproteins are becoming more and more available, almost routinely.
5. The study of synthetic analogues for the active sites in metalloproteins; design, synthesis, structure and spectroscopy is of great interest for many "low-molecular-weight" chemists.
6. The study of the catalytic reactions performed with synthetic analogues for the active sites in metalloproteins allows new applications.
7. Metal-containing drugs to prevent or cure diseases; synthesis of new compounds and mechanism of action are increasingly studied by chemists.
8. Removal and transport of metal ions and metal compounds to and from living systems (detoxification studies).

From a coordination point of view the interaction between metal ions and biomacromolecules is essential, and the focus of this chapter. The metal and its surrounding neighbours (called ligands) are of importance for the structure, stability and processes regulated or distorted by the metal species. The study of metal ions and their interactions with ligands is usually called *coordination chemistry*, or more specifically *biocoordination chemistry*, when either the metal or the ligand has a biological role. The interactions between metals and ligands in biological systems play a key role in almost every important event that takes place during biological processes, both natural processes and human induced processes (e.g. drugs, pollution, cleaning). The metals (and metal ions) are kept fixed in their position through coordination bonds with the ligands. The ligands (neutral or ionic) are usually oriented, polarized, activated, etc. through specific interaction with certain metal ions.

Unwanted excesses of (natural or unnatural) metals can often be removed by chelation with natural or specifically designed chelating ligands. Certain (xenobiotic) ligands may interact with metals in living tissue, thereby blocking coordination sites and hampering normal reactions. Some of the toxic metals may substitute non-toxic metal ions in metalloproteins or may bind to ligands *in vivo* (such as protein side groups and nucleic acids), changing the reactivities or fixation of certain conformations.

Finally, metal compounds may be used as drugs with a specific purpose of introducing either the metal or the ligand, or both, to certain positions in the living system (metal-containing drugs; metal-containing antibodies metallo-radiopharmaceuticals). For a better understanding of some of these processes, a classification of metals, ligands and their mutual interactions is needed. This is described in Sect. 3.2 of this chapter.

3.2 Overview of Ligands Present in Living Systems

3.2.1 General Coordination Behaviour of Ligands

Early attempts to distinguish between the coordination behaviour of the several metals, the "covalently binding" metal ions, have been considered as one category, and "ionic binding" metal ions, as the other extreme. This is, however, too much of a simplification. First of all the use of "covalent" binding is only valid from a kinetic point of view, i.e. slow breakage of the M–L bond, and not from a thermodynamic point of view (M–L bond strengths are always significantly weaker than typical covalent bonds like C–C and C–H bonds).

In the early days the use of the so-called HSAB theory [4, 5] was propagated to discriminate between preferences of certain metals for certain ligands, and this approach is still very useful. In fact, it is sufficient to know that ions such as Ca, Mg, Na, K and Mn belong to the ionic (class A) group, whereas other metal ions, such as Pt, Hg, Cd and Pb, belong to the more covalent group (class B). Transition metal ions such as Ni, Cu and Zn are usually considered as intermediate between A and B.

More or less similarly, ligands are classified according to their donor sites (i.e. metal binding atoms). Ionic or polar ligands are those with an oxygen-donor group (carboxylate, alcohol), and so-called covalent ligands (also called "soft ligands") are those with sulphur or phosphorus donor atoms (thioethers, thiolates, phosphanes). It should be noted again that the bonds as such are much weaker than typical covalent bonds. Nitrogen-donor ligands are usually considered as ligands with intermediate properties.

In the HSAB theory the basic rule is:

1. Ionic metals (class A) preferentially bind with ionic (or hard) ligands.
2. So-called covalent or soft metal ions (class B) preferentially bind with soft ligands.

This does not imply, however, that the HSAB theorem determines stability. It is only a *correction* term on the intrinsic acid and base properties of metals and ligands!

In biological systems the several metals (metal ions) are known to be able to coordinate to a variety of biomolecules; mentioned specifically are the following categories:

1. Proteins, and in particular to their side chains (N,O,S); however, also binding to the peptide bond (either to the $C=O$ or dehydronated N–H) is possible
2. Nucleic acids, both at phosphates and at base N-donor atoms
3. Carbohydrates, lipids (C–O; P–O groups)
4. Solids (bone, teeth, deposits such as kidney stones, magnetite)

3.2.2 Systematic Overview of Ligands that Bind to Metals in Biological Systems

Based on the classification above, it is possible to subdivide the ligands into biological systems in a systematic way. Frequently found metal-binding ligands in such systems are:

1. A variety of protein side groups and peptide-chain atoms, such as thioether (Met), imidazole (His), alcohol (Thr), thiolate (Cys) and carboxylates (Asp, Glu). Some of these are redrawn in Fig. 3.1. In several cases first dehydro-

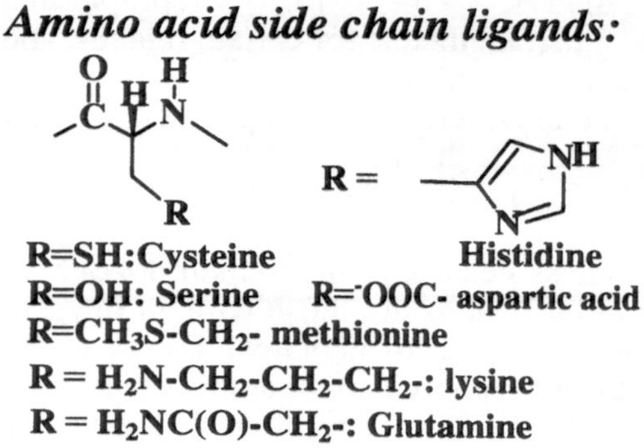

Amino acid side chain ligands:

R=SH:Cysteine   Histidine
R=OH: Serine   R=⁻OOC- aspartic acid
R=CH_3S-CH_2- methionine
R = H_2N-CH_2-CH_2-CH_2-: lysine
R = H_2NC(O)-CH_2-: Glutamine

Fig. 3.1. Overview of the structure of some important amino acids with coordination side chains, often found in metalloproteins

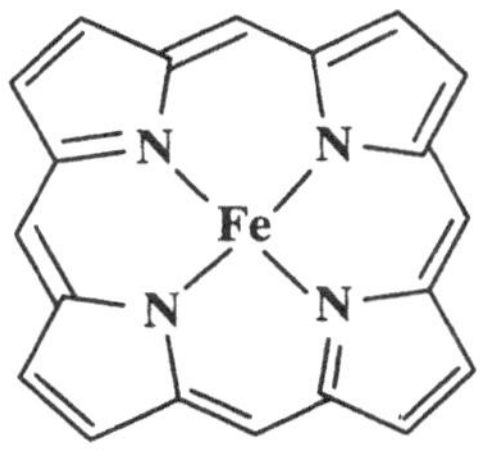

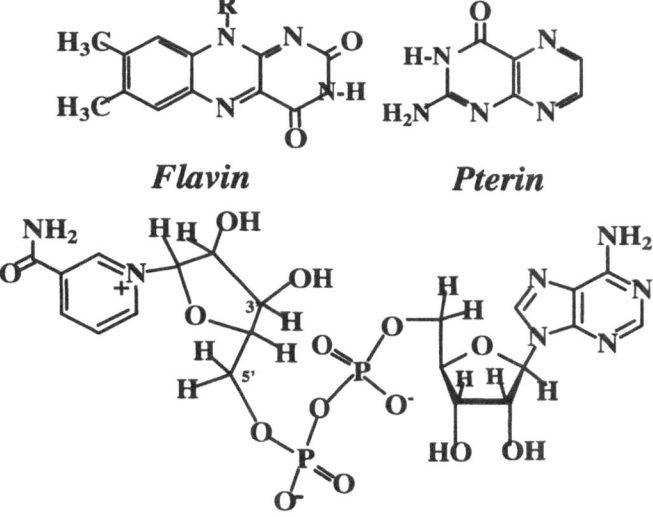

Porphyrin 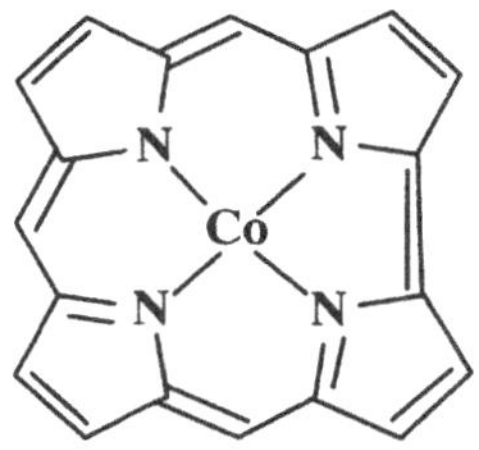   **Corrin**

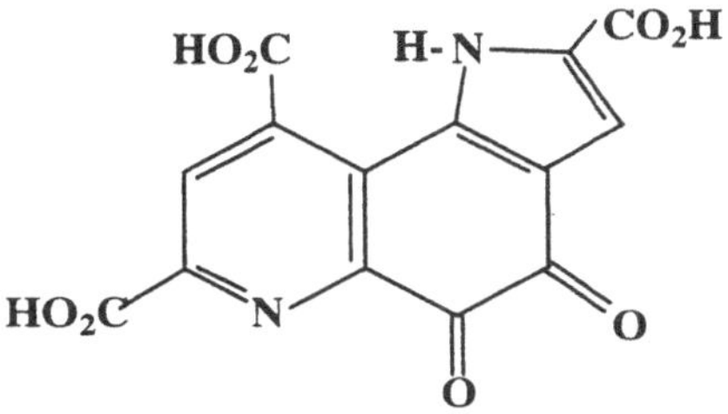

Pyrroloquinoline quinone, PQQ

Fig. 3.2. Overview of the structure of some important prosthetic groups (porphyrin, corrin) and a co-factor, pyrroloquinoline quinone (PQQ), often found in metalloproteins

Flavin   *Pterin*

Nicotinamide adenine dinucleotide (NAD⁺)

Fig. 3.3. Schematic structure of the co-factor groups Pterin, Flavon and NAD⁺, often found in metalloprotein

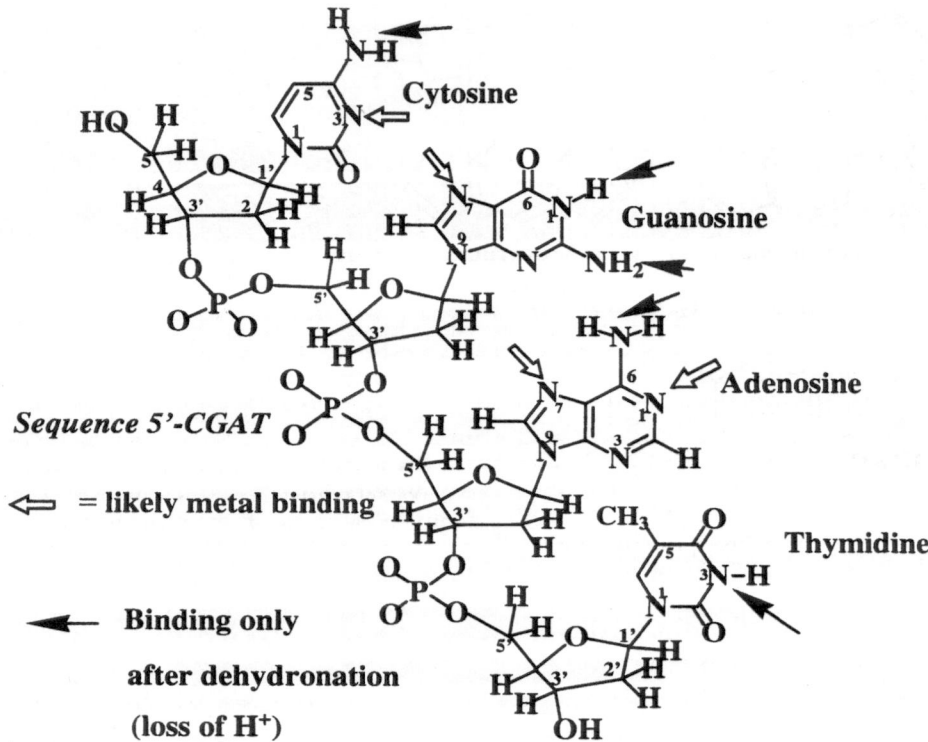

Fig. 3.4. The four major bases in DNA connected in a single-strand chain. The most important metal-binding sites are indicated with *arrows*. It should be noted that certain sites only binds metals at high pH, i.e. after loss of H⁺

nation (i.e. loss of a hydron, H⁺) takes place before binding to a metal ion, even at pH = 7.

2. Some prosthetic groups, such as tetrapyrroles and co-factors, such as PQQ and FAD (in many cases these co-factors are covalently attached to the protein, and therefore form a part of the biopolymer). Some examples, including the above and NADH, pterin, and a few other important metal ion binders, are given in Figs. 3.2 and 3.3. Some co-factors contain a metal intrinsically such as Moco, FeMoco in the porphyrins.

3. The N or O atoms from nucleic acid bases such as guanine. (In Fig. 3.4 the structures of four bases, linked in a DNA chain, and their possible metal-binding sites are depicted.)

4. Phosphates and diol groups in nucleic acids (also indicated in Fig. 3.4).

5. Finally, in biopolymers that bind metal ions, usually a variety of small inorganic ions can bind simultaneously to the metal such as OH⁻, O₂⁻ (a radical,

Table 3.2. Important bioligands and their presence in natural systems

Ligand group	Metal	Substance in which detected or proven
$=O$	Fe	P-450 enzymes
-OH	Fe, Zn	Carbonic anhydrase
H_2O	Fe, Zn, Ca	Many proteins; additional ligand
O_2/O_2^{2-}	Fe, Cu	Haemoglobin, haemocyanin, haemerythrin
O_2^-	Cu, Fe	Superoxide dismutase
$-OOH^-$	Fe	Haemerythrin
Tyrosine	Fe	Oxidases
Glutamate (and Asp)	Fe	Haemerythrin, ribonucleotide reductase
OPO_2R	Ca, Mg	Nucleic acids; ATP
NO_3^-, SO_3^{2-}	Mo	Several reductases
$-Cl^-$	Mn	Mn cluster in photosynthesis
$-S^{2-}$	Fe, Mo	Ferredoxin; nitrogenase
$-SR^-$ (cysteine)	Fe, Cu	Ferredoxin, plastocyanin, P-450, azurin
Me-S-R (methionine)	Cu, Fe	Plastocyanin, cytochromes, azurins
Imidazole	Cu, Zn, Fe, Mn	Plastocyanin, insulin
Benzimidazole	Co	Vitamin B-12
$(N<)^-$ (peptide)	Cu	Albumin
Tetrapyrroles	Fe, Co, Ni, Mg	Prosthetic groups; haemoglobin
CO	Fe	Toxic for myoglobin; cytochrome oxidase
$(CH_2-R)^-$	Co	Vitamin B-12

Table 3.3. Overview of most important ligands in biological systems (Y proven binding site; P probable binding site)

Ligand:	Mg	Ca	Mn	Fe	Co	Ni	Cu	Zn	Mo
O^{2-}			P	Y					P
OH^-			Y	Y			P	Y	
OOH^-			P	P			P		
O_2^{2-}			P	Y			Y		
O_2^-			P	Y			Y		
O_2			P	P			P		
H_2O	Y	Y	Y	Y			Y	Y	
H_2						Y			
N_2				P					P
RNH_2(lys)				Y					
Imidazole	P	Y	Y	Y	Y	P	Y	Y	Y
Porphyrin	Y			Y			P		
Corrin					Y				
Corphin						Y			
$RCOO^-$	P	Y	Y	Y				P	Y
SO_3^{2-}									P
NO_3^-									P
RPO_3^-	P	P							
S^{2-}				Y		P			Y
RS^- (cys)				Y		P	Y	Y	Y
$RSCH_3$(met)				Y		P	Y		P
O^-(tyr)			P	Y			Y		
OH(ser)	Y	Y							
OH(thre)	Y	Y							
$C=O$ (amide)	Y	Y	Y				Y		

so a better formula would be $O_2^{\cdot-}$), S_2^{2-}, $OOH^{\cdot-}$, CO_3^{2-} and small molecules (such as O_2, H_2O).

To indicate which ligands prefer which metal ion, in Table 3.2 a short overview of the ligands is given with a few representative examples of their biological presence in enzymes.

A more elaborate, but not necessary, complete classification of ligands and metal ions in biological systems is given in Table 3.3, where most of the known cases are tabulated in a matrix form. From this table it is clear that the imidazole group from histidine side chains is probably the most abundant ligand.

3.3 Metal Ions Occurring in Biological Systems

3.3.1 General Aspects

The metal ions and the ligands occurring in biological systems cannot be considered as just ligands and metals. When they undergo reactions *in vivo* they are an essential part of a complex system of equilibira, transports and storage. In a discussion about their presence, kinetics and thermodynamics have to be considered. Both kinetics and thermodynamics play a key role in all biological reactions involving metals.

The simple presence of a metal and a ligand in a biological system does not necessarily result in a reaction. The coordination process can either be far too slow (such as the case of the introduction of a metal into a tetrapyrrole system), or else the number of competing reactions with other ions or ligands may lead to a variety of different products. In fact, the kinetics of biological reactions dealing with Fe, Cu and Co are beginning to be somewhat understood. However, reactions with most other metal ions are far less well known. An example of a biomedical reaction where the first steps in the reaction are reasonably well understood is the mechanism of action of platinum antitumour compounds. Starting with drug administration and transport through the body, the drug enters the cell, is hydrolyzed and subsequently bind – mainly, but not exclusively – to DNA. Details are given in Sect. 3.4 of the present chapter.

3.3.2 Valency and Coordination Numbers of Metal Ions

In discussinc coordination compounds both the primary valence and the secondary valence ("coordination") of the metal need to be considered. In this regard it should be mentioned that valency often can be treated as a formality, i.e., whether a metal ion, such as copper, is Cu(I) or Cu(II) can usually be determined unambiguously. On the other hand, the coordination number of a metal ion, which is usually significantly larger than the valency, is not always determined unambiguously. Especially for metal ions such as Cu(II) the ligands are often at unequal distances, and for this metal ion structural chemists have introduced the indication "semi-coordination" to describe borderline cases. More importantly, the coordination number is not determined by the metal ion only: The ligand (charge, size, shape, bonding properties) greatly influence the observed coordination geometry around the metal ion.

Despite the uncertainties in coordination numbers, and the occasional ambiguity of valence states, several empirical observations have been made over the past decades that now allow a clear description of metal ion sites in biopolymers. The constraints exerted by the polymer are subsequently considered as a distortion effect.

The constraints exposed by the biological ligand (i.e. the polymer with its ligands; see also Chap. 2) can be large, and in a number of cases a very unusual geometry has been found for metal ions in proteins, often rare and in some cases even unprecedented, in low-molecular weight coordination compounds. Well-known examples of such unusual cases are:

1. The so-called Cu_A-site in cytochrome oxidase and nitrous reductase
2. The variety of Fe–S clusters in electron transfer proteins and enzymes
3. The so-called blue Cu site in electron transfer proteins

Blue Copper protein: plastocyanin

N=histidine ligand

The oxyhemocyanin structure, with O_2^{2-} and Cu(II)

Relevant distances:

Cu....Cu = 350 pm; O..O = 130 pm; Cu..N = 205 pm

Schematic structure of lactoferrin with coordinated carbonate

.... **hydrogen bonds**

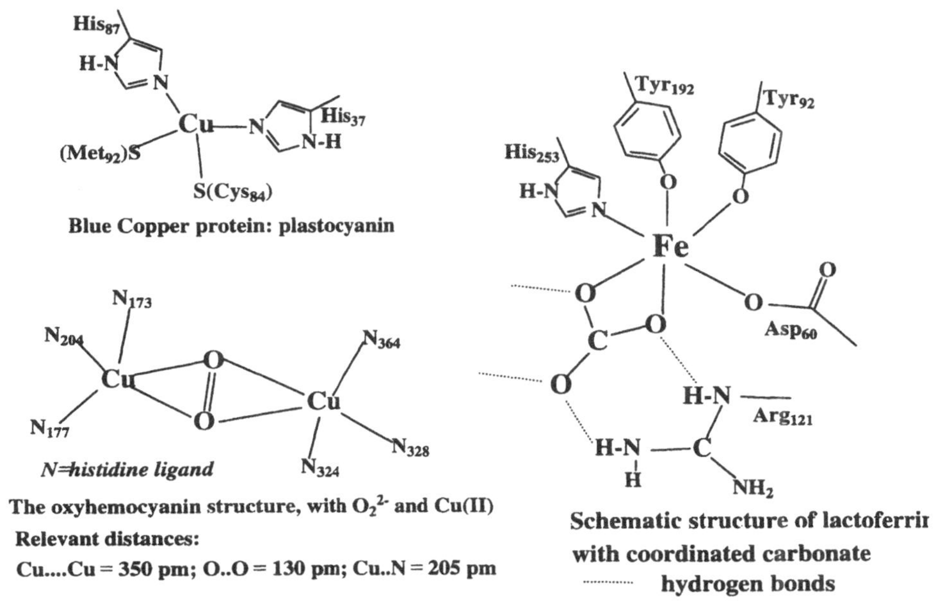

Fig. 3.5. Schematic structure of the active sites of some metalloproteins with a focus on the coordinating ligand to the metal [6–8]. The blue copper protein Plastocyanin, with a type-1 site. The dinuclear type-3 site in oxyhemocyanin. The metal-bound site in lactoferrin, in which H-bonding of the hydrogen carbonate is indicated as well

Some of these sites are discussed herein in more detail and in relation to their biological function.

Perhaps the most unusual geometries in nature are found for the copper proteins. According to their spectroscopic properties, the coordination units ("sites") have been classified as:

1. Type 1 copper sites (such as in plastocyanin and azurins). In the Cu(II) state this site has a characteristic EPR signal (with a low value of $A_{\parallel}$) and a strong

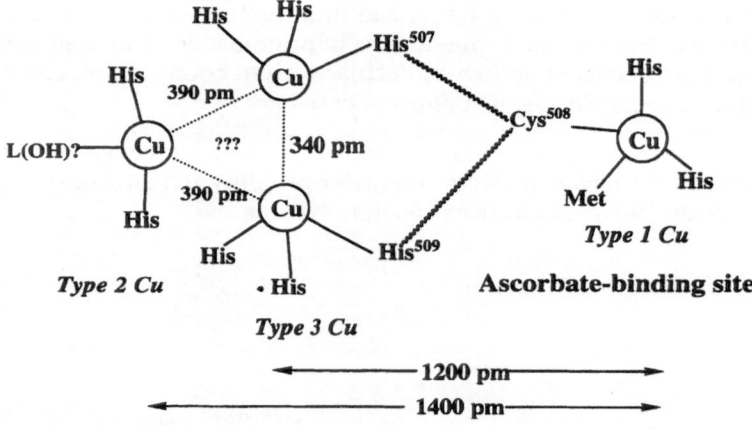

The active site of ascorbate oxidase

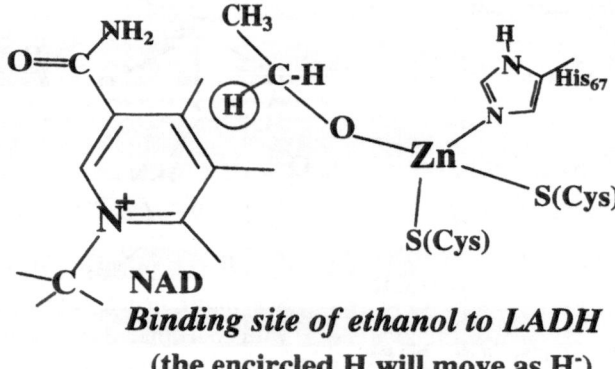

Binding site of ethanol to LADH

(the encircled H will move as H⁻)

Fig. 3.6. Schematic structure of the active sites of some metalloenzymes with a focus on the coordinating ligand to the metal [6–8], ascorbate oxidase (with type-1, type-2 and type-3 copper sites); liver alcohol dehydrogenase (LADH) showing the Zn site, the bound ethanol and the NAD⁺ ready to accept the H⁻

visible absorption (near 600 nm, molar extinction 3000–6000). The Cu_A site mentioned above has similar characteristics (see Sect. 3.3).

2. Type 2 copper sites (such as in superoxide dismutase). This site has normal spectroscopic properties when in the Cu(II) state.
3. Type 3 copper sites (such as the dinuclear site in haemocyanin). This site is EPR silent in the Cu(II) state, which appears to be related to a strong anti-ferromagnetic coupling between two Cu(II) ions.

In fact, many enzymes contain several copper ions usually comprising more than one spectroscopic type. A well-known example is **ascorbate oxidase** (see Sect. 3.3), which contains types 1, 2 and 3 Cu sites. In fact, types 2 and 3 sites are so close together that they may even be considered as forming a new type (type 4 has been suggested).

Iron proteins have an even greater variety of sites, and are usually classified according to their chromophore. The following are mentioned herein:

1. Iron prophyrins (role is either dioxygen transport, electron transport or catalysis).
2. Iron sulfur clusters (role is usually electron transport, but some enzymes have been found as well).
3. Non-haem iron proteins. These can be dinuclear, such as a haemerythrin, ribonucleotide reductase (RNR), Methane monooxygenase (MMO) or mononuclear, such as lipoxygenase and some dioxygenases such as pyro-catechuate 3,4-dioxygenase.

A few arbitrarily chosen, interesting structures are given schematically in Figs. 3.5 and 3.6, and in Sect. 3.3. A complete description of the possible sites and coordination of all metal proteins goes beyond the scope of this review. A brief introduction is to be found in two recent textbooks [6, 7] and a monograph on bioinorganic catalysis [8].

3.4 Applications of Metal Complexes in Living Systems

3.4.1 Transport and Storage of Metal Ions

In a large number of cases the role of a metal ion in a biological system is limited to a structural function, i.e. holding together certain (parts of) biomolecules in a more or less rigid structure. Classical, well-known examples are Ca^{2+} in thermolysin (a protein that has Zn at the active site), Zn in superoxide dismutase and alcohol dehydrogenase (both at non-catalytic sites), and in several solid structures such as skeletons, bones and teeth.

The structural role of magnetite (Fe_3O_4) found in bacteria, honey bees and, more recently, also in pigeons and higher organisms such as humans. It appears that single crystals of Fe_3O_4 are used – probably in combination with other objects – for orientation using the earth's magnetic field.

The very important role of zinc fingers in gene expression is the result of Zn-induced protein folding. This topic has been growing exponentially during the past few years, and several high-resolution NMR structures and even a few X-ray structures have become available for the so-called zinc fingers and their adducts with pieces of DNA.

Living systems using metal ions in several locations require efficient transport of metal ions. Well-known examples are transferrin (a hydrogencarbonate-dependent iron-binding protein) for iron transport in humans and albumin for copper transport. Ferritin is well known for iron storage. Apart from these natural transporting proteins, nature also uses efficient systems to remove excesses of toxic metal ions. Examples are the metallothioneines (cysteine-rich peptides) that take care of highly toxic metal ions, such as cadmium, and excesses of zinc and copper [3]. One such protein may bind up to seven metal ions.

3.4.2 Transport and Storage of Small Molecules by Macromolecules

The most important transport in nature is probably the transport of dioxygen by so-called dioxygen carriers. The best-known example is the haemoglobin molecule, which consists of iron-haem complexes held by a biopolymer to which the dioxygen can bind as an axial ligand.

Other examples are haemerythrin, a dinuclear iron-site-containing protein, to which the dioxygen is bound on one iron as an end-on peroxo ligand, and haemocyanin (a dinuclear Cu-containing site, to which dioxygen is bound symmetrically, bridging two Cu ions). The structures of these sites have only recently become available, and are currently being studied by a variety of "low-molecular weight" chemists to better understand the kind of binding and the reversibility.

Also, the transport of carbon dioxide, although not bound to a metal ion, takes place upon binding to a protein. The amine end of the protein chain, after binding with CO_2, is converted into a (labile) carbamate, thereby acting as a CO_2 carrier.

3.4.3 Catalysis by Metal Ions in Biomacromolecules

In many cases the metal ion and the binding of the ligands play an important catalytic role, and the study of this role is the main theme of the present monograph. This catalytic role can be in electron transfer only. It can also be a complicated redox reaction or a relatively simple acid–base reaction. The presence of the metal at the particular site usually results in activation reactions of substrates by the metal and the surrounding ligands.

All enzymes have been classified for many years into six major groups [9], a classification that also hold for the many metal-containing enzymes:

1. Oxidoreductases (catalyzing oxidation–reduction reactions). In this group one usually distinguishes between dehydrogeneases (all used dioxygen or hydrogen peroxide is converted into water), oxygenase (one of the oxygen atoms is introduced into the substrate) and dioxygenases (two oxygen atoms are used to be built into the substrate).
2. Transferases (transferring an organic group from a donor molecule to an acceptor molecule).
3. Hydrolases (hydrolytically cleaving of C–O or C–N bonds).
4. Lyases (cleaving C–N, C–O or C–C bonds by elimination).
5. Isomerases (geometrical or structural change within a molecule).
6. Ligases (these join together two molecules or fragments; often coupled with pyrophosphate hydrolysis).

Enzymes are usually classified by a so-called EC number, consisting of three or four digits separated by a dot, to indicate subgroups and sub-subgroups. Well known examples are bovine superoxide dismutase (SOD, EC 1.15.1.1), urease (EC 3.5.1.5), horseradish peroxidase (EC 1.11.1.7), nitrogenase (EC 1.18.6.1), and diol dehydratase (EC 4.2.1.28).

A. *Proposed dinuclear structure for Cu$_A$*

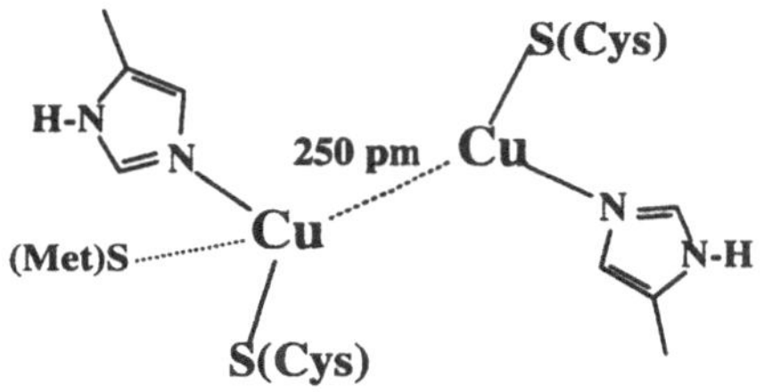

B. *Structure of Ribonucleotide reductase, RNR*

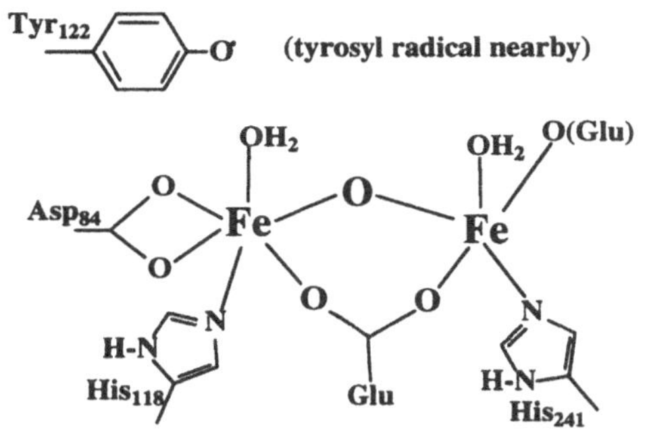

Fig. 3.7. Some unusual structures of metalloproteins; A proposed site for Cu$_A$ in cytochrome oxidase [11] with a short Cu–Cu contact and Cu in the mixed-valence state; **B** structure of the active site in RNR ribonucleotide reductase with a tyrosinyl radical nearly the dinuclear iron site [12]

It is evident from the lists of classified enzymes [9] (and expecially for groups 1 and 4) that very often a metal is involved, either directly at the active centre, or indirectly at another place, for instance, in the electron transfer process [10]. In fact, when Nature has to perform a very difficult chemical job, in almost all cases it uses metals coordinated in a biopolymer [8].

It is impossible to list even a small compilation of metal-containing enzymes (metalloproteins with a catalytic function). The reader is referred to the textbooks mentioned previously [3, 6, 7]. In some cases the metal ion is present in an enzyme when it has only a structural function, unrelated to the enzymatic reaction, such as the Zn ion in superoxide dismutase (SOD). In other cases the metal ion is at the heart of the chemical reaction centre and plays a role in the catalysis. This role can be either that of a Lewis acid or of a redox reaction (electron transfer involved). In some cases the metal ion is at the active site, and the reaction is a redox reaction, but the metal ion does not change valence states. The most well-known examples are the Zn ion in alcohol dehydrogenases (redox reaction done by $NADH/NAD^+$; see Fig. 3.6) and the ion in pyrocatechuate 3,4-dioxygenase. (In this enzyme the iron remains Fe(III) and dioxygen undertakes a direct attack on the substrate, with iron acting as a Lewis acid only.)

This section concludes with a few examples of unusual enzyme reaction centres, namely RNR and Cytochrome oxidase (Fig. 3.7). First of all, only very recently [11] has evidence been obtained that the Cu_A site in cyt c oxidase, with a very unusual Cu hyperfine splitting in the EPR, is due to a mixed-valence Cu(I)/Cu(II) pair, which acts as an electron transfer protein. All data agree with the structure drawn in Fig. 3.7A.

Another unusual structure was found for the so-called B2 protein of RNR (EC 1.17.4.1) from bacteria, where a dinuclear iron site with a μ-oxo bridge, resembling somewhat haemerythrin, is found to operate in connection with a tyrosinyl radical at a distance of less than 1 nm (Fig. 3.7B). This protein is essential for the synthesis of DNA in all organisms.

Given the enormously increased knowledge of the active-site structures in biopolymers, and the powerful possibilities of peptide synthesis, it is expected that within the next few years *de novo* design of proteins with special sites will become routine in many laboratories.

3.4.4 Medicinal Functions of Added Metal Ions In Vivo

As discussed previously, metal ions play an important role in structure, transport or catalysis. However, metal ions may also have deleterious effects as a toxic metal or play an important role as a metal-containing drug. Examples of M-containing drugs are:

1. Gold-thiolate and copper carboxylate compounds in the treatment of arthritis
2. Bismuth salts in the treatment of ulcers
3. Lithium salts to suppress symptoms of manic depression

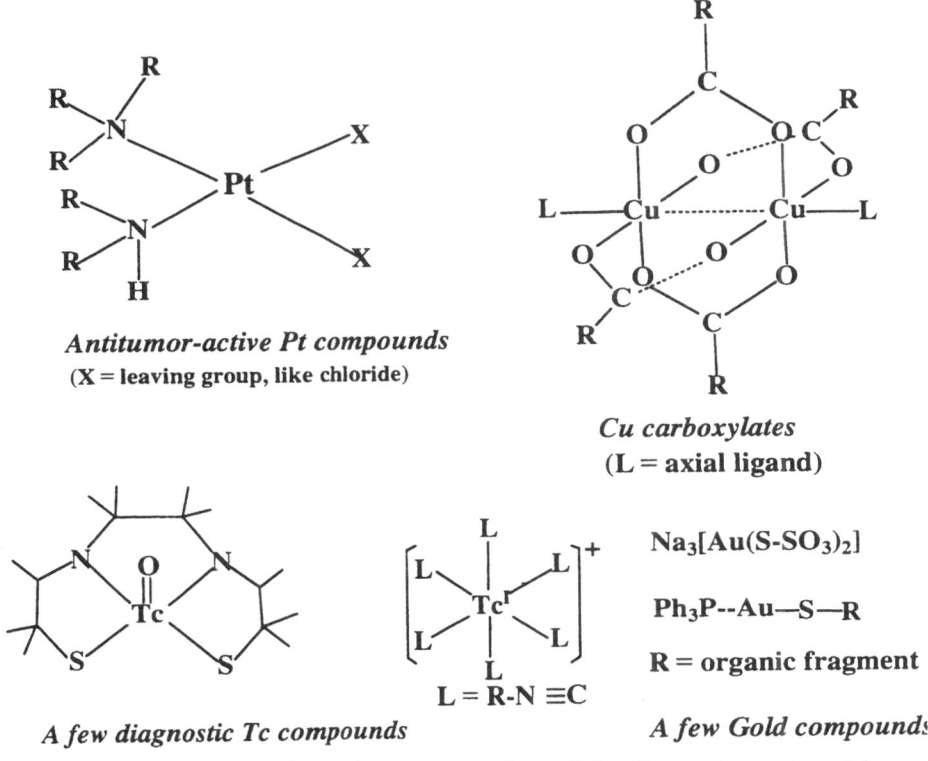

Antitumor-active Pt compounds
(X = leaving group, like chloride)

Cu carboxylates
(L = axial ligand)

$Na_3[Au(S-SO_3)_2]$

$Ph_3P--Au—S—R$

R = organic fragment

L = R-N ≡ C

A few diagnostic Tc compounds   *A few Gold compounds*

Fig. 3.8. Some structures of some important metal-containing diagnostic agents and drugs

4. Platinum-amine compounds as anti-tumour drugs
5. Radiopharmaceutical imaging with technetium coordination compounds

A few such compounds are schematically presented in Fig. 3.8.

Most well understood is the mechanism of action of platinum-anti-tumour compounds, and particularly their selective binding to DNA. The detailed distortion of DNA after Pt binding and its biological consequences are presented briefly.

The compound cis-$PtCl_2(NH_3)_2$, which has been known for 150 years and which is clinically known as "cisplatin", is a very successful anti-tumour drug, especially in combination with other drugs such as vinblastine and bleomycin. It is a very successful reagent in the treatment of testicular and ovarian cancer, and is also increasingly used against cervical, bladder and head/neck tumours. Second-generation derivatives, and particularly the complexes belonging to the cis-$PtCl_2(amine)_2$ structural class, show equal biological activity, although still relatively few Pt compounds have been tested in preclinical tumour screens.

More recently new structural classes of Pt anti-tumour drugs have been reported. These compounds possess chemical and biological properties very different from those of cisplatin, and such compounds are expected to be of great use in the clinic to overcome the problems of resistance that certain patients develop against cisplatin treatment. Among the recently discovered examples of new compounds that may provide a basis for further expanding the spectrum of activity of Pt-based drugs are the so-called asymmetrical Pt compounds (with chiral ligands) and compounds with larger leaving groups. A small selection of these novel Pt compounds is depicted in Fig. 3.9. Most recently even a class of dinuclear Pt compounds has been reported to show significant activity.

Detailed studies of specific interaction of these platinum compounds with cellular components have been performed with most attention being given to nucleic acid binding. The N7 atom of guanine appears to be much better (both faster and more stable) in binding to metal ions. It is now generally accepted that the kinetics of metal binding (assuming a metal-ammine/aqua ion) to the DNA at guanine-A7 is significantly enhanced by the H-bond interaction with O_6, whereas in the case of adenine-N7 such an H-bond attraction does not occur, because the NH_2 group is an H-bond donor (see also Fig. 3.4). After binding of Pt to purine-N7 sites, the resulting species will additionally be stabilized through H-bonding in the case of guanine, but not so in the case of adenine.

Many research groups currently study the synthesis of new Pt compounds and their specific interaction with nucleic acids. The structural changes of DNA (fragments) after the binding of a metal have received special attention.

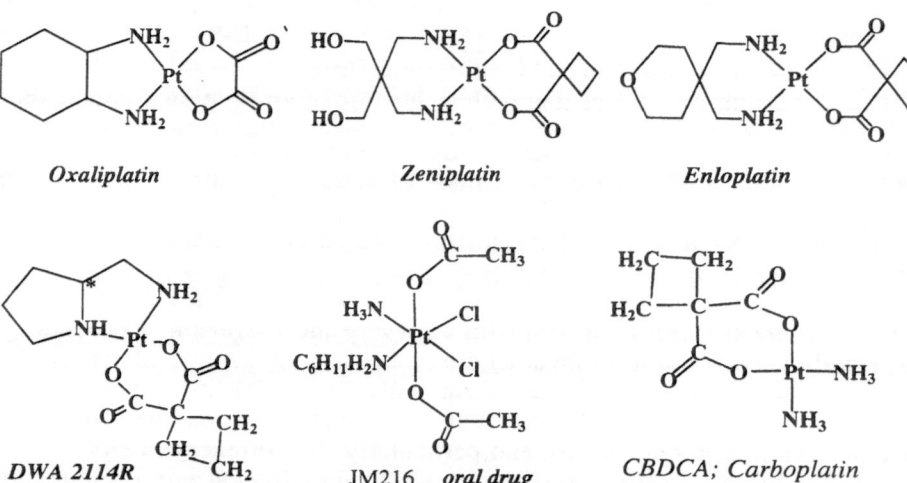

Fig. 3.9. A few recently introduced Pt-antitumor compounds with high activity and used in clinical trials

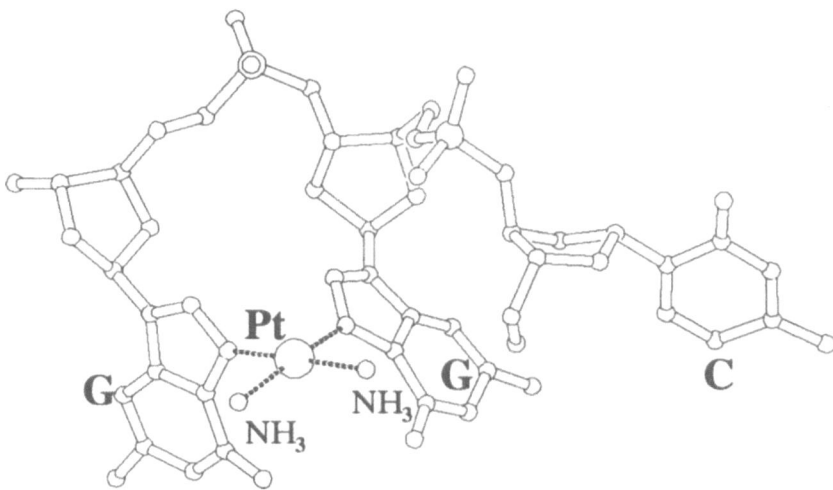

Fig. 3.10. Projection of the X-ray structure of the DNA chelate CGG-cisplatin. Pt-N7 and Pt-NH$_3$ bonds are dashed. Similar structure units are believed to be present in larger double-strand DNA based on nuclear magnetic resonance spectroscopy

Because Pt DNA binding is generally accepted to be the biologial target leading to cell killing, the distortion of DNA is summarized below (an overview is found insome of our regular work [13–15]:

1. In the DNA *cis*-platinum adducts with single-strand AG and GG chains (A adenosyl; G guanosyl), the two base are coordinated through N7 to *cis*-Pt in a "head-to-head" orientation. The conformational characteristics of the DNA hardly change upon platination. In fact, only the deoxyribose moiety of the 5'-guanine has adopted an almost pure N-type conformation compared with the S-type conformation in free d(GpG). Rotation about the Pt-GN7 bond remains possible after the first step. An X-ray structure of the GG-chelate structure with cisplatin at atomic resolution is presented in Fig. 3.10.
2. Although the degree of distortion of double-strand DNA by *cis*-Pt was initially expected to be very large given the structure presented in Fig. 3.10, most surprisingly the GG-platinated non-selfcomplementary decanucleotide d(TCTCGGTCTC) forms a duplex with its complementary strand d(GAGACCGAGA) [13]. Detailed spectroscopic analysis using proton and phosphorus nuclear magnetic resonance (NMR) and consideration of the circular dichroism (CD) spectra led to the conclusion that the double helix is somewhat destabilized after the platination, as reflected by the decrease in the melting temperature of the duplex by 10–20 °C with concentrations as used in NMR (3 mM).
3. Detailed analyses of the chemical shifts and coupling constants, using 2D-NMR spectra, followed by conformational analysis, suggests a small, but

significant, distortion of the double helix of the larger oligonucleotides upon platination, which has been described as a "kink" of approximately 40–70° in the helical axis at, or around, the GG lesion. Similar distortions have been reported by other workers using different sequences, and even for the AG lesion [16, 17]. Base-pair formation by hydrogen bonding still appears possible after the platination, as reflected by the appearance of imino-proton resonances. For the central GG sequence, these signals are only observed at low temperature, although shifted to lower field and broadened.

4. Comparison of CD spectra and ^{31}P NMR spectra of both GG platinated DNA from several sources and the platinated double-strand decanucleotide strongly suggests similar distortions for both cases.

5. The interactions with the 5'-phosphate group in the GG-chelated DNA seem to be important, i.e. it is involved in a hydrogen bond with an NH$_3$ ligand of platinum. This could be the origin for the observation that active platinum anti-tumour drugs need an acid N–H group. The phosphate–ammonia interaction might well induce and/or stabilize DNA distortions, thereby interfering with the replication process.

6. Earlier studies had shown that also GNG lesions might occur in DNA (N=A,T,C); most recently it has been shown to van Houte and van Garderen [18] that even in this case a double-helical DNA structure remains possible.

7. The inactive analogue of cisplatin, $trans$-PtCl$_2$(NH$_3$)$_2$, has been known to bind to DNA in a related way. Work by Dalbies et al. [19] has shown that even in this case a double-helical structure remains possible, albeit with base-pair disruptions near the Pt site.

3.4.5 Non-Biology Applications of Bioinorganic Principles in Polymers

During the past 2 decades it has been a challenge for many scientists to make use of the principles of metal ion binding by biological macromolecules. A few examples are briefly mentioned below; others are found in Chap. 2.

The first example deals with selective ion exchangers, in which chelating groups specific for, e.g. Cu(II), are covalently attached to polymers. Our early work was directed to hydrophobic polystyrene-type polymers [20]. These resins had as a disadvantage the poor swellability in water. More recently hydrophilic polymers have been synthesized and successfully used for selective removal of copper from waste water [21]. By making use of imidazole groups – excellent ligands for copper ions as shown by biological system – large selectivities have been observed for Cu(II) over zinc, cobalt, nickel and cadmium.

A second example deals with the possibility to bind certain proteins or enzymes covalently to an electrode. The covalent link is made up by attaching an imidazole ligand to a surface of glassy carbon. This ligand then coordinates to a genetically engineered protein (e. g. a blue-copper rotein) in which a ligand histidine close to the surface of the protein has been mutated to a glycine. The surface alkyl imidazole ligands not only binds to the Cu ion in the protein, but also links the protein to the electrode as a molecular wire. This modification allows the detection of a redox reaction in solution by following the reaction amperometrically [22].

The last example to be mentioned deals with the application of coordination compounds attached to polymers and their use as "immobilized catalysts". This technique has been used for a long time in organometallic catalysts, and even conferences dealing with immobilized catalysts have been held for a few decades. In fact, similar reactions can be performed with biomimetic catalysts such as with Cu(II) compounds that mimic the oxidases. This topic has not yet been studied in great detail, but is very promising. A review of polymeric copper imidazole complexes used in oxidative phenol coupling, to synthesize the engineering plastic PPE is available [23].

In the near future it is expected that catalysis by polymers consisting of modified antibodies will grow significantly. Antibodies elicited in such a way that they carry catalytically active groups will soon become a routine technique, as predicted by many recent studies. Catalytic antibodies in fact are only laboratory creations based on transition-state stabilization. Detailed treatment of this fascinating group of compounds goes beyond the scope of this chapter.

3.5 Conclusion

The great variety of exciting results about metal ion binding to biopolymers, unravelled during the past decades, has raised and partly answered the question: Can we possibly make use of what Nature is teaching us through bioinorganic systems? To what degree can be use smaller – non-polymeric – fragments, synthetically prepared, to perform similar reactions? What can we learn about electron transfer, transport of dioxygen, combustion and consumption of dioxygen, dinitrogen fixation, conversion of solar energy and communication between cells with electric signals, and how can we apply this knowledge? Part of these questions have been addressed in Sect. 3.4.3 and 3.4.5.

However, before such developments will become practical utility, understanding has to come first. For better understanding, mimicking of biological systems is important; after that recreation of the building blocks and reaction steps and their application on a large scale will become within reach. The first signs look promising!

In this regard an important question will become: How simple or how complicated will our models have to be to accomplish *in vitro* what Nature does *in vivo*? These studies are most likely to be stimulated from several directions.

Catalysis and energy research will most probably by strong stimulators (small-molecule activation and photoconversion). However, many other areas will also stimulate this type of research; one could think of medical research, where better chelating agents for selective removal of excesses of toxic metal ions will remain of great importance; one could also think of better understanding of biomineralization processes (surface coordination), of improved diagnostic reagents (monoclonals with an attached metal), environmental cleaning processes (metallothionein-like chelating ligands) and mutation research (role of certain metal ions on mutation and repair mechanisms). Polymeric ligands, either natural or synthetic, indeed have great potential in this regard.

Acknowledgements. The author is indebted to his several co-worker whose names are mentioned in the references. Continuous support from Leiden University and from the Foundation of Chemical Research in the Netherlands (SON) is gratefully acknowledged. Johnson & Matthey (Reading, UK) is thanked for a generous loan scheme for platinum. Financial support by the European Union, allowing regular exchange of preliminary results with several European colleagues (under contract ERBCHRXCT920016 and 920014) is thankfully acknowledged. The author is indebted to the EU for a grant as Host Institute in the EU Programme Human Capital and Mobility (1994–1997). Sponsorship concerted by COST Action D1-92/002 (Biocoordination Chemistry) is kindly acknowledged. Part of this work has been performed under the auspices of the joint BIOMAC Research Graduate School of Leiden University and Delft University of Technology.

References

1. Frieden E (ed) (1984) Biochemistry of the essential ultratrace elements. Plenum, New York
2. Frieden E (1985) J Chem Educ 62:917
3. Fraústo da Silva JJR, Williams RJP (1991) The biological chemistry of the elements. Oxford
4. Pearson RG (1963) J Am Chem Soc 85:3533
5. Pearson RG (1966) Science 151:172
6. Lippard SJ, Berg JM (1993) Principles of bioinorganic chemistry. University Science Books, Mill Valley, California
7. Kaim W, Schwederski B (1994) Bioinorganic chemistry: inorganic elements in the chemistry of life. Wiley, Chichester
8. Reedijk J (1993) Bioinorganic catalysis. Marcel Dekker, New York
9. Webb E (ed) (1984) Enzyme nomenclature (IUB Recommendations 1984). Academic Press, New York
10. Palmer G, Reedijk J (1991) Nomenclature of electron-transfer proteins (IUB Recommendations 1989). Eur J Biochem 200:599
11. Blackburn NJ, Barr ME, Woodruff WH, Van der Oost J, Vries S de (1994) Biochemistry 33:10401
12. Bollinger JM, Edmondson DE, Hyunh BH, Filley J, Norton JR, Stubbe J (1991) Science 253:292
13. Reedijk J (1992) Inorg Chim Acta 198–200:873
14. Admiraal G, Alink M, Altona C, Dijt FJ, van Garderen CJ, Graaff RAG de, Reedijk J (1992) J Am Chem Soc 114:930

15. Bloemink MJ, Dorenbos JP, Heetebrij RJ, Keppler BK, Redijk J, Zahn H (1994) Inorg Chem 33:1127
16. Herman F, Kozelka J, Stoven V, Guittet E, Girault JP, Huynh-Din T, Igolen J, Lallemand JY, Chottard JC (1990) Eur J Biochem 194:119
17. Fouchet MH, Gauthier C, Guittet E, Girault JP, Igolen J, Chottard JC (1992) Biochem Biophys Res Commun 182:855
18. van Garderen CJ, van Houte LPA (1994) Eur J Biochem 225:1169
19. Dalbies R, Payet D, Leng M (1994) Proc Natl Acad Sci USA 91:8147
20. Sahni SK, Driessen WL, Reedijk J (1988) Inorg Chim Acta 154:141
21. Verweij PD, van der Geest JSN, Driessen WL, Reedijk J, Sherrington DC, (1992) React Polym 18:191
22. den Blaauwen T, van de Kamp M, Canters GW (1991) J Am Chem Soc 113:5050
23. Challa G, Chen W, Reedijk J (1992) Macromol Chem, Macromol Symp 59:59

4 Electronic Processes in Macromolecular Metal Complexes

4.1 Ion-Conductive Materials of Macromolecules

E. Tsuchida and S. Takeoka

4.1.1 Introduction

The development of electronic devices is largely due to the emergence of microdevices, which can assemble many functions on microchips. However, it is not easy to reduce the size and weight of the energy sources, batteries. Because electrical energy is generated by the chemical reactions, electrolyte solutions are essentially required. The necessity of a compartment to encase the solution has prevented the construction of small batteries. The utilization of a flexible solid electrolyte film is very important to solve this kind of packaging problem and to manufacture small batteries and microdevices.

The idea of a "solid electrolyte" was realized firstly with inorganic crystals for super ion-conduction of AgI crystals above the $\beta \to \alpha$ transition (149 °C) [1]. Many efforts have been concentrated on the construction of similar crystalline states and super ion conductors such as $Rb_4Cu_{16}I_7C_{13}$ [2], β-alumina [3], and NASICON [4]. These super ion conductors were designed crystallographically and synthesized to provide 10^{-3}–10^0 S/cm even at ambient temperatures [5]. However, the lack of processibility of inorganic crystals is a serious drawback for engineering materials, whereas macromolecules have many attractive properties such as light weight, flexibility as thin films, good physical contact with electrodes, a variety of chemical modifications, etc. Since Wright reported in 1975 that poly(oxyethylene) (POE) dissolved inorganic salts and showed ionic conductivities of 10^{-8}–10^{-9} S/cm at ambient temperatues [6], the developments of ion-conductive macromolecules (solid macromolecular electrolytes) have greatly increased.

At present, the ionic conductivity of solid macromolecular electrolytes is 10^{-4} S/cm at ambient temperatures [7, 8], which is 10^4-fold higher than that reported by Wright, but still insufficient in comparison with inorganic solid electrolyte systems. If the ionic conductivity could be increased tenfold or 10^2-fold, many advantages of macromolecular materials, such as light weight, thin film formation, flexibility, etc., would lead to new types of electrochemical devices and batteries.

The purpose of this review is to introduce recent developments in ion-conductive macromolecules from basic studies of salt dissociation and ion conduction to the application to batteries and devices.

4.1.2 Advancement in Ion-Conductive Macromolecules

4.1.2.1 Macromolecules and Liquid Electrolytes

In lithium secondary batteries, inorganic salts having low dissociation energy ($LiClO_4$) and organic solvents having both high boiling points and high dielectric constants, such as propylene carbonate (PC, 64.4) and ethylene carbonate (EC, 89.0), are used. The electrolytes were incorporated into macromolecules having high dielectric constants, such as polyvinylidene fluoride (PVdF, 9.2) and polyacrylonitrile (PAN, 8.0), and were used as separators between electrodes. Ionic conductivities of ca. 10^{-6} S/cm were observed in those systems [9]. Recently, ionic conductivity has improved significantly by the use of mixed solvents of EC and PC in PAN [10, 11]. The conductivity of the "gel electrolyte" system under optimal conditions is $1.8 \cdot 10^{-3}$ S/cm at 20 °C. Bohnke et al. reported that the gel electrolytes of PMMA-$LiClO_4$-PC showed the ionic conductivity in the range of $5 \cdot 10^{-3}$ to $5 \cdot 10^{-5}$ S/cm depending on the amount of PMMA [12]. Based on the result that the ionic conductivity increased when a solvent with low viscosity was added, the effect of the solvent is due not only to the enhancement of salt-dissociation, but also to a plasticizing effect of low levels of solvent addition, and to the introduction of ion-conductive pathways in high levels.

Perfluorosulfonate ionomers, such as Nafion (E.I. du Pont de Nemours & Co., Inc.) or Flemion (Asahi Glass Co. Ltd.), have microsphase-separated structures of ionic clusters. The oligo (oxyethylene) COOE/$LiClO_4$ was incorporated into cylindrical columns (10–50 A in diameter) in order to construct ion-conductive pathways. Ionic conductivities of 10^{-5}–10^{-6} S/cm were obtained at 25 °C under conditions without leakage of OOE [13]. Recently, a casting method from a mixed solution of Nafion/OOE/DMF was found to be an effective method of preparing a thin film while maintaining a high incorporation ratio (55 %) of OOE [14].

4.1.2.2 Poly(oxyethylene)-Based Electrolytes

The development of ion-conductive macromolecules without organic solvent started extensively in the 1980s after the report of Wright [6]. The study on the (POE/inorganic salt) composite systems started with the analysis of the structures of complex crystalline materials. A helical POE chain was considered to surround a cation to provide a specific pathway for cation transport [15]. Chatani and Okamura determined the crystalline structure of $P(OE)_3NaI$ in detail in which sodium ions were located along the axis of a helix [16].

On the other hand, Armand et al. [17] studied the temperature dependence of the ionic conductivity of $P(OE)_nMSCN$, and reported that on Arrhenius plot of $P(OE)_{4.5}NaSCN$ and $P(OE)_{4.5}KSCN$ showed a bend at the melting point of POE. In contrast, Arrhenius plots of $P(OE)_8CsSCN$ composites showed an upper curved line [17]. Studies were focused on the phase diagrams of

(POE/inorganic salt) complex systems [18] (see Sect 4.1.4.2). Consequently, the development of ion-conductive macromolecules has been focused on the synthesis of amorphous macromolecules at ambient temperatures. Some amorphous polyether derivatives are listed in Fig. 4.1. The temperature dependence of ionic conductivities for these POE derivatives is shown in Fig. 4.2.

The normally atactic POP (Scheme 1 in Fig. 4.1) complexes are completely amorphous and show an ionic conductivity of approximately 10^{-6} S/cm, which is ten fold higher than that for the crystalline POE system [17]. However, the ionic conductivities of a POP composite become lower above the melting point of POE, because POP, having a low dielectric constant and steric hindrance due to the methyl groups for coordination, has a lower solvating property than POE. Oxymethylene-linked OOE (Scheme 2 in Fig. 4.1) with a molecular weight of more than 10^5 has a melting point of $15\,°C$ [19], and an amorphous composite with $LiCF_3SO_3$ ([O]/[Li] = 25) shows a high ionic conductivity of $5 \cdot 10^{-5}$ S/cm. One of the methods of synthesizing the amorphous ion-conductive macromolecules is to produce segmented block comacromolecules in which chain regularity is disrupted, and hence, the crystallization is inhibited. Nagaoka and coworkers linked short POE chains with dimethyl siloxane units (Scheme 3 in Fig. 4.1) [20]. The macromolecules show a low glass transition temperature ($-123\,°C$) and a conductivity of $2.0 \cdot 10^{-4}$ S/cm.

Comb-branched macromolecules have oligo(oxyethylene) side chains attached to a macromolecular backbone. If the length of these side chains is restricted to prevent crystallization, totally amorphous systems can be produced. Thus, systems based on substituted methacrylate macromolecules have been described by Kobayashi et al. [21] (Scheme 4 in Fig. 4.1) ($PMEO_n$). A similar system, but with a higher POE side-chain density, based on poly-(itaconic acid) (Scheme 5 in Fig. 4.1) [22] and (Scheme 6 in Fig. 4.1) [23] were synthesized. By using a backbone, which has itself a low glass transition temperature, a more flexible comb-branched system can be realized, which is likely to result in an electrolyte of superior conductivity. Hall et al. [24] and Fish et al. [25] produced liquid comb-branched siloxane macromolecules (Scheme 7 in Fig. 4.1). A conductivity of $1.6 \cdot 10^{-4}$ S/cm at $25\,°C$ was reported for the $LiClO_4$ electrolyte ($n = 12$). However, the chemical and mechanical stabilities of the macromolecule are low. Fish et al. enhanced the chemical stability of the comb-branched siloxane macromolecules by connecting the POE branch with an Si–C bond (Scheme 8 in Fig. 4.1) [26]. Blonsky and coworkers studied systems based on phosphazene macromolecules[poly(bis-(methoxy ethoxy ethoxide) phosphazene (MEEP)] (Scheme 9 in Fig. 4.1) [27]. This material has a Tg of $-80\,°C$ and all systems are completely amorphous. A conductivity of 10^{-5} S/cm was also realized for an Li-salt system.

Recently, Cowie and Sadaghianizadeh [7] synthesized cross-linked comb-branch macromolecules from macromer (Scheme 10 in Fig. 4.1): $CH_2=CHO(CH_2CH_2O)_5CH_3$ and cross-linking agent; $CH_2=CHO(CH_2CH_2O)_3$-$CH=CH_2$. An ionic conductivity of $1.0 \cdot 10^{-4}$ S/cm, ($LiClO_4$, [Li]/[O]=0.05) was observed when the cross-linking density is ca. 5%.

Cheradame and coworkers have concentrated on making chemically cross-linked macromolecular electrolytes that resist crystallization and have good

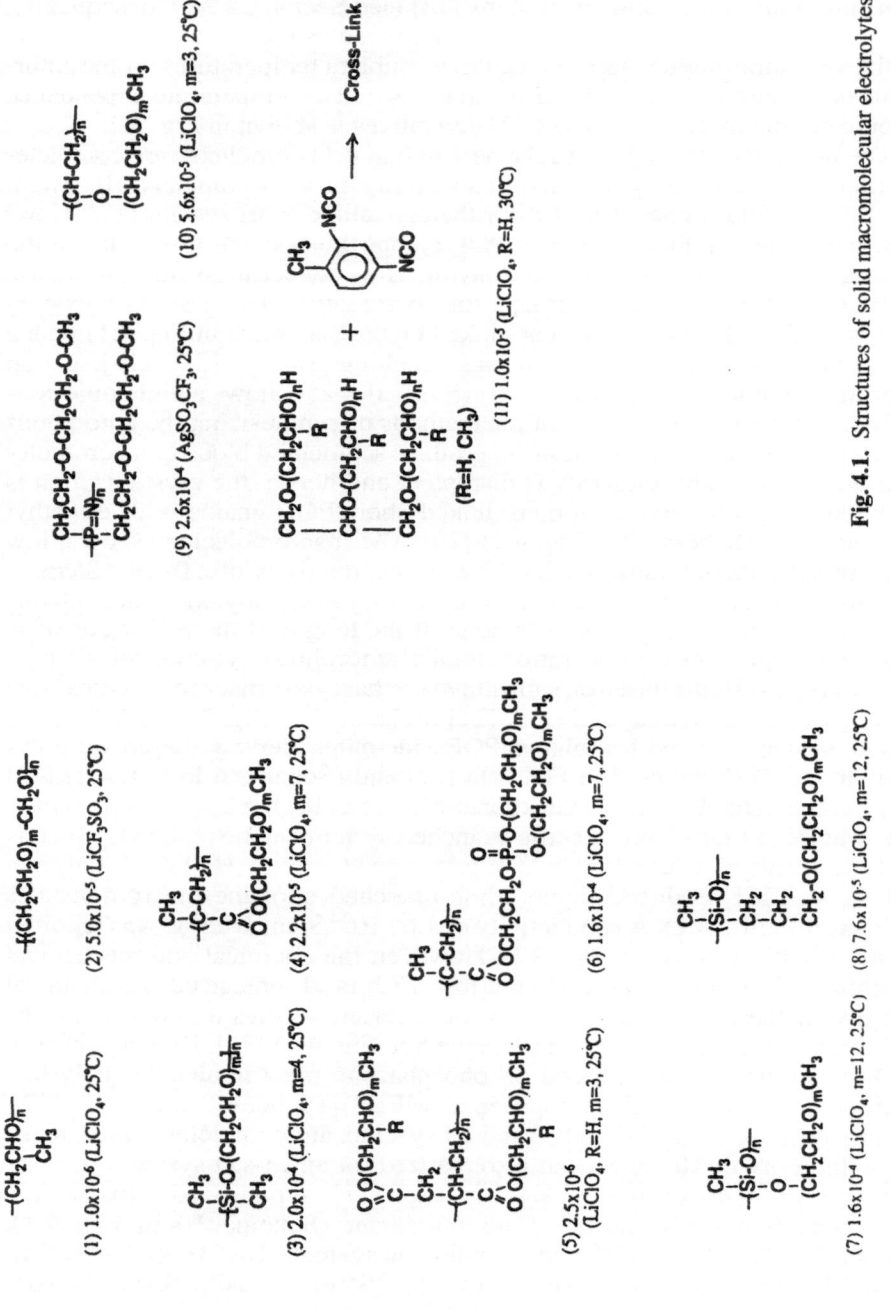

Fig. 4.1. Structures of solid macromolecular electrolytes

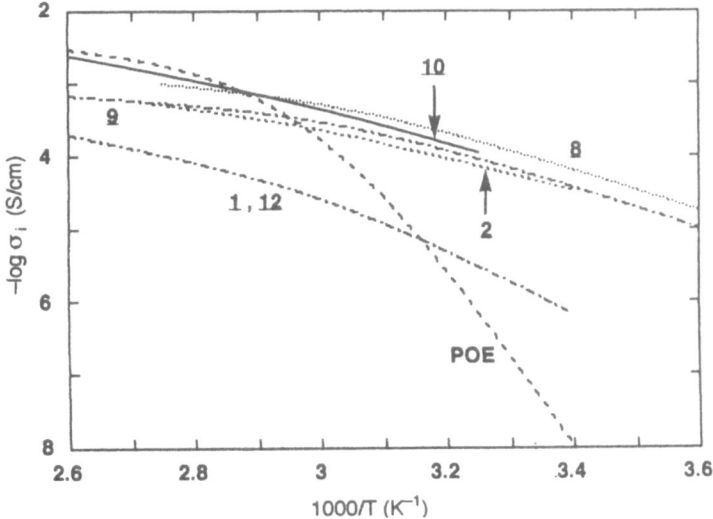

Fig. 4.2. Temperature dependence of ionic conductivity for typical macromolecular electrolytes shown in Fig. 4.1

mechanical properties [28]. Cross-linking density affects the ionic conductivity by restricting the mobility of the chain segments [29]. Network macromolecular electrolytes based on POE linked by isocyanates (Scheme 11 in Fig. 4.1) have also been studied by Watanabe and coworkers [30].

4.1.2.3 Single Ion-Conductors

In ordinary solid macromolecular electrolyte systems, a significant decrease in ionic conductivity occurs under d.c. polarization, because active electrodes are generally utilized for cations, not for anions, and the transference numbers of cations are usually less than 0.5. There are essential requirements, therefore, for the solid macromolecular electrolyte with only cation migration (the transference number of cation is unity). Several kinds of cation-conductive solid macromolecular electrolyte have hitherto been proposed [31–40]. There are, in principle, three approaches to designing single ion-conductive systems:

1. Blend of polyelectrolyte and ion-conductive macromolecules
2. Copolymerization of corresponding monomers for ion conduction and for a carrier source
3. Polymerization of ion-conductive macromer having a carrier ion source.

The single ion conductors are shown in Fig. 4.3. A blend of a polyelectrolyte salt with an ion-conductive macromolecular matrix, such as poly(oxyethylene), is a typical and simple approach. Bannister et al. first reported this kind of single ion conductor [31]. The blend of poly(2-sulfoethyl methacrylate lithium salt) and poly[2-(4-carboxyhexafluorobutanoyloxy)ethyl methacrylate lithium salt]

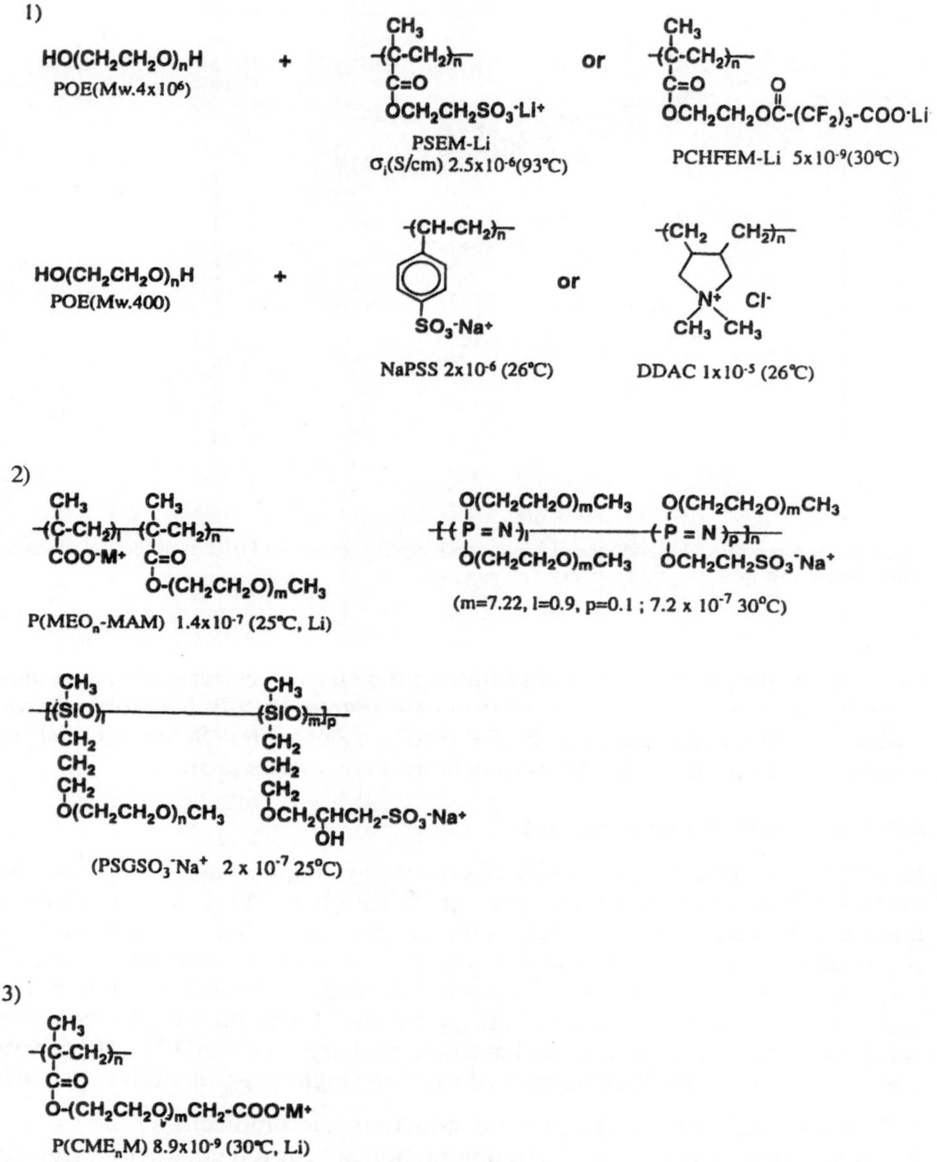

1)

$HO(CH_2CH_2O)_nH$
POE(Mw.4x10^6)

+

$-(\overset{\underset{\displaystyle CH_3}{|}}{\underset{\underset{\displaystyle OCH_2CH_2SO_3^-Li^+}{|}}{\underset{\displaystyle C=O}{|}}}C-CH_2)_n-$

PSEM-Li
$\sigma_i(S/cm)$ 2.5x10^{-6}(93°C)

or

$-(\overset{\underset{\displaystyle CH_3}{|}}{\underset{\underset{\displaystyle OCH_2CH_2O\overset{\displaystyle O}{\overset{||}{C}}-(CF_2)_3-COO^-Li}{|}}{\underset{\displaystyle C=O}{|}}}C-CH_2)_n-$

PCHFEM-Li 5x10^{-9}(30°C)

$HO(CH_2CH_2O)_nH$
POE(Mw.400)

+

$-(CH-CH_2)_n-$
![benzene ring]
$SO_3^-Na^+$

NaPSS 2x10^{-6} (26°C)

or

$-(CH_2 \quad CH_2)_n-$
N^+ Cl$^-$
CH_3 CH_3

DDAC 1x10^{-5} (26°C)

2)

$-(\overset{\underset{\displaystyle CH_3}{|}}{\underset{\displaystyle COO^-M^+}{|}}C-CH_2)_l-(\overset{\underset{\displaystyle CH_3}{|}}{\underset{\underset{\displaystyle O-(CH_2CH_2O)_mCH_3}{|}}{\underset{\displaystyle C=O}{|}}}C-CH_2)_n-$

P(MEO$_n$-MAM) 1.4x10^{-7} (25°C, Li)

$-((\overset{\underset{\displaystyle O(CH_2CH_2O)_mCH_3}{|}}{\underset{\displaystyle O(CH_2CH_2O)_mCH_3}{|}}P=N)_l-(\overset{\underset{\displaystyle O(CH_2CH_2O)_mCH_3}{|}}{\underset{\displaystyle OCH_2CH_2SO_3^-Na^+}{|}}P=N)_p)_n-$

(m=7.22, l=0.9, p=0.1 ; 7.2 x 10^{-7} 30°C)

$-((\overset{\underset{\displaystyle CH_3}{|}}{\underset{\underset{\underset{\underset{\displaystyle O(CH_2CH_2O)_nCH_3}{|}}{\displaystyle CH_2}}{\underset{\displaystyle CH_2}{|}}}{\underset{\displaystyle CH_2}{|}}}SIO)_l-(\overset{\underset{\displaystyle CH_3}{|}}{\underset{\underset{\underset{\underset{\displaystyle OCH_2CHCH_2-SO_3^-Na^+}{|}}{\displaystyle CH_2}}{\underset{\displaystyle CH_2}{|}}}{\underset{\displaystyle CH_2}{|}}}SIO)_m)_p-$
OH

(PSGSO$_3^-$Na$^+$ 2 x 10^{-7} 25°C)

3)

$-(\overset{\underset{\displaystyle CH_3}{|}}{\underset{\underset{\displaystyle O-(CH_2CH_2O)_mCH_2-COO^-M^+}{|}}{\underset{\displaystyle C=O}{|}}}C-CH_2)_n-$

P(CME$_n$M) 8.9x10^{-9} (30°C, Li)

Fig. 4.3. Structures of single ion-conductive macromolecular electrolytes

showed $2.5 \cdot 10^{-6}$ S/cm at 66 °C, but only $4 \cdot 10^{-9}$ S/cm at 30 °C. Hardy and Shriver reported the sodium ionic conductivity of $2 \cdot 10^{-6}$ S/cm at 26 °C in a poly(styrene sulfonate sodium salt) and polyethylene glycol mixed system [32].

Comacromolecules of oligo(oxyethylene) methacrylate and alkali-metal methacrylates showed the conductivities on the order of 10^{-7} S/cm at room temperature [33, 34]. Other comacromolecular systems using poly(phosphazene) [35] or poly(siloxane) branched macromolecular derivatives [36], of which the carrier sources were $-CH_2SO_3^-$, were reported recently to have Na^+ conductivities of ca. 10^{-7} S/cm at 30 °C. A polyether network with an ionizable cross-linking agent is the other approach, developed by both Le Nest et al. [37] and Watanabe et al. [38]. They reported ionic conductivities of 10^{-7}–10^{-8} S/cm for Li^+.

The sophisticated system is a single ion-conductor consisting of macromer that serves three different roles such as flexible structure, ion-conductive pathway, and the source of carrier ion in one repeating unit [39, 40]. However, Na^+ conductivity was low ($6.5 \cdot 10^{-8}$ S/cm at 30 °C) because of the low dissociation ability of $-COO^-$.

4.1.3 Dissociation of Inorganic Salts in Macromolecules

4.1.3.1 Ion-Dipole Interaction in Macromolecules

In order to facilitate the dissociation of inorganic salts in macromolecular solids, the following requisites must be considered: (a) smaller lattice energy of the inorganic salt, (b) larger solvation energy (ion-dipole interaction) of the macromolecule, and (c) higher dielectric constant of the macromolecule. The thermodynamics of salt-dissociation into poly(oxyethylene) has been considered in detail by Papke et al. [41]. A pair of a cation (hard acid) with a large diameter and an anion the charge of which is largely delocalized (soft base), have a low lattice energy. Table 4.1 shows the relationship between lattice energies of inorganic salts and the solubility of the salts in high molecular weight POE. There are critical lattice energies for the solvation by the poly(oxyethylene). The critical lattice energy increases with decreasing size of the cation, indicating the high solvating ability of POE for a smaller cation.

Table 4.1. Lattice energy (kJ/mol) of alkali-metal salts and solubility in POE

Cation/anion	Li^+	Na^+	K^+	Rb^+	Cs^+
F^-	1036 (I)	923 (I)	821 (I)	785 (I)	740 (I)
Cl^-	853 (S)	786 (I)	715 (I)	689 (I)	659 (I)
Br^-	807 (S)	747 (S)	682 (I)	660 (I)	631 (I)
I^-	757 (S)	704 (S)	644 (S)	630 (S)	604 (S)
SCN^-	807 (S)	682 (S)	619 (S)	616 (S)	568 (S)
$CF_3So_3^-$	725 (S)	650 (S)	605 (S)	585 (S)	550 (S)
ClO_4^-	723 (S)	648 (S)	602 (S)	582 (S)	542 (S)

S soluble; I insoluble.

The solvation energy of POE for cations and anions relates to the donor and acceptor number of macromolecule, respectively. Polyethers, such as poly(oxyethylene), have a small acceptor number (10.8, glyme) and a high donor number (22, glyme) that is higher than that of water (16.4). Therefore, polyether solvates a cation more preferentially than an anion. The best structure of the polyether is up to now restricted to polyethers having a repeating unit -C-C-O-, where the configurations of -C-O-, -O-C, and -C-C- are *trans, trans,* and *gauche,* respectively, because the sequence makes several ether oxygens in the same macromolecule coordinate a cation for the effective dissociation of the salt. Therefore, the solvation energy is the sum of the coordination energy of plural ether oxygens, leading to the relatively high degree of dissociation despite the low dielectric constant.

4.1.3.2 Dielectric Constant for Ion Dissociation

The dielectric constant of a macromolecular matrix influences the dissociation energy of the inorganic salt. Saito [42] and Barker [43] found that the logarithm of the ionic conductivity of many conventional macromolecules was proportional to the inverse of the dielectric constant (ε) of the macromolecules. In those cases carriers were considered to be ionic impurities. The number of ionic carriers (n) is expressed in Eq. (1):

$$n = n_0 \exp(-W/2\varepsilon kT), \tag{1}$$

where n_0 and W are the constant number and the dissociation energy of the salt, respectively.

The dielectric constant of poly(oxyethylene), where a crystalline phase and an amorphous phase coexist, is 4 at 25 °C, and that of the amorphus phase is 8 [44], and it increases with increasing the salt concentration [45]. This relatively high dielectric constant of amorphous POE in comparison with those of conventional macromolecules is effective as a macromolecular matrix from the point of salt dissociation. However, the dielectric constants of the macromolecule are considerably low in comparison with those of organic solvents such as propylene carbonate (64.4). The formation of ionic clusters is a serious problem in such a low dielectric matrix [41, 46].

4.1.3.3 Association States of Ions in Macromolecular Solids

Ionic clusters exist as singlets, pairs, triplets, and further-aggregated states. The total charge number and the ionic mobility of the ionic clusters relate to the ionic conductivity of the entire system. A higher association number leads to a lower ionic mobility. Torell and Schantz developed a novel method by applying Raman spectroscopy to the systematic analyses of the dissociated states of inorganic salts [46]. As shown in Fig. 4.4, the symmetric stretching vibration of anions separated into a few peaks corresponding to the associating states such as single ions, ion pairs, and ion triplets (ion aggregates). With increasing salt concentration, the ratio of single ions decreases. In an $LiClO_4$ system the ratio of single ions is high in comparison with an $NaCF_3SO_3$ system, and ion aggregates do not exist even in a high concentration of the salt (O/Li = 5), indicating the higher ionic conductivity of POE/LiClO$_4$ composites [47]. The temperature dependence of the states of ionic clusters was also reported using this method [48]. The ratio of free ions decreases significantly with temperature. Very recently, the study of ion association in poly(oxyethylene) was extended to trivalent cations, such as Eu^{3+} and Ce^{3+}, coupled with luminescence studies [49].

4.1.3.4 Development of New Salts

Thus far, the only useful salts with the high electrochemical stability in solid electrolytes for Li batteries are limited to LiClO$_4$ or LiCF$_3$SO$_3$. The noble lithium salt of bis(trifluoromethanesulfone)imide, [LiN(CF$_3$SO$_2$)$_2$], was recently synthesized and characterized [50]. The values of conductivity are excellent with an ionic conductivity of $5 \cdot 10^{-5}$ (S/cm) for (P(OE)$_8$/Li(CF$_3$SO$_2$)$_2$N at 25 °C) [69]. The anion of this salt can be considered to confer a "plasticizing" effect because of the low rotational barrier around the S−N bond [51]. All the sulfur, nitrogen, and oxygen atoms, as well as the terminal -CF$_3$ groups, can participate in dispersing the charge on the imide anion, which promotes dissociation. Furthermore, a delocalized salt Li$^+$[(CF$_3$SO$_2$)$_3$C]$^-$ was also recently synthesized [52]. As a reference the degree of dissociation in organic solvent (1:1 PC:DME) appears to be in the following sequence [53]:

$$LiPF_6 = LiAsF_6 > LiClO_4 = Li(CF_3SO_2)_2N$$
$$= Li-SO_2(CF_2)_4SO_2N >> LiBF_4 >> LiCF_3SO_3$$

The ionic conductivities of single ion-conductive macromolecules are generally approximately 10^{-2} times that of ordinary amorphous ion-conductive macromolecules because of the low dissociation of carboxylate or sulfonate salts in comparison with MClO$_4$ or MCF$_3$SO$_3$. Taking this into consideration, a perfluorosulfonate ionomer should be an excellent candidate for a macromolecular single ion conductor with high ionic conductivity. However, the relatively high ratio of phase-separated perfluoroethylene regions in the perfluorosulfonate-type ionomer results in a small number of carriers and ion-conductive pathways [14]. The development of a perfluorosulfonate-type ionomer with the low equivalent molecular weight becomes necessary.

In another method for enhancing single ionic conductivity, cryptands or crown ethers, which form complexes with cations and separate them from counter anions, were added to the single-ion conductive macromolecules. The ionic conductivity was enhanced more than ten fold because of the enhancement of dissociation of the sulfonate group [54]. Furthermore, sterically hindered phenol groups bound covalently to poly(siloxane-g-ethylene oxide) backbones were synthesized. In this system, t-butyl substituents at the 2,6-positions of phenoxide serve in separating alkali-metal ions, such as K$^+$ and Na$^+$, from the phenoxide [55]. The maximum single ionic conductivities were $5 \cdot 10^{-6}$ S/cm for Li at 30 °C.

4.1.3.5 Composite Systems (Molten Salts/Macromolecules)

Certain pyridinium or imidazolium salts form molten salts at ambient temperatures when mixed with AlCl$_3$. Watanabe et al. studied the complexation of

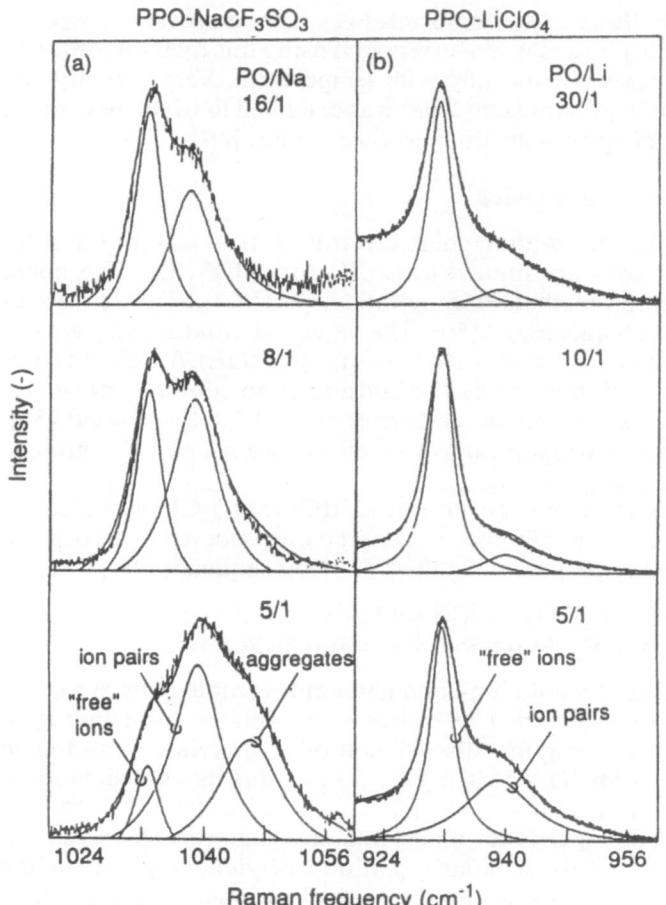

Fig. 4.4. The anion symmetric stretching mode of Raman spectra for different salt concentration in POP

the molten salts with poly(1-butyl-4-vinyl pyridinium halide) [56]. This new class of macromolecular electrolyte has a high ionic conductivity of 10^{-3} S/cm at room temperature. In this case the carrier ion may be $AlCl_4^-$. Angell et al. studied various mixtures of lithium salts (e.g., $(LiI)_{0.5}$, $(LiOAc)_{0.3}$, and $(LiClO4)_{0.2}$), which have a T_g far lower than room temperature [57, 58]. To such molten salts small amounts of macromolecules, such as POP or polyvinyl lithium sulfonate (PVSLi), were added to form Li^+-conductive ionic rubbers. The Li^+ conductivity of 18.3-mol% PVSLi in 81.7 mol% $(LiClO_4)_{0.6-}$ $(LiNO_3)_{0.4}$ was 10^{-5} S/cm at 25 °C [58].

4.1.4 Ion-Conduction Mechanism in Macromolecular Systems

4.1.4.1 Role of Segmental Motion of Macromolecules for Ion Conduction

The ionic conductivity (σ) is known to be given simply by

$$\sigma = ne\mu, \tag{2}$$

where n, e, and μ are the total number of carrier ions, the elementary electric charge, and the carrier mobility, respectively. To design ion-conductive macromolecules with a high ionic conductivity, an effort must be made to increase the factors (n and μ). Because even a proton, the smallest ion, has a weight of ca. 2000-fold that of an electron, physical space is necessary for ion conduction. Miyamoto and Shibayama expressed the ion conduction in macromolecules based on the free volume theory [59]:

$$\sigma = \sigma_0 \exp\left[\{(-\gamma V_i^*/V_f)-(E_j + W/2\varepsilon)/kT\}\right], \tag{3}$$

where E_j is the activation energy for ion transfer in the macromolecule, W is the dissociation energy of a salt in the macromolecule, ε is the dielectric constant of macromolecule, γ is a numerical factor that takes into account the possible overlap of free volume, and V_i^* is the minimum hole size necessary for ion transfer, arising from thermal fluctuations of the free volume. V_f is the average free volume per ion at a temperature (T) above the glass transition temperature (T_g):

$$V_f = V_g[f_g + \alpha(T-T_g)], \tag{4}$$

where V_g, f_g, and α are the relative volume at T_g, the average free volume fraction per diffusing unit at T_g, and the thermal expansion coefficient, respectively. Based on Eqs. (3) and (4) when a high dielectric environment and a sufficient free volume are provided in a matrix of macromolecules containing inorganic salts, the ionic conductivities should be ignored.

Killis et al. [60] studied the viscoelastic behavior of networks based on POE or POP containing various salts and estabished master curves for the William-Landel-Ferry (WLF) equation correlating the ionic conductivity with frequency and temperature using the WLF equation [61]:

$$\log[\sigma(T)/\sigma(T_g)] = C_1(T - T_g)/(C_2 + (T - T_g)), \tag{5}$$

where $\sigma(T_g)$ is the conductivity of the relevant ionic species at T_g, and C_1 and C_2 are the WLF parameters of the free-volume equation for ionic transport. They calculated the WLF parameters of $C_1(10\pm2)$ and $C_2(50\pm5\,K)$, which were similar to the values obtained from a WLF treatment of the storage modules vs temperature data. Watanabe et al. [62] studied the ionic conductivity for amorphous POE networks with different alkali-metal salts. Although the conductivity at the same salt concentration was different by at most two orders of magnitude, the WLF plots were represented by one master curve ($C_1 = 9.6$, $C_2 = 45.6\,K$), irrespective of the salt species as shown in Fig. 4.5. That means that ion conduction in a macromolecular solid strongly relates to the segmen-

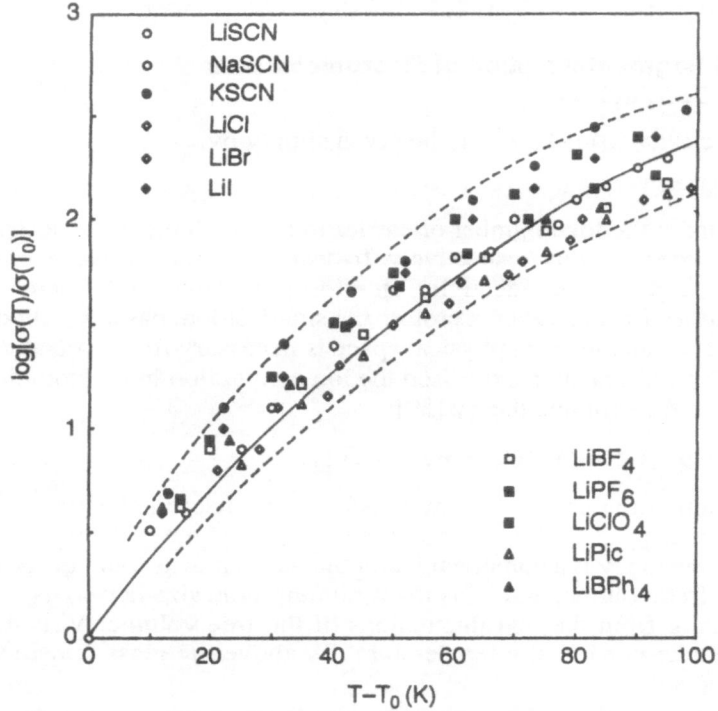

Fig. 4.5. William-Landel-Ferry plots of cross-linked POE containing various inorganic salts

tal motion of macromolecular chains. The WLF equation was conducted based on Eq. (3) proposed by Miyamoto et al. when the term of $(Eb' + W/2\varepsilon)$-$(T-Tg)/TTg$, which is derived from the second part of the equation, was ignored. C_1 and C_2 correspond to $\gamma v_2^*/2.3 vgfg$ and fg/α, respectively.

The salt-concentration dependence of the ionic conductivities generally reveals its highest level as shown in Fig. 4.6 [63]. The conductivity maxima of $PMEO_n$ ($n = 3, 7, 12$, and 17) appear at lower concentration of $LiClO_4$ when $P(MEO)_n$ with a larger n is employed. This is attributed to the compensation of two factors, namely that the initial increase in the ionic conductivity can be explained from the increase in the carrier number as the $LiClO_4$ concentration increases, and that the following decrease is due to the lower segmental motion by a T_g increase due to the coordination of the dissociated ion to the macromolecules [63]. A decrease in segmental motion would be more extensive in $P(MEO)_n$ with a larger n because of the stronger interaction between the lithium ion and the longer side chains. The optimum number (n) of oxyethylene units is seven in $P(MEO)_n$.

4.1.4.2 Phase Diagrams of (Macromolecule/Salt) Composites

Phase diagrams of (POE/inorganic salt) composites were made using thermal analyses [64], X-ray diffraction [18], ionic conductivity measurements [18],

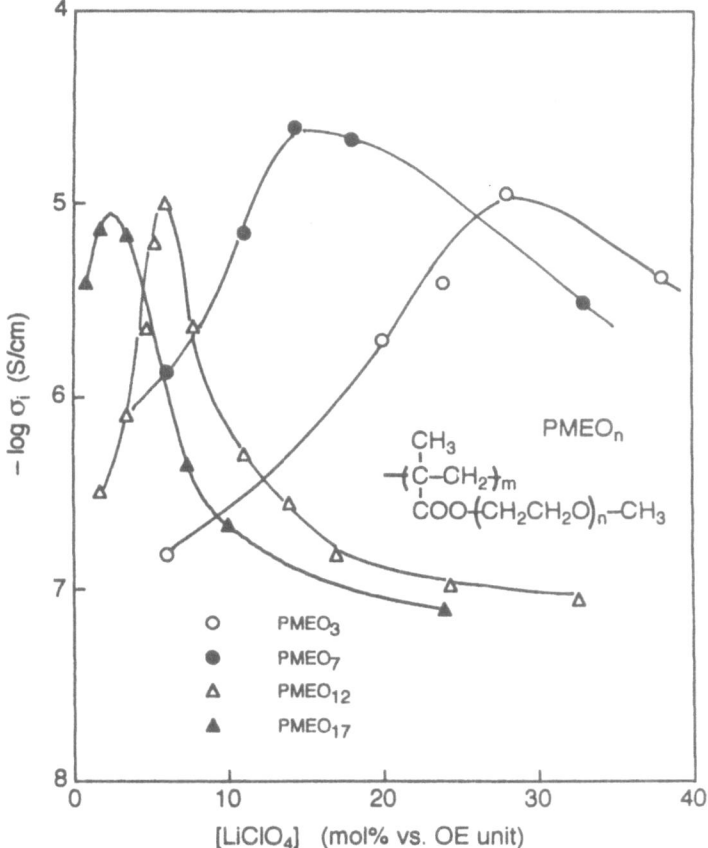

Fig. 4.6. Salt-content dependence of the ac (1 V) ionic conductivity for PMEO$_n$ hybrid films

and optical microscopy [18]. The resulting phase diagrams are generally very complicated, having a (POE/inorganic salt) complex crystalline phase with a high melting point, a POE crystalline phase with a melting point of 60°C, and a POE amorphous phase in which the inorganic salts are soluble. Furthermore, the composition of these phases changes with temperature. Fig. 4.7 shows the phase diagram of a POE/LiClO$_4$ system and the temperature dependence of ionic conductivity of each composition [18], respectively. The turning point of the temperature dependence corresponds to the boundary of the phase change. The behavior of the temperature dependence of the ionic conductivity was explained by a single microstructural model [65]. Ohno and Sasayama developed a new apparatus for the dynamic measurement of the ionic conductivity by raising or lowering the temperature at constant rates (0.1–10°C/min) [66]. With this method the properties and phase states of (POE/inorganic salt) systems can be studied dynamically in detail in a short time (typically within 20 min).

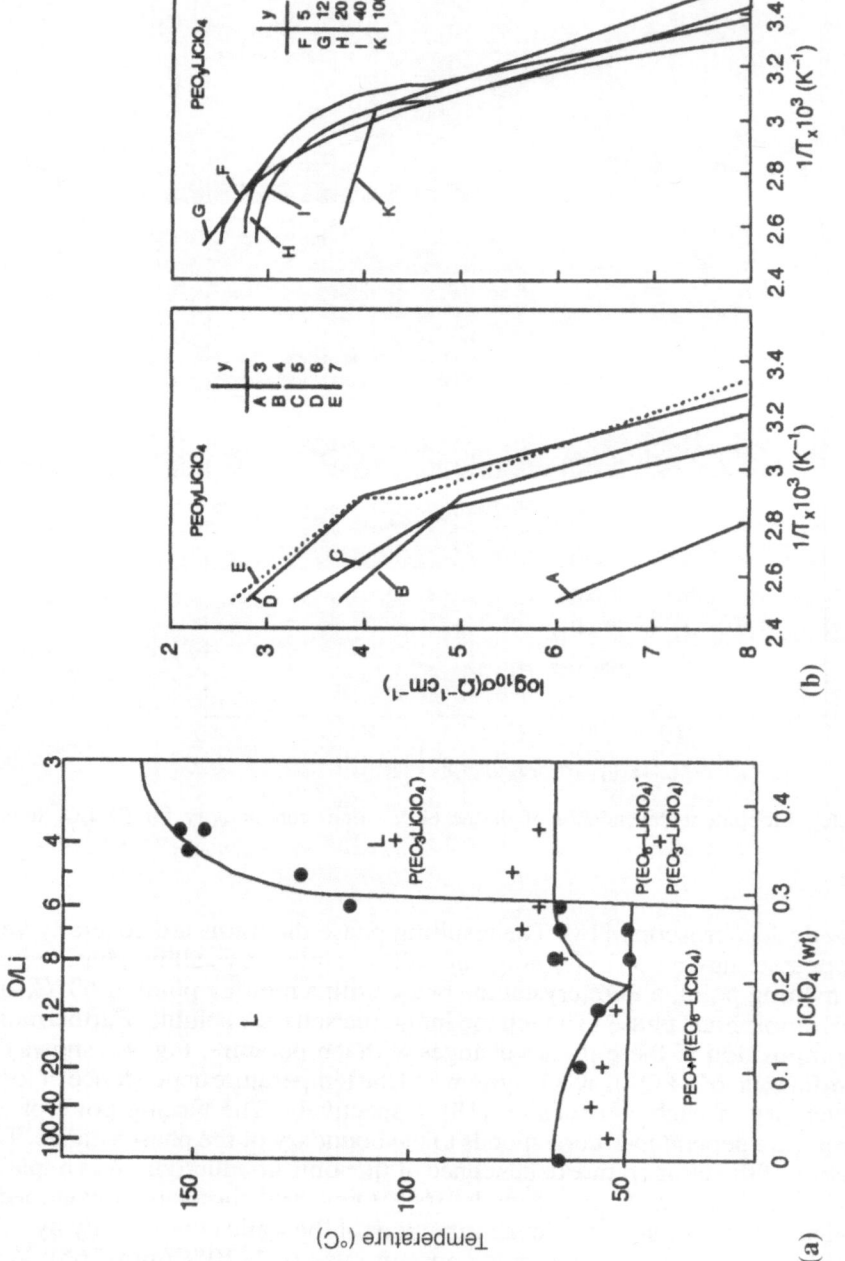

Fig. 4.7. Relation between **a** phase diagram and **b** temperature dependence of ionic conductivity

4.1.4.3 Ions with Different Size and Ion–Oligoether Interaction

In order to study the effect of the polyether structure on the ion-dipole interaction, various oligo(oxyethylene) derivatives were incorporated into the ionic
clusters of M^+-Nafion films [67]. In the Nafion/POE system, the dissociation
of the cation-perfluorosulfonate pair can be studied easily by observing the
peak shift of the $-SO_3$ symmetric stretching mode (V_{SO_3}-) in the IR trans-
mittance spectral [14]. The V_{SO_3}- of Li^+, Na^+, and K^+-Nafion not containing
ethers appeared at 1075, 1066, and 1060 cm^{-1}, respectively. This means that
the larger cation causes a smaller degree of polarization of the S–O dipole because of weaker ion–ion interaction. The rate of the peak shift can be calculated
according to Eq. (6):

$$\text{Peak shift } (\%) = \frac{V_0 - V_1 \cdot 100}{V_0 - 1051} \tag{6}$$

where V_0 and V_1 are the V_{SO_3}- of Nafion films without and with ethers, respectively, and 1051 cm^{-1} is the estiamted value at 100% dissociation.
Linar oligo(oxyethylene) and cyclic oligo(oxyethylene) (crown ethers; 15C5,
18C6, and 21C7) were incorporated into the micropores of a M^+-Nafion thin
film in an amount equimolar to the perfluorosulfonate groups. Figure 4.8
shows the rate of the peak shift of a linear oligo(oxyethylene) system: it increases with increasing cation diameter, meaning that the order of salt disso-

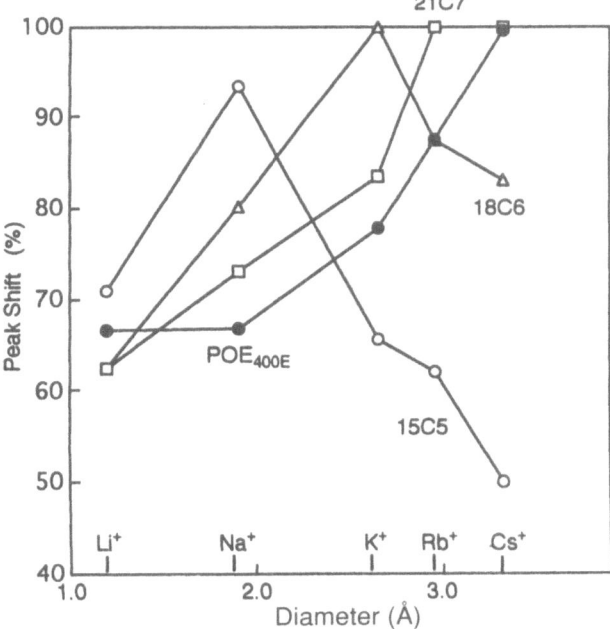

Fig. 4.8. Influence of cation diameter on the peak shift of $v(SO_3^-)$

ciation also increases with cation diameter [68]. In the case of M⁺-Nafion/-cyclic oligo(oxyethylene) composite films, the high degree of dissociation is formed preferentially when the size of the cation fits the cavity size of the cyclic oligo(oxyethylene) [69]. In the case of 15C5, 18C6, and 21C7 the highest values were confirmed for Na⁺, K⁺, and Rb⁺ (Cs⁺), respectively.

M⁺-Nafion/linear oligo(oxyethylene) clarified that larger cations can move faster than smaller cations. These results were also confirmed in a series of alkali-metal salts of poly{α-carboxymethyl-ω-methacryloyloxyoligo(oxyethylene)] [70]. The difference in the mobility should originate from the different interaction force between the alkali-metal cation and ether oxygens. Larger ions have smaller surface charge density and therefore interact weakly with both an anion and ether oxygens. Accordingly, larger cations can move faster in macromolecular solids through the fast exchange between sites. On the other hand, the cation conductivities of M⁺-Nafion/cyclic oligo(oxyethylene) composite films are plotted in Fig. 4.9. The conductivity of M⁺-Nafion/18C6 increased in the order Li⁺>Na⁺>K⁺>Cs⁺ despite the highest salt-dissociation of K⁺. This means that the order of salt dissociation could not contribute to the order of ionic conductivity. The interaction between the cation

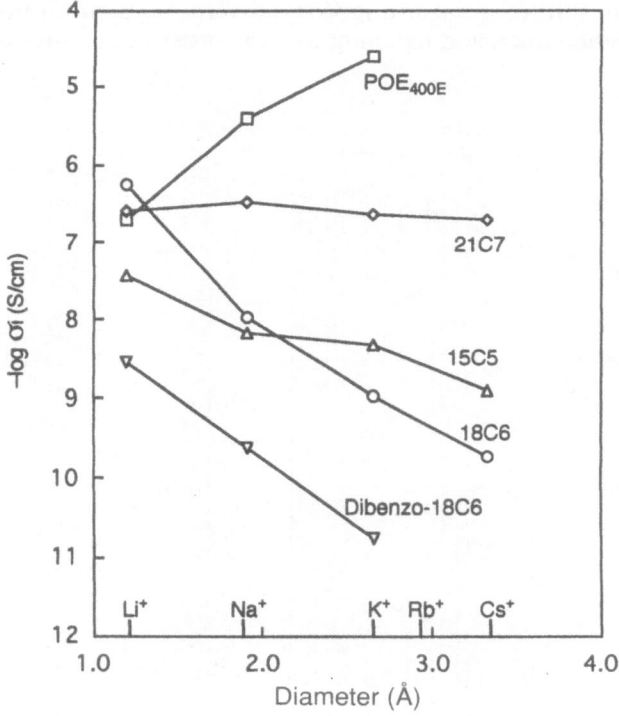

Fig. 4.9. Ionic conductivity of M⁺-Nafion/oligo(oxyethylene) composite films (at 100°C in film)

and ether oxygens increases when the size deviation between the cation and the cyclic oligo(oxyethylene) becomes small, and K^+ shows the highest interaction with ether oxygens. The high interaction decreases the migration of the cation.

Next, the modified oligo(oxyethylene)(OOE) was introduced into M^+($M =$ Li, Na K)-Nafion thin films [8]. One type of modification of OOE is the exchange of the terminal hydroxyl groups with methoxy or acetoxy groups, and the other is the replacement of the center ethylene with methylene or propylene. Modification of the center divides a series of ether oxygens, which coordinate a cation intramolecularly. At least four continuous oxyethylene units are necessary for dissociation of the Li salt, but a more continuous oxyethylene unit is better to provide the more intramolecular freedom for Li^+, and OOE with over nine oxyethylene units crystallize at ambient temperatures. Therefore, six or seven oxyethylene units are considered to be the best side chain length. In the modification of the end groups of OOE, the methyl group is superior to the acetyl group. The intermolecular mobility of the cation is high when the end group can move with greater freedom. Considering this knowledge, the macromer of [$CH_2 = CHCH_2O(CH_2CH_2O)_7CH_3$] was synthesized and showed an excellent ionic conductivity of $1.4 \cdot 10^{-4}$ S/cm at 25 °C ($LiClO_4$, Li/OE = 0.05) [71].

4.1.5 Performance of Ion-Conducting Macromolecules

4.1.5.1 Lithium Secondary Batteries

The application of solid macromolecular electrolytes (SME) to lithium secondary batteries for automobiles, which have high power and energy densities, are flexible for rolling and are lightweight, would be the goal for many researchers in these fields. Features expected for solid macromolecular electrolytes are follows:

1. A thin-film configuration, which enables a roll-type cell
2. A large surface-to-thickness ratio (a film with a large area) that permits low-current density, leading to high energy efficiency and cycling efficiency
3. Ease of assembling and packaging of an all-solid-state device
4. A high flexibility and high mechanical stability
5. A thin separator to avid Li dendrites
6. An elastomeric binder for the composite cathode to enhance the cycling efficiency and processibility.

The SME Li secondary battery combines the use of a thin film of Li^+-solid macromolecular electrolyte with Li^+-reversible electrodes. This type of battery has thin-film configurations of the type illustrated in Fig. 4.10 [72, 73], where the negative electrode is a lithium metal foil, the macromolecular electrolytes are linear poly(oxyethylene) containing $LiClO_4$ or $LiCF_3SO_3$, and the positive electrode is a composite mixture of V_6O_{13} or TiS_2 as active materials, macromolecular electrolyte for ion transport, and acetylene black as an electronic conductor. At discharge, lithium metal is oxidized to lithium ion at the interface between the negative electrode and the electrolyte, transfers through

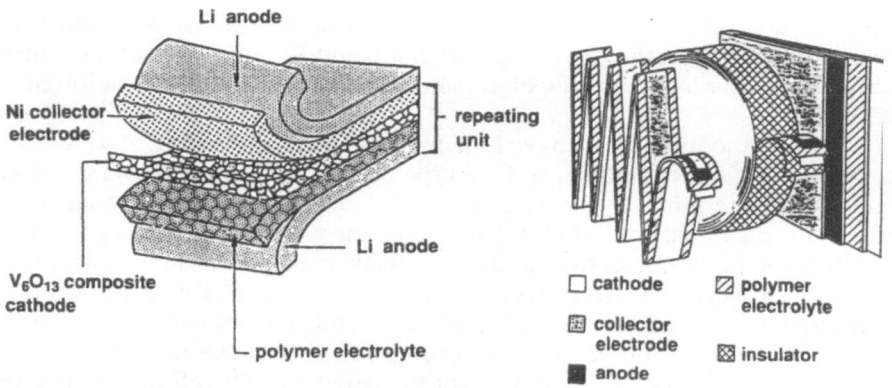

Fig. 4.10. Structure of lithium secondary battery with solid macromolecular electrolytes

the electrolyte and intercalates between layers of the active material at the positive electrode. The reverse occurs on charging. Because the carrier is only lithium ion in such a system, the electrolyte functions as a separator to prevent short circuits, and is not needed to serve as a pool for the carriers. The advantage of a macromolecular electrolyte to form a thin film is one of its excellent merits, leading to a high energy density of the batteries.

Recently, lithium metal was used for a negative electrode as a nonmetal compound, $Li_yM_nY_m$ [74]. The electrochemical process at this electrode is the uptake of lithium ions during charge and the release during discharge. If the positive electrode is made of an Li^+-intercalation compound, A_zB_w, the lithium ion "rock" from one side to the other during charge and discharge processes accordig to Eq. (7). This type of cell is sometimes called a "rocking-chair battery" [74].

$$Li_yM_nY_m + A_zB_w \underset{charge}{\overset{discharge}{\rightleftharpoons}} Li_{y-x}M_nY_m + Li_xA_zB_w, \qquad (7)$$

where $Li_yM_nY_m$ is Li_xC_6, $LiWO_2$, or $Li_6Fe_2O_3$, and A_zB_w is TiS_2, V_2O_5, WO_3, and NbS_2.

Table 4.2 summarizes the properties of several laboratory batteries [75–84]. The research on the lithium-secondary battery using solid macromolecular electrolytes has been promoted vigorously in the ACEP (French acronym for macromolecular electrolyte battery) project [75, 81]. The ACEP reported two types of batteries, the so-called first and second generations [79]. The second generation (1987–1988) of solid macromolecular electrolytes (amorphous SME based on oligo(oxyethylene) side chains, and the new plasticizing lihium salt) is superior to the first generation (1985) of solid macromolecular electrolyte (low crystallinity POE) used in the ACEP project.

Linear poly(oxyethylene)-based macromolecular electrolytes provide batteries that generate a current density of only a few $\mu A/cm^2$ at ambient tem-

Table 4.2. Performance of lithium secondary batteries using macromolecular electrolytes

Cell configuration	Temperature (°C)	Discharge voltage (V)	Current density (mA/cm²)	Cathode utilization (%)	Charge/discharge cycles	Reference
Li/P(OE)$_8$LiClO$_4$/TiS$_2$	100	2.2	0.49	85	250	75
Li/P(OE)$_8$LiCF$_3$SO$_3$/TiS$_2$	130	2.2	0.25	70	–	75
Li/P(OE)$_8$LiClO$_4$/V$_6$O$_{13}$	100	2.4	0.50	60	–	75
Li/P(OE)$_4$LiCF$_3$SO$_3$/V$_6$O$_{13}$	140	2.4	0.8	40	–	76
Li/P(OE)$_4$LiCF$_3$SO$_3$/Li$_{1+x}$ V$_6$O$_6$	90	2.5	0.25	85	–	77
Li/P(OE)$_8$LiClO$_4$-γLiAlO$_2$/LiV$_3$O$_{13}$	120	2.9	0.2	40	20	78
Li/(50:50 MEEP-P(OE)$_8$)LiClO$_4$/TiS$_2$	100	2.2	2.0	50	1	79
Li/P(OE)$_8$LiN(SO$_2$CF$_3$)$_2$/X	80	3.0	0.125	75	83	80
Li/P(OE)$_8$LiClO$_4$/MoO$_2$	20	1.6	0.008	40	300	75
Li/ACPE(II)/TiS$_2$	25	2.2	0.09	60	450	81
Li/(87:13 MEEP-PVP)$_4$LiClO$_4$/TiS$_2$	20	2.2	0.05	17	1	82
Li/P(OE)$_8$LiN(SO$_2$CF$_3$)$_2$/X	35	3.0	0.25	17	1	80
Li/PAN-EC/PC-LiAsF$_6$/LiMn$_2$O$_4$	r.t.	3.0	0.1	67	138	83
LiTiS$_2$/PAN-EC/PC-LiClO$_4$/Li$_{1-x}$CoO$_2$	r.t.	1.8	0.02	55	26	84

X-poly(1,3,4-thiadiazole disulfide).

peratures, because the ionic conductivity of the macromolecular electrolytes is less than 10^{-7} S/cm [75]. In contrast, oligo(oxyethylene) (OOE, mol. wt. 400) generated $250 \mu A/cm^2$, because its ionic conductivity was $7 \cdot 10^{-4}$ S/cm [82]. An ionic conductivity of more than 10^{-4} S/cm should be necessary for batteries to generate a current density of $0.5-5$ mA/cm^2. Therefore, in the case of linear POE, working temperatures are usually ca. $80-100 °C$. In an (Li/(POE)$_8$LiClO$_4$/TiS$_2$) battery, the cathode utilization, remained at 80%, even after 250 discharge–charge cycles at $100 °C$ [75]. When the temperature was lowered to $80 °C$, the utilization decreased to 60%. When LiCF$_3$SO$_3$ was used instead of LiClO$_4$, it decreased to 37% because of the low ionic conductivity of LiCF$_3$SO$_3$ in comparison with LiClO$_4$ [75]. In the second generation the power characteristics of the SPE battery at room temperature have been significantly upgraded; however, the current density is still low, ca. $0.05-0.5$ mA/cm^2 [79, 80]. The battery performance of the gel electrolyte systems [83, 84] and a rocking-chair battery [84] are also listed in Table 4.2. However, further advances are still needed in order to improve the conductivity behaviour at low temperatures and attain the cationic conductivities of 10^{-4} S/cm at $25 °C$.

The cathode utilization of TiS$_2$ was superior to that of V$_6$O$_{13}$ [75]. That indicates that lithium ion diffusion in the intercalation process should be considered in constructing a lithium secondary battery of higher quality. Lithium secondary batteries using electrodes containing poly(disulfide)s, namely solid redox polymerization electrodes (SRPEs), overcome some of the limitations of using intercalation compounds as the positive electrode. Liu et al. applied these poly(disulfide)s as active materials for cathodes [80]. As indicated in Eq. (8), Li$^+$ ions insert into the cathode as counter ions for the depolymerized fragment anions, -SRS-, on discharge, and are released from the electrode on charge, the polymerization by disulfide bonding:

$$-(-SRS-)n- + 2nLi + 2ne \underset{\text{charge}}{\overset{\text{discharge}}{\rightleftarrows}} n(Li-SRS-Li). \qquad (8)$$

Theoretical energy densities of batteries based on poly(disulfide)s are higher than those of intercalation compounds, such as TiS$_2$, due to the low equivalent weight. The results of performances of a cell utilizing Li/(POE)$_8$LiN(SO$_2$CF$_3$)$_2$/poly(1,3,4-thiadiazoledisulfide) composite cathode are also summarized in Table 4.2 [77]. At $100 °C$ the cell achieved a power density of 1700 W/kg (current density 10 mA/cm^2) at an energy density of 140 Wh/kg with 96% utilization of the poly(disulfide).

The potential advantages of both electrically and ionically conductive macromolecules have led to several attempts to build all-macromolecular batteries. Chiang reported a polyacetylene battery with POE-based macromolecular electrolytes [85], but the poor interpenetration of cations and anions into the macromolecules limited the performance of this battery. In 1983 Fouletier et al. measured lithium insertion into polyacetylene combined with a POE-based macromolecular electrolyte [86]. Nagatomo et al. reported a rechargeable all-macromolecular battery fabricated using a polyacetylene film as the solid electrolyte [87]. Nova'k et al. studied an all-macromolecular battery having a lithium anode and a poly(propylene) and poly(oxyethylene) composite cathode [88].

4.1.5.2 Electrochromic Displays

A variety of materials have been employed as electrochromic materials such as WO_3 [89], $Ir(OH)_x$ [90] and Prussian Blue (PB) [91], bipyridinium compounds [92], and electric conducting macromolecules [93, 94]. In most cases, however, electrochromic displays (ECDs) have been constructed as solution-containing systems. The leakage of the electrolyte solution and low stabiity are therefore inevitable drawbacks for these wet electrochromic devices. Because all solid-state electrochromic displays based on solid macromolecular electrolytes should have advantages, such as thin films, light weight, processibility, etc., macromolecular electrolytes can be expected to eliminate these drawbacks [95].

All solid-state WO_3/PB based on electrochromic devices were prepared usig solid macromolecular electrolytes. The electrochromic redox reactions of the PB and WO_3 films are shown in the following equations [89, 91]:

$$Fe_4^{3+}[Fe^{II}(CN)_6]_3 + 4e^- + 4Li^+ \underset{\text{coloring}}{\overset{\text{bleaching}}{\rightleftharpoons}} Li_4Fe_4^{2+}[Fe^{II}(CN)_6]_3 \qquad (9)$$

$$WO_3 + xe^- + xLi^+ \underset{\text{bleaching}}{\overset{\text{coloring}}{\rightleftharpoons}} Li_xWO_3. \qquad (10)$$

The oxidized and reduced states of a WO_3 film are coloress and blue, respectively, whereas a PB film is blue and colorless in the oxidized and reduced states, respectively. Consequently, the optical density change between coloring and bleaching was large in a WO_3/PB system compared with the case where either of them is employed [95]. In ordinary wet ECDs the cation species was limited to K^+, because reversible intercalation into and deintercalation from the PB lattice requires a smaller hydrated ion size than the cavity (3.2 A in diameter) in the lattice. Hydrated Li^+ and Na^+ are not effective because of their larger size than hydrated K^+. In macromolecular electrolytes, Li^+, Na^+, and K^+ can migrate as naked ions; therefore, all of these ions are useful for solid-state ECD [96].

4.1.5.3 Molecular Devices

An amorphous macromolecular electrolyte can dissolve various functional molecules and provide a solid reactive matrix for them. Macromolecular electrolytes enable the transport of counter ions to compensate the change in the charge number of redox-active molecules after charge transfer. This strongly suggested that electrode redox reactions could be carried out in the macromolecular electrolyte. Therefore, various kinds of redox-active molecules, such as TCNQ [97], ferrocene [98], hemin [99], etc., have been studied for oxidation and reduction electrochemically even in macromolecular solids. In such applications, these macromolecular electrolytes can be called as "macromolecular solvents" [100, 101]. That will open a new field in solid-state ionics, and can be applied to new types of electrical sensors made of thin films.

However, the low ionic conductivities of macromolecular electrolytes (maximum 10^{-5}–10^{-4} S/cm in comparison with those of organic solvents 10^{-2}–10^{-3} S/cm) used for electrochemical analyses causes the high IR drop. Therefore, the usual voltammetry has been considered impossible. The combination with ultramicroelectrodes and miniaturized electrochemical cells can compensate for the problem of the typically low ionic conductivity in macromolecular electrolytes. It is now possible to perform quantitative voltammetry of molecular solutes and to discuss electron transfer and mass-transport phenomena in macromolecular solids. This development offers important new approaches in solid-state chemistry. Quantitative solid-state voltammetric investigations in poly(oxyethylene) also provide the basis for macromolecular electrolytes for mass transport rates [102], electron-transfer dynamics [103], and formal potentials of monomer solutes dissolved in the macromolecular solid. Solid-state voltammetry also probes the dynamics of the macromolecule itself [104].

Various proteins, such as hemoglobin [105, 106] and myoglobin [107], can also be incorporated into (polyoxyethylene/inorganic salt) composite films by modification of proteins with POE in order to become soluble in macromolecular solids. Although proteins had not been considered to work in the absence of water, modification of the proteins preserved the funtions of the proteins in organic media and led to excellent stability in the macromolecular media. Murray et al. reported the POE-covered cytochrome-C-coated electrode [100]. Actually, cytochrome C was confirmed to work as an electron transfer site even in solid POE.

In macromolecular electrolytes containing redox-active molecules, electron transfer generally occurs by electron hopping between the redox-active molecules to (from) the electrode and results in low efficiency. If molecular wires for electron transfer, redox-active molecules, and macromolecular electrolytes are combined at a molecular level by the molecular assembling system, these systems will be significantly developed as well as the cathodes for batteries containing redox-active molecules [108].

4.1.6 Conclusion

Recent developments concerning ion-conductive macromolecules have been reviewed. Ion-conductive macromolecules have the potential ability and possibility to convert wet-type devices (or method) into dry. These applications always require advancement in the performance of materials. The advancement, on the other hand, leads to novel branches in the academic research field. We hope that this review will be of some help in breaking the limit of ionic conductivity, molecular design, and application.

4.2 Transport Phenomena and Separation of Small Molecules

H. Nishide and E. Tsuchida

4.2.1 Facilitated Transport with Metal Complexes as a Carrier

Gas separation performed on the largest scale is that of nitrogen and oxygen from air. Nitrogen and oxygen gas is the second and the third largest commodity chemical in the United States, and the third and the first in Japan, respectively. Nitrogen is used for purging gases in explosion protection, in food storage, and in manufacture of fine electronics and pharmaceuticals, and oxygen for combustions, chemical oxidations, respiratory therapy, refreshment, aquaculture, waste-water treatment, etc. Whereas the air separation is mainly performed cryogenically presently, separation by a polymer membrane, such as a thin polydimethylsiloxane membrane, is increasing from the point of view of low equipment costs and running costs, and of simple and compact device for the separation [109].

Besides air separation, there are several industrial applications of polymer membranes to gas separation processes, e.g., polyimide membranes for the recovery of carbon dioxide in mining natural gas and of hydrogen from refinery streams. Permeability of a small or gaseous molecule through these polymer membranes is represented by the product of solubility in and diffusivity through the dense polymers, which constitute active layers of the membranes (Fig. 4.11a). Slight difference in both the solubility and the diffusivity brings about permselectivity of the gases through the membranes [110]. On the other hand, microporous membranes involving 10~100 Å pore-size are also often effective in separating gaseous molecules with different diameter or different molecular weight (Fig. 4.11b). This process is available to sepa-

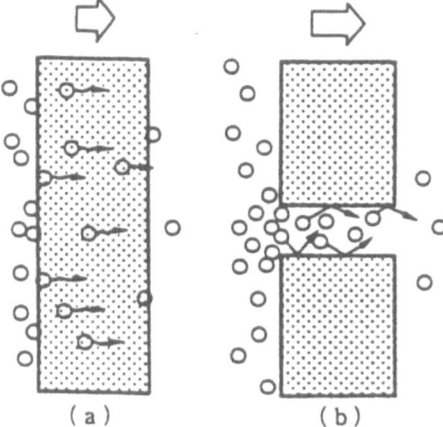

(a)   (b)

Fig. 4.11. Gas separation by a dense polymer membrane (**a**) and a microporous membrane (**b**)

rate, e.g., the mixture of helium and argon, and of hydrogen and carbon monoxide. However, presently only limited cases, such as small-size nitrogen production from air, are economical using the membranes compared with competing separation processes, such as cryogenic distillation and pressure swing adsorption, due to restricted permeability and selectivity of presently existing polymer membranes. To overcome the competitive processes, the requisites for the gas-separation membrane in the next generation are both high permeability and high permselectivity, e.g., as the air-separation membrane high permeability more than $10^{-8} cm^3(STP)cm\ cm^{-2}s^{-1}cmHg^{-1}$ and high oxygen/nitrogen selectivity over 10.

One of the characteristic chemical functions of metal ions and their complexes is the specific and reversible binding of small or gaseous molecules. A typical example is the efficient oxygen carriage by hemoglobin in a living body. Metalloporphyrins have been studied as the carrier of oxygen, and some metal complexes of chemically modified porphyrins bind molecular oxygen rapidly and reversibly in response to partial oxygen pressure [111].

The metal complexes attached to synthetic polymer matrices or macromolecular metal complexes (MMCs) often show specific behaviors in the binding reaction of small molecules, because the reactions are affeted by the polymers that surround the complex moieties [112]. Immobilizing and orienting of the complex moieties and physical properties of the polymer matrics strongly reflect on the kinetic, equilibrium, and lifetime profile of the molecule-binding reactions. For example, synthetic and macromolecular cobalt–porphyrin (CoP) complexes can specifically, rapidly, and reversibly bind oxygen from air with working lifetime over 3 months [113].

Complexation-mediated transport or facilitated transport membranes are expected to be a promising candidate to produce the high performance membrane, and has often been studied for the oxygen/nitrogen separation with liquid membranes [114] as represented by Fig. 4.12a. A metal complex is dissolved as a mobile carrier of oxygen in a liquid membrane, i.e., a complex solution is retained in a porous supporting membrane, which separates the upstream of feed-air from the downstream or product stream. The complex (carrier) acts as a shuttle, picking up oxygen selectivity from air at the upstream side, diffusing across the membrane, as the oxygenated complex in response to its concentration gradient, releasing oxygen to the downstream side, because the oxygen concentration is low by being evacuated or with sweep gas. The carrier complex diffuses back to the upstream side and repeats the process of Fig. 4.12a. Cobalt Schiff's base complexes have been applied to the oxygen-transporting fluid of an oxygen-selective liquid membrane [115]. They were also investigated for reversible oxygen binding in the solid state [116].

Although the liquid membranes holding the metal complexes for facilitated transport provided an elegant method for improving both the selectivity and flux, there remained the unresolved issues listed in Table 4.3. To overcome these problems with liquid membranes, an absolutely solvent-free, solid-state membrane containing the complex as a fixed-carrier is expected to be a promising membrane.

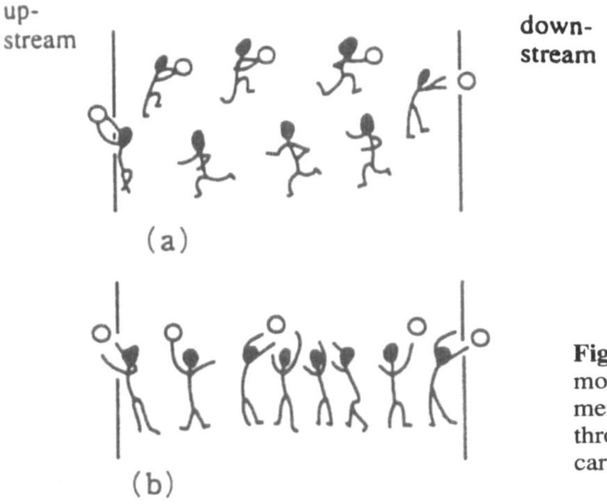

up-
stream

down-
stream

(a)

(b)

Fig. 4.12. Facilitated transport of small molecule (*circle*) through the liquid membrane with a mobile carrier (**a**) and through the solid membrane with a fixed carrier (**b**)

Table 4.3. Facilitated transport for gas separation

Mobile carrier/liquid membrane	Fixed carrier/solid membrane
Low diffusivity of carrier	Just a diffusion of small, gaseous molecule
Chemical instability of carrier	Strong suppression of carrier deactivation
Evaporation loss of liquid	Stable solid state
Limited carrier solubility in membrane	High carrier incorporation into membrane
Low temperature operation	Wide temperature window in operation
Membrane thickness	Thin membrane

Figure 4.12 b schematically represents the facilitated transport of a small or gaseous molecule *via* a fixed carrier (complex) in a solid membrane [117]. For the example of oxygen, the fixed carrier picks up oxygen specifically from air at the upstream–membrane interface. The oxygen taken up into the membrane is transferred by the fixed carriers from the upstream to the downstream side in response to the concentration gradient of oxygen across the membrane. At the membrane–downstream interface the carrier releases oxygen to the downstream side. If the passage of oxygen by the fixed carrier is efficient and rapid, oxygen transport is enhanced.

The fixed carrier and solid membrane provides the following advantages for gas separation (Table 4.3):

1. The carrier complex is fixed in the solid membrane and just a small and gaseous molecule diffuses across the membrane.
2. Inactivation of the carrier complex is strongly suppressed by fixing the complex in the solid matrix.
3. The dry membrane is mechanically tough against gas pressure and adequate for a huge amount of gas passing.

4. Incorporation amount of the complex in the membrane in increased.
5. Temperature window for the operation is relatively wide.
6. Thin membrane is available.

In this section, specific and reversible bindings of small molecules from their gaseous mixtures to the MMCs in the solid membrane states, and selective transport or permeation of the small molecules across MMC membranes, are described by especially discussing the binding reactions and the transport phenomena of oxygen/nitrogen from air. Physiochemical aspects and application potentials of the MMC membranes for gas separation are also described.

4.2.2 Chemically Specific and Reversible Binding of Small Molecules in Solid MMCs

One of the major difficulties encountered in attempts to prepare 1/1 metal/gaseous molecule-coordianted complexes that can reversibly bind and dissociated the molecule is the strong driving force toward the irreversible formation of the molecule-bridged metal dimer as represented by using the example of oxygen coordination (oxygenation):

$$M(II) + O_2 \rightleftharpoons M-O_2 \rightarrow M-O_2-M \rightarrow M(III)-O-M(III). \qquad (1)$$

The oxygenated complex rapidly reacts with another deoxy M(II) complex [M(II) = lower valence metal ion] and the binuclear dioxygen-bridged complex is formed. This binuclear complex is irreversibly oxidized to produce the oxo-bridged M(III) dimer [M(III) = higher valence metal ion]. In other words, the first requisite of reversible gaseous molecule-binding is how to inhibit this dimerization reaction. One approach to satisfy this requisite is attaching the complexes to a rigid polymer chain (MMC formation) so as to prevent two complexes from approaching each other closely enough to lead to dimerization. A solid state of the polymer-attached metal complexes additionally immobilizes the complex moiety and considerably inhibits the irreversible dimerization. By these means a reversible oxygen binding to metalloporphyrins (Schemes 4.1–4.5) and cobalt Schiff's base (Scheme 4.6) complexes, and a reversible nitrogen binding to cyclopentadienylmanganese (Scheme 4.7) and benzenechromium (Scheme 4.8) complexes have been achieved with a feasible working lifetime under air atmosphere at room temperature [118–124].

Selective and reversible oxygen binding to the fixed complexes was confirmed, e.g., with a microvolumetric measurement (Fig. 4.13) [125]. For example, the membrane in Scheme 4.1 containing 30% CoP complex moiety sorbed ca. 7 ml oxygen/g polymer, which is more than 500 times greater than that of physically dissolved nitrogen. This extraordinarily large amount of dissolved oxygen in the polymer membrane is based on the chemically specific and reversible oxygen-binding to the CoP complex in the polymer. It is one of the advantages of solid MMCs that the large amount of CoP can be incorporated into the membrane, due to small molecular size (> 10 Å) of cobalt com-

Schemes 4.1–4.8

plexes and through the coordination of a nitrogenous residue of polymer. The sorption isotherm for the membrane containing 7% CoP indicates that the amount of dissolved oxygen corresponds to the CoP concentration in the membrane. Figure 4.13 also indicates that the oxygen sorption is in response to an atmospheric oxygen pressure, and is according to Henry's law for the physical sorption and Langmuir isotherm for the chemical dissolution to the complex, to give a physical solubility coefficient k_D of the gaseous molecule and an equilibrium constant K for the binding reaction (Eq. (2), respectively:

$$\text{ImPCO} + O_2 \underset{k_{off}}{\overset{k_{on}}{\rightleftharpoons}} \text{ImPCO} - O_2, \qquad K = k_n / k_{iff}. \tag{2}$$

Here, PCo and Im represent CoP and imidazolyl residue, respectively. Examples of the K values are 3.0, 7.4, 4.3, 4.0, and $0.51 \cdot 10^3$ M^{-1} for Schemes 4.1 – 4.5, respectively: This mans that the oxygen-binding affinity of the CoP complex can be controlled with the chemical structure of the complex [126]. These oxygen-bindig affinities are adequate to oxygen separation from air.

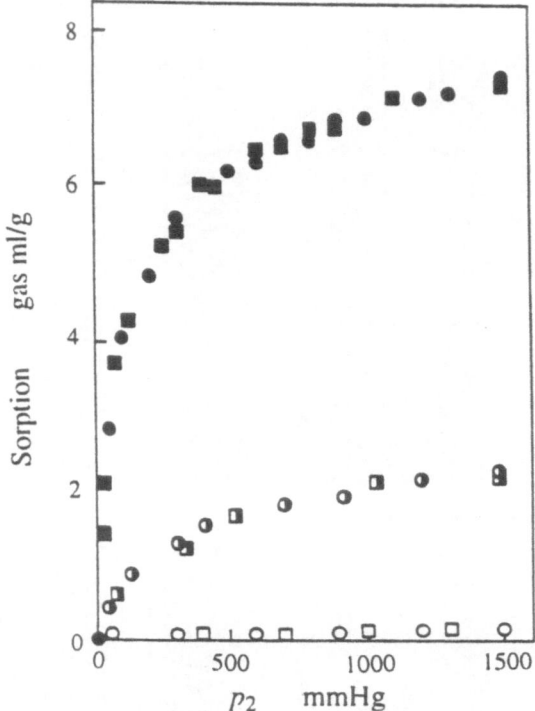

Fig. 4.13. Oxygen sorption into the cobalt-porphyrin (CoP) membrane (Scheme 4.1). Sorption amount of oxygen (*closed symbol* and *half-closed symbol*) and of nitrogen (*open symbol*); sorbed gas volume per membrane unit weight, p_2: atmospheric gas pressure, the CoP concentration in the membrane 7 and 30% for *half-closed symbol* and *closed symbol*, respectively; square symbols: repeated data

The colors of the membranes change upon exposure to the oxygen atmosphere, from dark red to brilliant red for the CoP membranes and from orange to deep violet for the cobalt Schiff's base (Scheme 4.6) membrane. These color changes of the membranes also become a good probe of oxygen-binding, and can be monitored easily with visible absorption spectrometry. The oxy-deoxy spectral change was reversible in response to partial oxygen pressure with isosbestic points [127]. This is crucial evidence that the cobalt complex acts as an effective oxygen-binding site from the equilibrium point of view, even after fixing as a solid MMC membrane.

It is known that metal-coordinated gaseous molecules are photodissociated under flash irradiation, and that their rapid binding reactions can be analyzed. The photodissociation and recombination of the gaseous molecules in the solid MMC membrane was observed by improving pulse ad laser flash spectroscopic techniques [126]. The oxygen-binding rate constant k_{on} and dissociation rate constant k_{off} in Eq. (2) were estimate by second-order kinetics: k_{on} ~$10^7 M^{-1} S^{-1}$ and k_{off} ~ $10^{3-5} s^{-1}$ for the CoP complexes in the membranes, and

$k_{on} = 4.8 \cdot 10^5$ and $k_{off} = 44$ for the cobalt Schiff's base complex (Scheme 4.6) in the membrane [128, 129]. These k_{on} and k_{off} values of the complexes in the membranes are similar to those of the complexes in toluene solution. This means that the cobalt complexes are also kinetically active for oxygen-binding even after fixing as dry, solid MMC membranes. The CoP complexes with cavity-like stereostructures, such as the complexes shown in Schemes 4.1 and 4.3−4.5, maintain their oxygen-binding site vacant in the solid state and yield a very rapid gaseous molecule-binding reactivity. The details are discussed in the following section. From the kinetic point of view in the oxygen-binding reaction, the CoP complexes are of great promise for facilitated transport of oxygen through the membrane.

An MMC to which molecular nitrogen coodinates rapidly and reversibly is expected to absorb and/or transport nitrogen selectively, and to offer the possibility of a membrane for enriching nitrogen from air. Cyclopentadienyl-dicarbonylmanganese (CpMn) and benzendicarbonylchromium complex attached to polymers (Schemes 4.7 and 4.8) have one unsaturated coordination site to bind molecular nitrogen reversibly [130]:

$$CpMn + N_2 \rightleftharpoons CpMn-N_2 \qquad (3)$$

The polymer matrix protects the nitrogen-coordinated complex even in air, probably because the polymer matrix fixes the complex to suppress irreversible dimerization of the complex and retards diffusion of water vapor to the complex moiety.

Sorption measurement of the membrane indicated that nitrogen sorption was selectively enhanced by chemical nitrogen binding to the CpMn complex moiety, and that the nitrogen binding equilibrium curve also obeyed a typical Langmuir isotherm [123]. The equilibrium constant in Eq. (3) $K = 980\ M^{-1}$, is a little smaller than thse for oxygen binding, but still practical for nitrogen separation from air. The color of the membrane depicted in Scheme 4.7 changed from yellow to brown upon exposure to nitrogen. Rapid and reversible nitrogen binding to the CpMn moiety in the membrane was also confirmed by laser flash photolysis of the nitrogen-coordinated complex [123]. The nitrogen binding rate constants ($k_{on} = 2.9 \cdot 10^5\ M^{-1}s^{-1}$ and $k_{off} = 3.0 \cdot 10^2 s^{-1}$) mean that the nitrogen-binding reaction possessing an organometallic character is still a rapid reaction, although the rate constants are 10^2 times smaller than those for oxygen binding.

In any case, when these solid MMC membranes are placed under a gaseous pressure gradient, the complexes are expected to contribute to the permeation flux across the membranes of a specific gaseous molecule.

4.2.3 Solid MMC Membranes for Facilitated Transport and Gas Separation

We first succeeded in the facilitated transport of a gaseous molecule through a solid MMC membrane containing kinetically active CoP for chemically selective oxygen transport. The oxygen permeability coefficient (P_{O_2}) for the membrane in Scheme 4.1 is shown in Fig. 4.14 [131]. P_{O_2} is larger than the

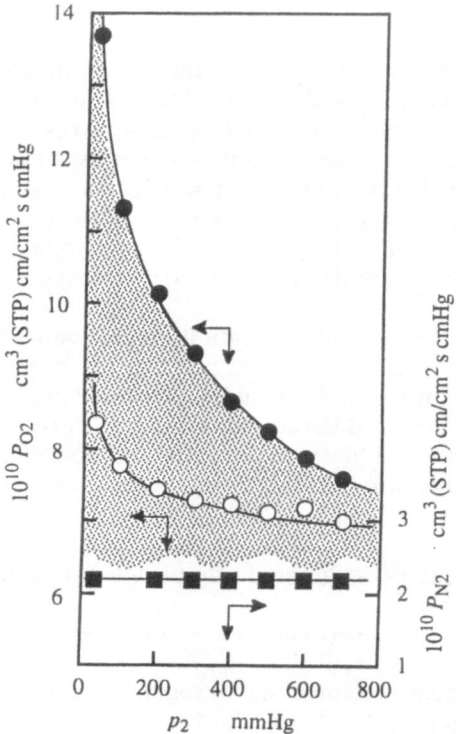

Fig. 4.14. Oxygen (P_{O_2}) and nitrogen (P_{N_2}) permeability coefficients for the CoP membrane (Scheme 4.1). p_2: upstream gas pressure, the CoP concentration in the membrane 1.2 and 4.5 % for *open symbol* and *closed symbol*, respectively, at 30 °C

nitrogen permeability coefficient (P_{N_2}) and steeply increases with a decrease in the oxygen upstream pressure [$p_2(O_2)$]. On the other hand, P_{N_2} is small and independent of the nitrogen upstream pressure [$p_2(N_2)$], because the fixed complex does not interact with nitrogen. P_{O_2} is also small and independent of $p_2(O_2)$ for the control membrane composed of the inert Co(III)P, which does not interact with oxygen, i.e., the active CoP complex fixed in the membrane facilitates oxygen transport in the membrane and enhances the oxygen permeation additionally, represented as the shadowed area in Fig. 4.14. P_{O_2} increases (as shown in Fig. 4.14) with the incorporated-CoP concentration in the membrane. The permeability ratios of oxygen/nitrogen (P_{O_2}/P_{N_2}) was 3.2, 5.7, and 12 for the membrane containing 0, 2.5, and 4.5 % CoP complex moiety, respectively. This result indicates the possibility of high permselectivity with an MMC membrane.

The time course of the permeation of gaseous molecules through membranes showed an induction period followed by permeation with a constant slope (steady state). The induction period (θ) for the oxygen permeation was longer than θ for nitrogen, and was prolonged with the decrease in $p_2(O_2)$ [131]. On the other hand, θ for nitrogen permeation was short and independent of $p_2(N_2)$. This behavior indicates that oxygen clearly interacts with the complex in the membrane, and that its diffusivity in the membrane was reduc-

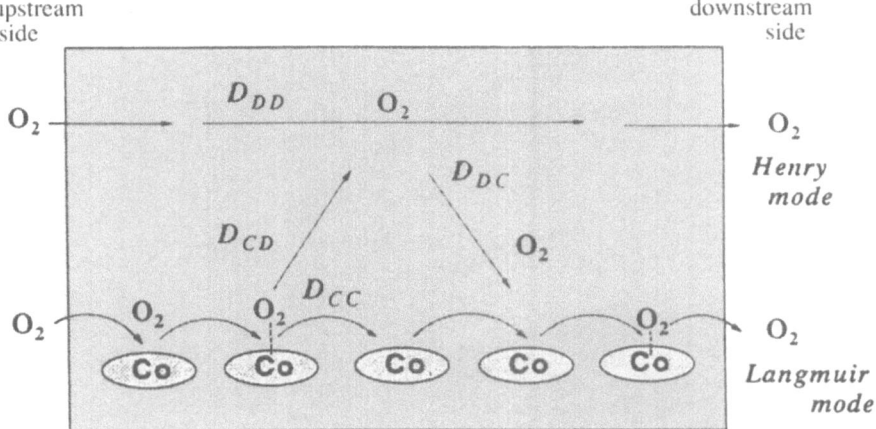

Fig. 4.15. Dual-mode oxygen transport in the solide MMC membrane containing CoP as fixed carrier.

ed by repeated binding and dissociation of oxygen to the fixed complex. θ_{O_2} was prolonged with the incorporated-CoP concentration in the membrane. θ_{O_2} and the $p_2(O_2)$ dependency of θ_{O_2} were enhanced at lower temperature, because the oxygen-binding rate constants k decreased, and the equilibrium constant K increased with decreasing temperature. θ is the best parameter to distinguish whether facilitation by the complex contributes to transport in the MMC membrane.

These results suggest that a dual-mode-transport theory [132] is mathematically applicable to the oxygen transport in MMC membranes. Figure 4.15 schematically shows the oxygen permeability in the MMC membrane that is governed by two modes, i.e., the oxygen permeation is equal to the sum of a first term that represents the physical Henry mode permeation, and a second term that represents the chemical Langmuir mode permeation. For the Henry mode (the uper permeation route in Fig. 4.15), oxygen physically dissolves in a polymer membrane according to Henry's law, and the dissolved oxygen diffuse physically. For the Langmuir mode (the lower permeation route), oxygen is specifically and chemically taken up into the membrane by the selective binding reaction to the complex moiety fixed in a membrane and diffuses *via* the fixed complex moiety by repeating the binding and dissociation reaction to and from the complex. Oxygen transport is accelerated by this Langmuir mode in addition to the physical Henry mode. The dual-mode transport model is mathematically given as [131, 133]:

$$P = k_D D_{DD} + \frac{C_C K D_{CC}}{(1 + K p_2)} \tag{4}$$

$$+ \left[\frac{C_C K D_{CD} - k_D D_{DC}}{(1 + K p_2)} + \frac{2 k_D D_{DC}}{K p_2} \ln (1 + K p_2) \right].$$

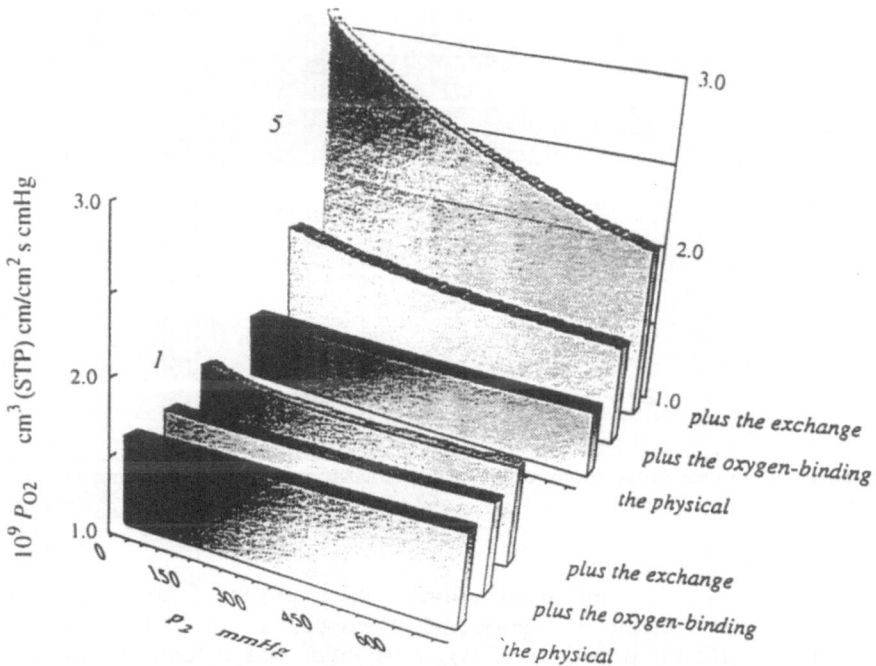

Fig. 4.16. Contribution of each mode in the dual-mode oxygen transport to the total permeability coefficients for the CoP membranes of Scheme 4.1 and 4.5 at 45 °C

Here k_D is the physical solubility coefficient for the Henry mode, determined e.g., in Fig. 4.13. D_{DD}, D_{DC}, D_{CD}, and D_{CC} are the diffusion coefficients for the physical Henry permeation, for the diffusion from the polymer matrix to the fixed carrier (complex), for the diffusion from the fixed carrier to the polymer matrix, and for hopping between the fixed carriers, respectively. C_C is the fixed carrier concentration in the membrane, determined also by the sorption experiment such as in Fig. 4.13. The total permeability coefficient (P) is equal to the sum of the first physical permeation term, which is given by the product of the solubility and the diffusivity, the second term, which represents the Langmuir mode attributed to a specific binding and diffusion of oxygen to and *via* the fixed carrier (complex) and is given by the product of the Langmuir isotherm and the diffusivity *via* the fixed carrier, and the third term, which is also the Langmuir mode attributed to the exchange between the first and second terms.

The experimental data in Fig. 4.14 were analyzed by Eq. (4), and the parameters for oxygen transport through the membranes given by the dual-mode transport model are listed in Table 4.4. D_{DD} was calculated to be 10^{-6} $cm^2 s^{-1}$, which agreed with the previously reported diffusion constants of oxygen in rubbery polymers [131] (glass transition temperature of the membranes of this experiment ca. 0 °C) and is almost one-tenth of those in liquids (e.g., $10^{-5} cm^2 s^{-1}$ for oxygen in water at 25 °C [134]. That suggests that segmental

Table 4.4. Diffusion constants of oxygen and oxygen-binding rate constants in the cobalt–porphyrin (CoP) polymer membranes

CoP Scheme	Diffusion constants (cm² s⁻¹)				$10^{-3} K$ (M⁻¹)	Rate constants	
	$10^6 D_{DD}$	$10^7 D_{DC}$	$10^8 D_{CD}$	$10^9 D_{CC}$		$10^{-7} k_{on}$ (M⁻¹s⁻¹)	10^{-3} k_{off}(s⁻¹)
4.1	2.2	0.97	0.87	3.1	3.0	0.98	3.2
4.4'	2.2	2.2	1.5	7.3	4.3	1.4	3.2
4.4	2.3	2.6	2.0	9.0	4.0	2.3	5.7
4.5	2.2	3.2	22	144	0.51	3.4	66

Scheme 4.4': mono(methacrylamido)-tris(pivalamido)-substituted porphyrin derivative; $10^4 k_D$ (cm³(STP) cm⁻³ cmHg⁻¹) = 7.2; C_C (cm³(STP) cm⁻³) = 0.2; the CoP concentration in the membranes = 1.3%; 45 °C.

moblity of the polymer matrix does not significantly retard oxygen diffusivity in the membrane. Furthermore, D_{DD} is not influenced by CoP species or the porphyrin structure.

On the other hand, D_{DC}, D_{CD}, and D_{CC} involving the oxygen diffusion *via* the fixed CoP complex are smaller than D_{DD} and depend on the CoP species or the oxygen-binding properties of CoP. It is noticed in Table 4.4 especially that D_{CD} and D_{CC} are enormously large for the CoP in Scheme 4.5. Comparison of the diffusion constants with the spectroscopically determined rate constants in Table 4.4 elucidated logarithmically linear correlation of the rate constants of oxygen binding and dissociation of CoPs with the diffusion constant *via* CoPs, i.e., k_{on} corresponds to D_{DC}, and k_{off} to D_{CD}. D_{CC} relates to both k_{on} and k_{off}; especially k_{off} clearly reflects on D_{CC}. The MMC membrane containing CoP with a large rate constant of oxygen binding and dissociation of CoP leads to a large diffusion constant to yield oxygen permselectivity or facilitated oxygen transport.

Figure 4.16 shows the contribution of each transport mode in Eq. (4) to the permeation coefficient [135], by using the example of the membrane in Scheme 4.1 and 4.5. The most important among these factors in the oxygen transport is the diffusion constants of oxygen *via* CoP. The diffusion constants, D_{DC} and D_{CD}, for the exchange third term in Eq. (4) augments P_{O_2} or the facilitated transport of oxygen at a wide range of $p_2(O_2)$. On the other hand, the binding affinity K enhances P_{O_2} at low $p_2(O_2)$, because the specific oxygen uptake into the membrane is effective at its dilute-feed condition.

Oxygen/nitrogen permselectivity (P_{O_2}/P_{N_2}) increases with the increase in the permeability coefficient P_{O_2}. The complexation-mediated or facilitated transport was successfully utilized in enhancing both permeability and permselectivity; their relationship had remained a reverse one for presently available polymers [110]. P_{O_2}/P_{N_2} is more than 10 for the membrane containing Scheme 4.5 at an upstream pressure of 5 mm Hg, despite its low concentration (1.3 wt%) in the membrane. Oxygen permselectivity for the membrane containing a large amount of Scheme 4.5 (10 wt.%) is calculated to be 30

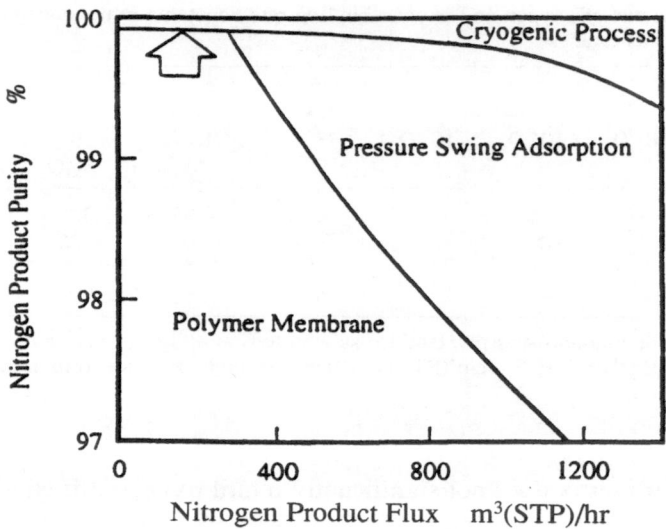

Fig. 4.17. Nitrogen production systems from air for different purity/flow combination

according to Eq. (4). Although the reduction in steric bulkiness of the substituent on the porphyrin plane causes the decrease in oxygen-binding affinity for the Scheme 4.5, its much larger D_{DC}, D_{CD}, and D_{CC} efficiently facilitate oxygen transport. The high diffusivity of oxygen *via* the fixed CoP complex contributes strongly to the transport, besides the oxygen binding affinity.

The specific oxygen uptake into the MMC membrane is extremely efficient at dilute oxygen concentration of the upstream or feed gas, and P_{O_2} or P_{O_2}/P_{N_2} is steeply enhanced with the decrease in $p_2(O_2)$, as shown in Fig. 4.14. Figure 4.17 shows the areas of most applicability for different production systems of nitrogen gas from air as a function of purity and product flow rate. Nitrogen production from air using membranes tends to be the optimum delivery system presently when product flows and/or required product purities are moderate. However, the cryogenic system still takes over in the high nitrogen purity range. The MMC membranes are now being tested as a system (the arrow in Fig. 4.17) that produces >99% pure nitrogen from the crude nitrogen supplied by the presently available membrane system or that cleans up trace amounts of oxygen in the crude nitrogen using the facilitated transport.

The cobalt Schiff's base complex in the solid MMC membrane state also acted as a fixed carrier for oxygen transport [122]. The oxygen/nitrogen selectivity increased with the cobalt complex concentration and was above 10 for the membrane containing 12% complex moiety. For the CpMn complex (Scheme 4.7) in MMC, nitrogen transport through its membrane was selectivity-augmented, due to rapid and reversible nitrogen binding to the fixed CpMn complex. The nitrogen permeability coefficient (P_{N_2}) increased with a decrease in $p_2(N_2)$, whereas oxygen permeability was independent of $p_2(O_2)$.

The CpMn complex in the membrane interacted specifically with nitrogen and not with oxygen, and nitrogen transport through the membrane was facilitated by the CpMn complex.

The binding-reaction character of the fixed metal complex with a gaseous molecule clearly relates to the selective transport of the gaseous molecule across the solid MMC membrane.

4.2.4 Surface Diffusion of Oxygen in Porous MMC Membranes

Facilitated oxygen transport has been demonstrated in dry or solvent-free polymer (dense) membranes containing CoPs and cobalt-Schiff's base complexes as fixed carriers of oxygen. However, an upper limit of the permeability coefficient of oxygen (P_{O_2}) of 10^{-8} cm^3(STP)cm cm^{-2}s^{-1}cm Hg^{-1} through the polymer dense membrane remained, despite the oxygen permselectivity attributed to the facilitated transport.

Oxygen permeates in the gas phase by Knudsen diffusion through a porous membrane (pore size < mean free path of oxygen) with a huge flux of $P_{O_2} = 10^{-6}$. However, the permselectivity for mixed gases is inversely proportional to the square root of the molecular weights of the permeating gases. For oxygen and nitrogen, which have almost the same molecular weight, $P_{O_2}/P_{N_2} = 0.94$. Surface diffusion of a gaseous permeate has been reported for a porous membrane of which the pore surface has an affinity for the gaseous molecule. However, surface diffusion was observed only for condensable gases such as carbon dioxide and hexane [136]. In this section a chemically specific surface diffusion of a noncondensable gaseous molecule, oxygen, is described by using a porous membrane modified on its pore surface with the oxygen-binding CoP [137].

The CoP complex was fixed on the pore surface in the chemically modified porous glass through coordination with the imidazolyl group (Scheme 4.9). The five-coordinated structure of deoxy-CoP, which can bind oxygen rapidly, was confirmed by spectroscopies, and the homogeneous introduction of CoP

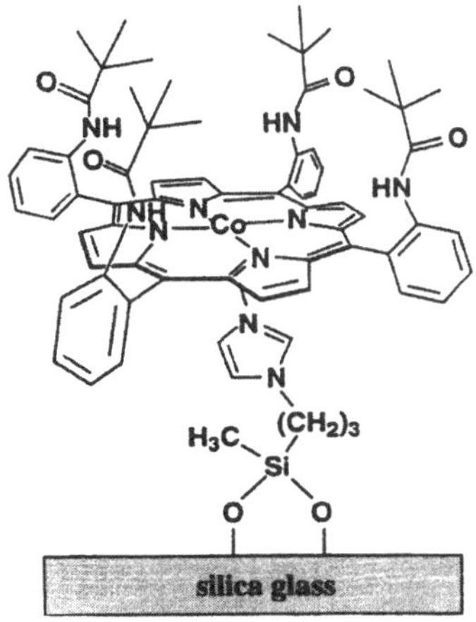

Scheme 4.9

into the membrane was confirmed by X-ray microanalysis. The oxygen-binding equilibrium constant of the fixed CoP was determined in Eq. (2) on the basis of Langmuir isotherm ($K = 0.18\,\text{cm}\,\text{Hg}^{-1}$), which agreed with those of the CoP cordinated with imidazole in toluene. The sorption isotherm of oxygen obeyed that of Langmuir, and the saturated amount of oxygen sorption was $0.80\,\text{cm}^3(\text{STP})\,\text{g}^{-1}$, which almost corresponded to the introduced amount of CoP and was enhanced at lower temperature. These results indicated that the CoP fixed in the membrane acts as a chemically specific and Langmuir-type oxygen-binding site.

The oxygen-binding reaction to the CoP fixed on the pore surface was very rapid, and the oxygen-binding rate constants ($k_{on} = 2.8 \cdot 10^6\,\text{M}^{-1}\,\text{s}^{-1}$; $k_{off} = 3.4 \cdot 10^3\,\text{s}^{-1}$) estimated by laser flash photolysis were comparable with those of the corresponding CoP in toluene solution. The oxygen-binding kinetics directly in response to the atmospheric oxygen pressure suggest that CoP is located in the inside surface of the micropores. CoP is kinetically active to bind oxygen even after fixation in the porous membrane and acts as an effective carrier for the passage of oxygen.

The pore structure of the membrane modified with CoP was studied by BET nitrogen adsorption measurement: pore volume $\varepsilon = 23\%$, tortuosity $\tau = 8.2\,\text{cm/cm}$, specific surface area $123\,\text{m}^2/\text{g}$, and man pore size $26 \pm 10\,\text{Å}$ (62%). These data also support the incorporation of CoP into the inside of the micropores of the membrane. The average pore size was still 26 Å to allow gas-phase diffusion of oxygen and nitrogen. Gas-phase diffusion through the CoP-modified porous membrane was confirmed by using a control membrane, i.e., the same membrane modified with inactive Co(III)P. As shown in Fig. 4.18, the permeability coefficients for the control membrane were independent of upstream pressure with $P_{O_2} = 8.8 \cdot 10^{-7}\text{cm}^3(\text{STP})\text{cm}\,\text{cm}^{-2}\text{s}^{-1}\text{cmHg}^{-1}$, $P_{N_2} = 9.4 \cdot 10^{-7}$, and $P_{O_2}/P_{N_2} = 0.93$ (theoretical 0.93): Gas-phase or Knudsen diffusion is predominant for the membrane.

P_{O_2} and P_{N_2} for the CoP-modified porous membrane are also shown in Fig. 4.18. P_{N_2} is larger than P_{O_2} at the higher region of upstream nitrogen pressure $p_2(_2)$ and is independent of $p_2(\text{N}_2)$, because nitrogen permeates through the membrane according to gas-phase diffusion and the fixed carrier CoP does not bind with nitrogen. On the other hand, P_{O_2} is small with $P_{O_2}/P_{N_2} = 0.94$ at the high $p_2(\text{O}_2)$ region because of the saturation of CoP with oxygen, but steeply increases with a decrease in $p_2(\text{O}_2)$: (P_{O_2}/P_{N_2}) reaches 1.4 at $p_2(\text{O}_2) = 7\,\text{mm Hg}$. These indicate that the CoP fixed on the pore surface interacts specifically with oxygen and augments the diffusion of oxygen in the membrane ascribed to chemically specific surface diffusion (Fig. 4.19).

In contrast to the temperature dependence of P_{N_2}, P_{O_2} has a minimum at ca. 35 °C, as shown in Fig. 4.18. This means that the surface diffusion of oxygen involves a chemical reaction with CoP in addition to its gas-phase physical diffusion. Total gas flux could be considered to be the sum of the surface diffusion and the gas-phase diffusion as given in Eq. (5):

$$P_{O_2} = P_S + P_g = \frac{1}{\tau}\,D_s\,\frac{C_S K}{(1 + K p_2)^2} + \frac{\varepsilon}{\tau}\,D_g\,\frac{1}{RT}\,, \qquad (5)$$

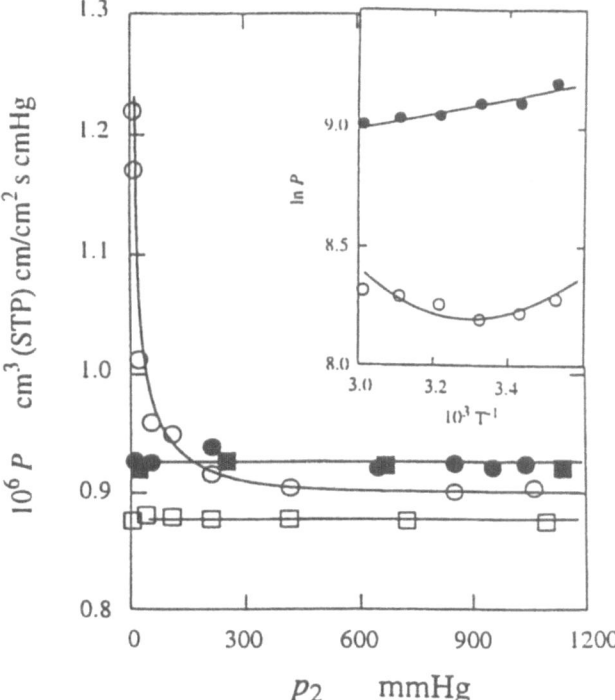

Fig. 4.18. Effect of upstream gas pressure (p_2) and temperature on the permeability coefficients for the CoP modified porous membrane at 25 °C. *open symbols:* oxygen; *closed symbols:* nitrogen, *square symbols:* inactive control membrane

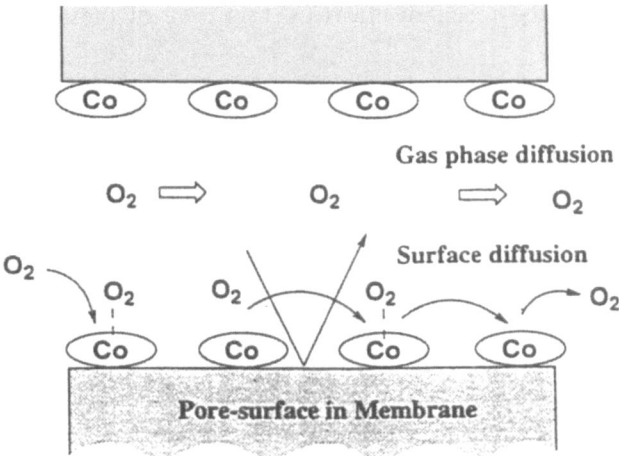

Fig. 4.19. Chemically specific surface diffusion of oxygen through the CoP-modified porous membrane

where D_S and D_g are diffusion constants for the surface and gas-phase diffusion, respectively. C_s is the concentration of active CoP in the membrane with the oxygen-binding equilibrium constant K (given in Eq. (2) and determined by sorption and spectroscopic measurements). The permeability coefficient of the surface diffusion (P_s) was obtained by subtracting the permeability coefficient of the gas-phase diffusion (P_g) measured for an ideal gas (e. g., helium) from the total gas diffusion.

Substituting the C_s, K, ε, τ, and p_2 values in Eq. (5) yields the gas-phase diffusion constant of oxygen, $D_g(O_2) = 2.5 \cdot 10^{-3} cm^2 s^{-1}$ at 25 °C, and that of nitrogen, $D_g(N_2) = 2.7 \cdot 10^{-3}$: their ratio. 0.94, also supports gas-phase diffusion. The surface diffusion constant, $D_s(N_2) = 2.9 \cdot 10^{-5}$, was 100 times smaller than $D_g(O_2)$. The activation energy of $D_g(N_2)$ was 0, which is also consistent with that of gas-phase diffusion for nitrogen. On the other hand, the activation energy of $D_s(O_2)$ for oxygen was 11 kcal mol^{-1}, which corresponds to the chemically specific surface diffusion and is comparable with the activation energy of the oxygen-binding reaction (10 kcal mol^{-1}).

In Eq. (5), C_s and K also contributed to P_{O_2} in addition to $D_S(O_2)$. The enormously large C_s of 0.7 cm^3 cm^{-3} is caused by the large surface area of the porous membrane and enhances P_s by overcoming the small D_S. The requisites for establishing a much higher P_{O_2} are a higher C_s, a moderate K, and a higher D_S. The third is connected with a larger k_{on} and k_{off} in the oxygen-binding reaction of the carrier CoP complex.

Specific combinations of binding reactions between smaller gaseous molecules and metal complexes have now been established in the field of inorganic chemistry. The introduction of these metal complexes to polymers, i. e., MMC, offer the possibility of membranes for highly selective and efficient gas separation. The practical advantages of the ease of membrane formation, membrane and carrier stabilities, and the feasibility of MMC membranes provide a great incentive for the research effort on this type of facilitated-transport membrane.

4.3 Assembled Porphyrins and Oxygen Coordination

E. Tsuchida and T. Komatsu

4.3.1 Introduction

One of the most exciting challenges in porphyrin chemistry is to construct functional multiporphyrin assemblies composed of a large number of self-organized porphyrinatometal complexes. In a biological system hemoproteins or membrane-bound enzymes include highly ordered metalloporphyrin arrangements as active sites and cooperatively allow sequential reactions through the porphyrin moieties. For example, the profile of the

multistep O_2-binding to hemoglobin (Hb) revealed the allosteric phenomenon [138, 139]. The first steps in photosynthesis are also mediated by
highly arranged porphyrinic complexes in the membrane-spanning protein
[140, 141]. The highly ordered geometrical arrangement of the redox
centers facilitates extremely fast electron transfer with a very high quantum
efficiency.

Such characteristics of oriented porphyrinic array should be mimicked
with synthetic porphyrins, which then may act as new molecular devices
whose functions are different from those of natural systems, i.e., as super
hemoproteins, multiphotosensitizers, superconductors, molecular wires, etc.
If a porphyrinatometal complex itself produces a highly organized aggregate,
a super molecular assembly, including an ordered reactive site with a high
density, is materialized. Consequently, the elucidation of the organization
process, microstructure, and electronic properties of the highly ordered porphyrin
assemblies is a topic of current interest in aqueus, organic, and solid systems. In
this section recent advances in the performance of assembled porphyrin systems
are discussed.

4.3.2 Assembled Porphyrin Systems in Solid State

4.3.2.1 Crystal Structure of Porphyrin

The most prominent natural porphyrin is the protoporphyrin IX (PPIX;
Scheme 4.10) iron complex. Although characteristic rhomb-shaped
Teichmann crystals of hemin have been known for 100 years, there exists no
fully solved crystal structure. Only the lath-shaped crystal $(2 \cdot 0.5 \cdot 0.05\,\mathrm{mm})$
has been shown to be composed of disordered chlorohemin [142]. The
iron atom of chlorohemin lies out of the plane of the four nitrogen atoms
toward the chloride atom. Whereas each of the pyrrole residues is planar,
the porphyrin ring is slightly puckered. It has obviously been difficult
to obtain good single crystals from asymmetrical porphyrins having
amphiphilic residues. On the other hand, a number of crystal structures
of symmetrically substituted porphyrins have been reported thus far [143,
144].

Tetraphenylporphyrin (TPP) molecules pack very efficiently into
two dimensions. Recently, TPP-based lattice clathrates, so-called
porphyrin sponges, have been developed by Byrn et al. [145]. From the
analysis of the crystal structure of over 100 TPP derivatives, the extent
to which van der Waals interactions between TPP molecules govern the
crystal packing has been revealed. In all these crystals, corrugated sheets
of tightly packed porphyrin molecules stack to form arrays of parallel
channels in which a remarkable variety of guest molecules are accommo-
dated (Fig. 4.20). Conservation of the host structure in the absence of
any covalent or hydrogen-bonding connection between host molecules
indicates an engineering strategy for the construction of porous molecular
solids.

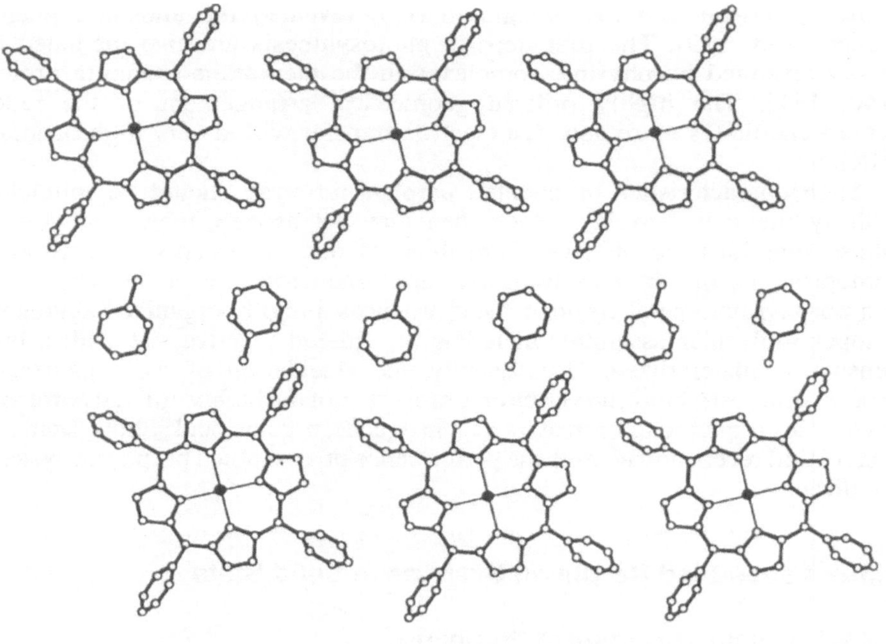

$$Zn^{2+} (Toluene)_2$$

Fig. 4.20. Porphyrin sponges consisting of Zn(II) tetraphenylporphyrin (TPP) containing toluene

4.3.2.2 Structure of Polyporphyrins and Electron Transfer

Performances of oligo- and polyporphyrins have been of interest for many years. Especially polyporphyrinatometals, which are formed by a bridging ligand, have been widely studied, and with very long π-electron systems, are of intense interest as organic semiconductor and nonlinear optical materials.

The most simple and classic examples of oligoporphyrinic molecules are the oxobridging complexes [146, 147]. The structure of the μ-oxo oligomer of Fe(III)TPP was determined by Hoffman et al. [148]. The Fe–O–Fe angle is 174.5° despite the strong antiferromagnetic coupling between the iron atoms [149]. In the case of phthalocyanines, efficient columnar crystals are formed [150]. Furthermore, large numbers of bridged coordinated Fe-, Ru-, and Co-phthalocyanine derivatives with unsaturated ligands (e. g., pyradine) are well known as "shish-kebab" polymers. Their structure and electronic conducting properties were clarified in detail by Hanack et al. [150, 151].

Ligand-bridged poly-octaethylporphinatometals (Fe, Ru, Os) have been synthesized and characterized by Collman et al. (Fig. 4.21) [152]. The Ru and Os polyporphyrins exhbit high electrical conductivities ($10^{-3}\sim10^{-2}$ $(\Omega cm)^{-1}$)

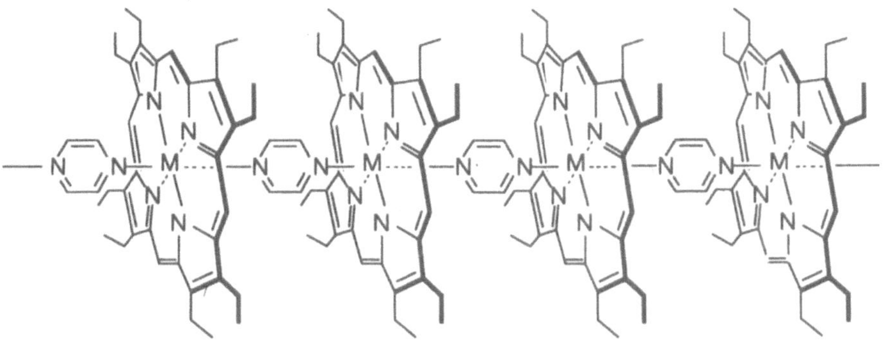

M: Fe, Ru, Os

Fig. 4.21. "Shish-kebab" polyoctaethylporphinatometal complex

when partially oxidized. The conductivity of these polymers depends on the extent of doping, the nature of the central metal, and the bridged ligands. The doped polyporphyrins exhibit strong infrared absorption due to mixed-valence transitions. The electron transport in these polyporphyrins proceeds exclusively along the metal–pyradine backbone, unlike most of the previously reported polyporphyrinic molecules in which the macrocycles support the conduction.

The porphyrin monomer having a covalently bound base group can itself form a polymer by the coordination of the peripheral substituent of the central metal of an adjacent porphyrin. The tetrakis(*o*-nicotinamidophenyl)porphinatoiron(III) formed polymeric chains [153]. The X-ray structure of this polyporphyrin indicated that polymerization was supported by the coordination of the nitrogen of the nicotinamide to the Fe of an adjacent porphyrin. The chloride ion occupies the six-coordination site inside the cavity of the four nicotinamide groups.

5-Pyridyl-10,15,20-triphenylporphinatozinc(II) formed a polymer in $CHCl_3$, which was characterized by visible ^1H nuclear magnetic resonance (NMR) and fluorescence spectroscopy [154]. The coordination of pyridine on the porphyrin periphery to the central zinc of an adjacent porphyrin produced polyporphyrin. The crystal structure of the polyporphyrin was found to be a long chain of *zigzag* conformation with an unusual 25° tilt to the pyridine ring (Fig. 4.22). The structure of the polymer in solution was shown to be similar to that in the solid state.

More recently, a conjugated "porphyrin ladder" has been prepared by Anderson [155]. Conjugated butadiene-linked porphinatozinc(II) dimers form a very stable 2:2 ladder complex with 1,4-diazabicyclo[2.2.2]octane (Fig. 4.23). The NMR ring current shifts and exciton coupling showed that the porphyrin is coplanar in the aggregate. The electronic spectra demonstrate overlap between the porphyrin π-systems.

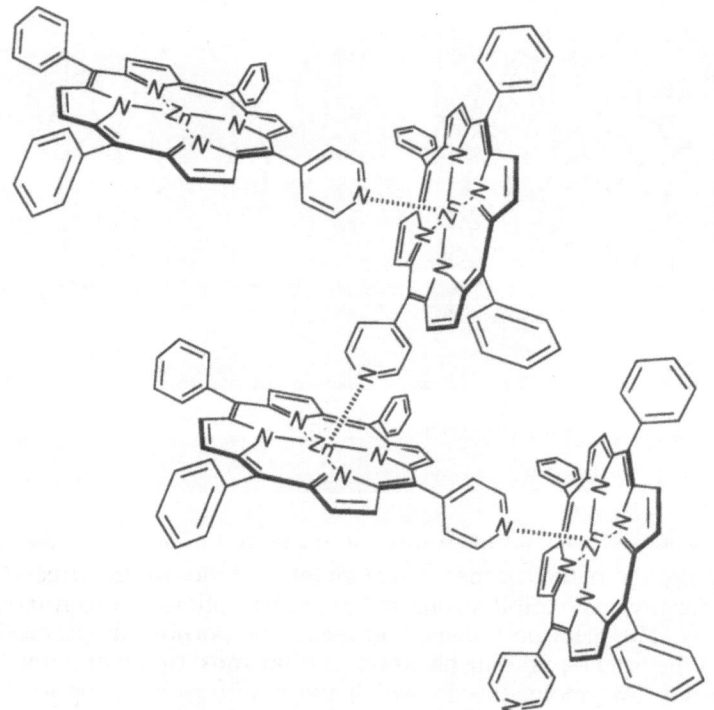

Fig. 4.22. Zigzag conformation of polyporphyrinatozinc(II) complexes

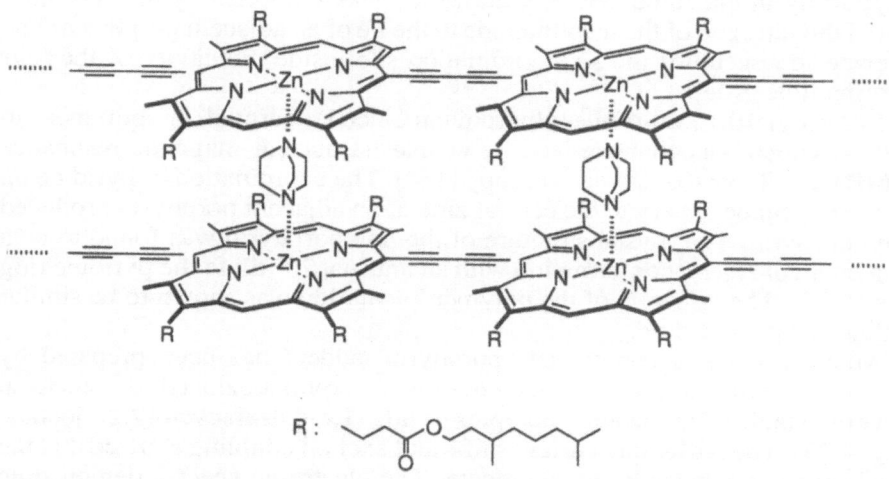

Fig. 4.23. Planar structure of polyporphyrin ladder

4.3.3 Self-Assembly of Amphiphilic Porphyrins in Aqueus Medium

4.3.3.1 Morphology and Structure of Porphyrin Assembly

The most important condition for self-organization of porphyrin derivative in aqueous medium is to introduce suitable hydrophilic groups to the structure. PPIX (Scheme 4.10) is a typical amphiphilic molecule and its aggregation behavior has been studied for several decades [156]. Alexander prepared a macromolecular layer of PPIX on the surface of water and proposed that the PPIX molecules are packed face-to-face and vertically oriented [157]. In this manner the polar carboxylic acid groups are in the water, whereas the vinyl groups are far from the water. This face-to-face stacking model has been the basic assumption in virtually all the studies of planar porphyrin aggregation. PPIX provided dramatic spectral changes with dilution, unlike the *meso*-substituted porphyrins [158]. Similar spectral changes were observed for hematoporphyrin (HP) under the same conditions, however, the spectral changes indicated that the dimerization constant of PPIX ($K > 10^5$ M^{-1}) was greater than that of HP ($> 10^4$ M^{-1}). Caughey et al. pointed out that in the presence of steric factors, electron-withdrawing substituents at the 2,4-positions enhance dimerization [159].

Recently, a few studies on assembly behavior of PPIX in aqueous solution have been reproted [160, 161]. PPIX is dissolved in water above pH 9. At pH 4.5, where about half of the carboxylate groups are protonated, a split Soret band with peaks at 360 and 460 nm is found. Although the PPIX indeed produces assemblies of high molecular weight in water, its assemblies in aqueous system were very difficult to handle especially for electron microscopy. Therefore, no electron micrograph of the defined PPIX assembly has been obtained thus far.

Bacteriochlorophyll *a* (BChl-*a*; Scheme 4.11) oligomers form cylinders with a diameter of 15 nm and an average length of 200 nm in 25 % dimethylformamide (DMF) aqueous solution [162]. The structure of these cylinders has been calculated by its giant CD bands. It was estimated that one fiber contains approximately 300 BChl-*a* dimers, and that the distance between the dimers is ca. 2 nm. Because the BChl-*a* molecule bears strong hydrogen-bonding groups, the assembly must be held by solvent molecules. The keto group

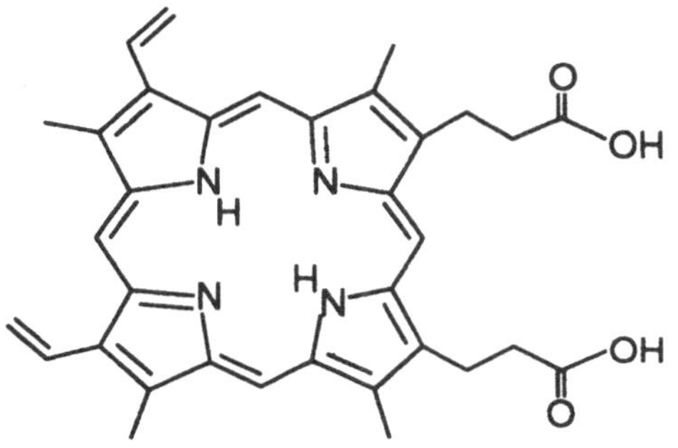

Scheme 4.10

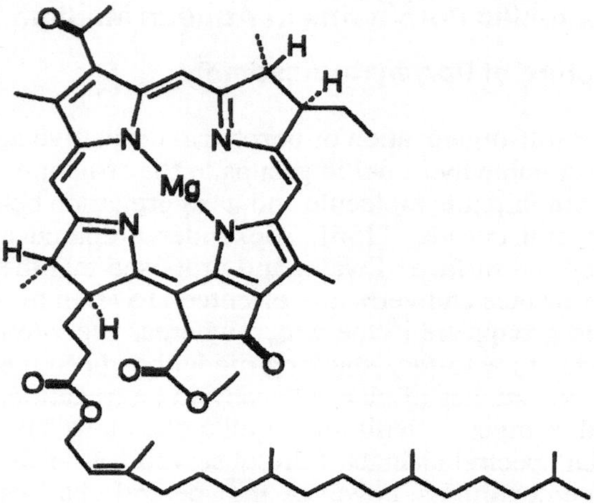

Scheme 4.11

acts as a proton acceptor, and the phytyl side chains play the role of the solvo-phobic part. As a result the whole assembly has the character of an inverse micellar strand with a helical twist.

In the case of artificial amphiphilic porphyrins, there have been several reports on the characterization of the morphology and microstructure of the self-assemblies in aqueous solution [163, 164]. Recently, Fuhrhop et al. showed the beautiful micellar fiber and vesicular tubules consisting of amphiphilic porphyrin derivatives [161, 165, 166]. A series of protoporphyrin derivatives having glucosamide groups (Scheme 4.12) produce short micellar ribbons of 4~6 nm width at pH 4−8 [161]. In the case of the racemic D, L-mannonamide derivative, longer fibers of 5 μm are formed. The molecular weight of the fibers

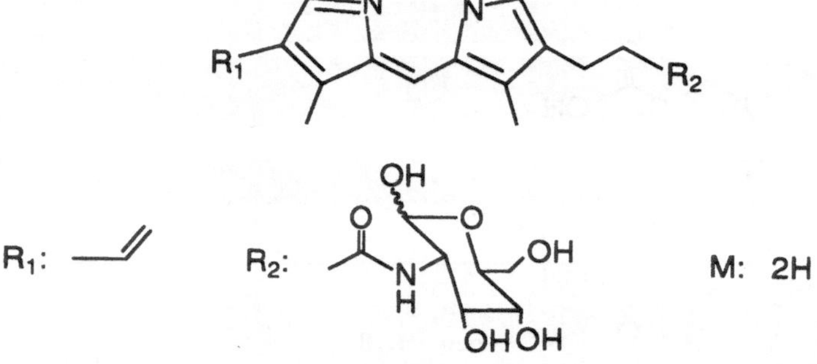

R₁: $\diagup\!\!/$ R₂: [structure]   M: 2H

Scheme 4.12

was determined to be $> 2 \cdot 10^6$ dalton by gel-column chromatography. The UV–visible spectra of the aqueus solution showed a Soret band, which is split by approximately 100 nm, by excition interactions (460 and 360 nm). The fibers made of pure enantiomers show strong CD bands. From both the absorption and CD spectra, the microstructure of porphyrin fiber was simulated based on excition calculations. Because the porphyrin macrocycle has a high density of π-electrons on the pyrrole ring and a positive hole in the center of the porphyrin, it dimerizes with a lateral shift; the electron-rich pyrrole ring of the porphyrin stacks on the center of a partner. Stable dimers are thus formed in solution and were postulated to connect to twisted ribbons by strong edge-to-edge interactions (Fig. 4.24).

Tin(IV) porphyrin attached to open-chain carbohydrate groups (Scheme 4.13) formed extremely long fibers with a diameter of 5 nm and a length of several μm at pH 0 [165]. Visible and CD spectra indicate a lateral arrangement of the facially protonated monomer. The cationic HCl pairs on both sides of the porphyrins act as a kind of facial head group and makes the hydration of the porphyrin plane possible, producing very thin porphyrin fibers.

The PPIX derivative having two phospholipid-like groups (lipidprotoporphyrin: Schemes 4.15–4.17) also formed highly organized fibers or an oval multilamellar vesicle in aqueous medium [167, 168]. The compound in Scheme 4.15 produced highly organized rod-like fibers with ca. 60 A widths (Fig. 4.25 a) [167]. The length of the fibers varied from 0.1 to 1.5 μm. Because

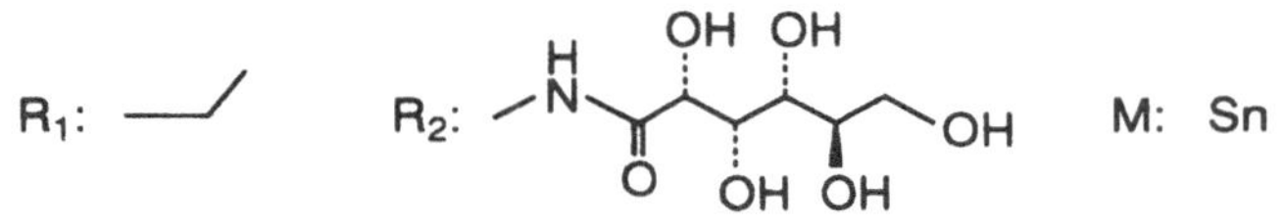

R: $-\overset{O}{\overset{\|}{C}}O(CH_2)_{11}O\overset{O}{\overset{\|}{P}}O(CH_2)_2- \overset{CH_3}{\overset{|}{N^+}}-CH_3$

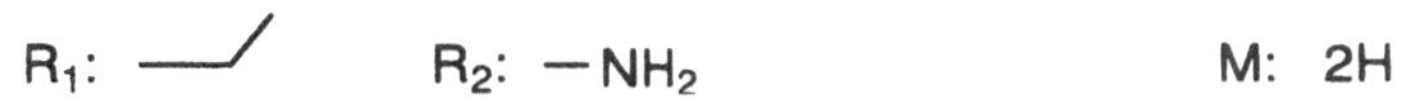

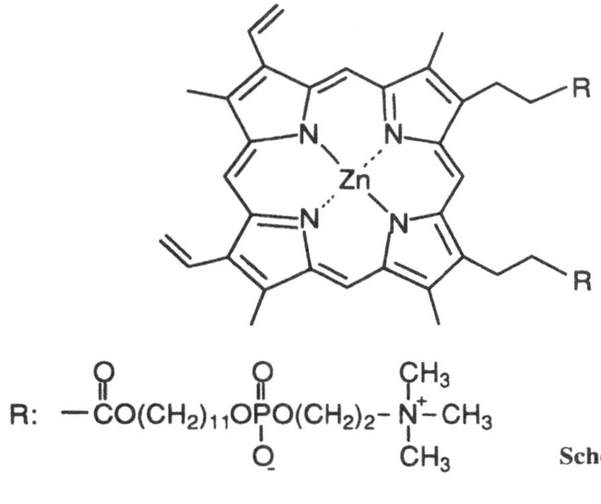

Scheme 4.15

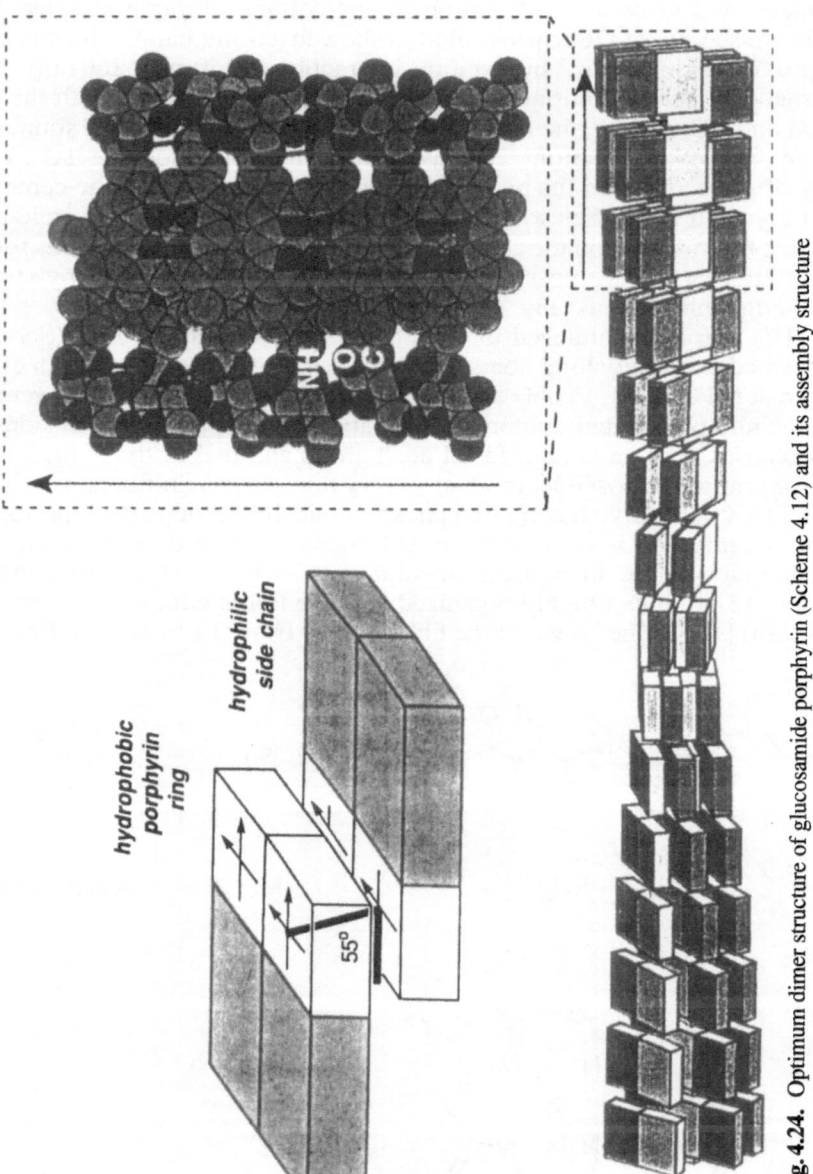

Fig. 4.24. Optimum dimer structure of glucosamide porphyrin (Scheme 4.12) and its assembly structure

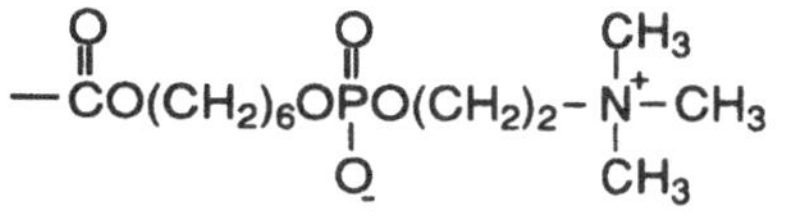

$$-\overset{\overset{\displaystyle O}{\|}}{C}O(CH_2)_6O\overset{\overset{\displaystyle O}{\|}}{P}O(CH_2)_2-\overset{\overset{\displaystyle CH_3}{|}}{\underset{\underset{\displaystyle CH_3}{|}}{N^+}}-CH_3$$

Scheme 4.16

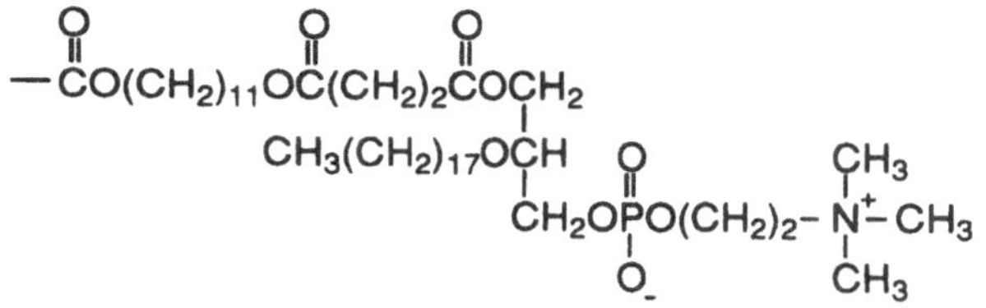

$$-\overset{\overset{\displaystyle O}{\|}}{C}O(CH_2)_{11}O\overset{\overset{\displaystyle O}{\|}}{C}(CH_2)_2\overset{\overset{\displaystyle O}{\|}}{C}OCH_2$$

$$CH_3(CH_2)_{17}O\overset{|}{C}H$$

$$CH_2O\overset{\overset{\displaystyle O}{\|}}{P}O(CH_2)_2-\overset{\overset{\displaystyle CH_3}{|}}{\underset{\underset{\displaystyle CH_3}{|}}{N^+}}-CH_3$$

Scheme 4.17

the molecular length of the compound in Scheme 4.15 is ca. 35 A, it might be presumed that the fibers arise from the stacked porphyrinatozinc(II) aggregate. On the other hand, the compound in Scheme 4.16 formed twisted ribbon-like fibers. Visible absorption spectra of the lipoporphyrinatozinc(II) fibers showed Soret bands at $\lambda_{max} = 389$ nm for the compound in Scheme 4.15, which were shifted toward the blue region compared with that of the micellar dispersion containing the monomeric form ($\lambda_{max} = 420$ nm). Fluorescnece of the aqueous dispersion of the lipidporphyrinatozinc(II) was not present, which also suggested formation of a $\pi-\pi$ aggregation. Therefore, it can be assumed that the compound in Scheme 4.15, having two long alkylphosphocholine groups, formed highly organized fibers.

The spectral features of the lipidprotoporphyrin fibers remained essentially unchanged after a year and were not influenced by the addition of NaCl (450 mmolL^{-1}). The fibrous aggregate does not seem to dissociate upon dilution or heating. These results showed that long-lived lipodprotoporphyrin fibers do not change their structure, even under physiological conditions.

Interestingly, the lipidprotoporphyrin fibers were spontaneously incorporated into the bilayer of the phospholipid [1,2-bis(parmytoyl)-sn-glycero-3-phosphocholine (DPPC)] vesicle (40 ~ 50 nm). The absorption spectrum of the mixed solution of the fiber in Scheme 4.16 (40 μmolL^{-1}) and DPPC vesicle (2 mmolL^{-1}) changed from 386 nm to 420 nm within 30 min at 25 °C. From a TEM of the mixed solution of the fiber in Scheme 4.16 and DPPC vesicle, only small unilamellar vesicles with diameters of ca. 50 nm were detected, i.e., the lipidprotoporphyrin fibers were incorporated into the vesicle and the porphyrin molecules were homogeneously embedded in the bilayer of the membrane. This spontaneous formation of hybrid assembly was induced by disequilibrium of the hydrophobic interaction between the porphyrin fibers and the phospholipid vesicle.

On the other hand, the lipidprotoporphyrin (Scheme 4.17) was easily homogenized in deionized water by sonication to give an oval multilamellar vesicle with a major axis of 200–250 nm (Fig. 4.25 b) [168]. The uniform thickness of each layer was ca. 7 nm. Because the molecular length of the compound in Scheme 4.17 is ca. 4 nm, it can be presumed that the vesicle arises

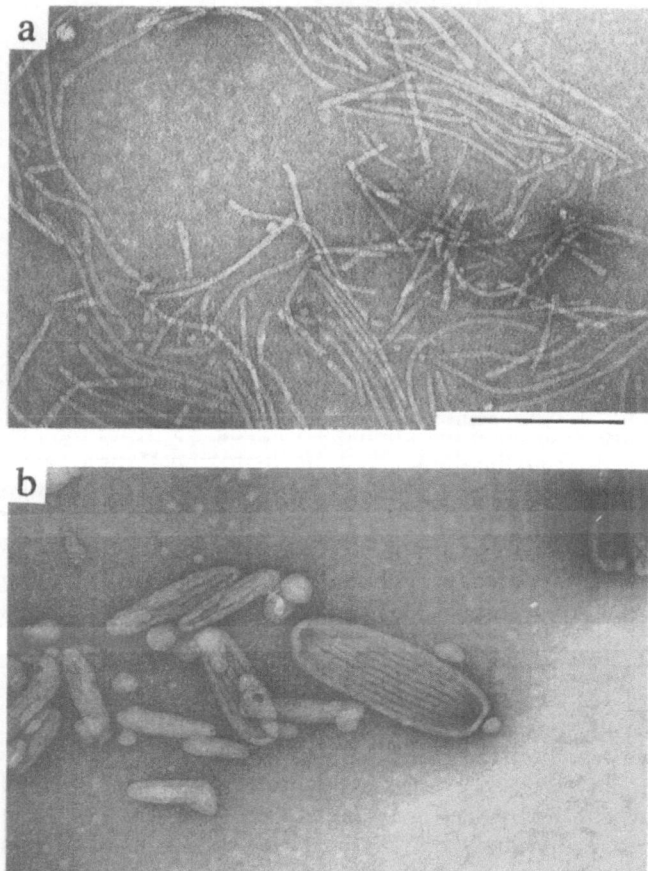

Fig. 4.25 a, b. Transmission electron micrographs of lipidprotoporphyrin (Scheme 4.15) fiber and mulitlamellar vesicle (Scheme 4.17; bar: 100 nm)

from bilayer aggregates of the porphyrin molecules. This is the first example of formation of a vesicle consisting of only PPIX derivatives. The Soret band of the aqueous dispersion of the compound in Scheme 4.16 (λ_{max}: 394 nm) was also shifted toward the blue region compared with that of a monomer (λ_{max}: 418 nm), suggesting formation of $\pi-\pi$ aggregation. The morphology of the assembly of the compound in Scheme 4.16 was characterized by its oval form and a large particle size compared with a spherical phospholipid (e. g., DPPC) vesicle (40 ~ 50 nm) prepared by sonication. Because the porphyrin moiety of Scheme 4.17 is a rigid and hydrophobic unit ($0.8 \cdot 0.8 \cdot 0.4\,nm^3$, determined using creatine phosphokinase models), stacking of the porphyrin ring results in the small curvature of the vesicle.

4.3.3.2 Electronic Process of Porphyrin Fiber

Aminoethylporphyrin (Scheme 4.14) is water-soluble in the pH range from 4 to 6 [166]. It shows a broadened Soret band with a peak at 370 nm. This blue-shifted Soret band comes from transition dipole interactions of the face-to-face stacked porphyrin configuration. The electron micrographs of the slightly acidic solutions (pH 6.5) show vesicular tubules with an outer diameter of ca. 30 nm and average length of 800 nm. At higher pH values the Soret band appears split at 445 nm and at 360 nm, and the morphology was transformed to very long micellar rods with diameters between 6 and 20 nm. The 445-nm band at high pH is consistent with a lateral ribbon stabilized by $NH_2...NH_3^+$ hydrogen bonds. Because the porphyrin rings are closely associated with each other in the highly ordered assembled porphyrin, the photo- and redox-chemistries are explored. Upon excitation of the 30% DMSO aqueous solution of the fibers in Scheme 4.14 with laser flashes, a charge separation occurred from the singlet state. The differences spectrum after the flashes showed a maximum at 835 nm and a minimum at 660 nm. This does not correspond to the triplet-state spectrum, which is very likely related to a charge separation within the fiber leading to an anion radical and a cation radical. The porphyrin–porphyrin distances within the fiber should be short enough for very fast electron transfer, which has to compete with a fast internal conversion.

4.3.3.3 Octopus-Porphyrin Assembly

Various metalloporphyrins have been synthesized as models for Hb and myoglobin (Mb), and their O_2-coordination behaviors were studied in detail in aprotic organic solvents [169–173]. In particular, both face-hindered tetraphenylporphinatoiron(II)s are considered to be more effective than the single face-hindered one, because the bulky substitutes on each side of the macrocycle impede, the formation of an intermolecular μ-oxo dimer leading to irreversible oxidation [174–177].

Amphiphilic tetraphenylporphins having four alkyl phosphocholine groups on each side of the ring plane (octopus-porphyrin; Scheme 4.18) were easily dispersed in deionized water by vortex mixing to give a transparent red solution [178]. The critical micelle concentration (cmc) of the octopus-porphyrins was ca. 1 μmolL^{-1} as estimated using the Wilhelmy method. Above this concentration, the compound in Scheme 4.18a formed fibrous aggregates with uniform widths of ca. 80 A. The fluorescene of an aqueus dispersion of the compound in Scheme 4.18 was present at the same intensity as that of its monomer dispersion; no quenching of fluorescence of the octopus-porphyrin fibers was observed. Furthermore, the fiber of the compound in Scheme 4.18 produced a triplet state after excitation at 532 nm [179]. In self-organized porphyrin aggregate, the stable triplet state cannot generally be observed, because the porphyrin rings, are stacked on each other. In the case of the octopus form, the bulky 2,2-dimethyl groups on both sides of the ring plane impede the completely stacking of the porphyrin macrocycle resulting in preventing the quenching of the excited state. The octopus-porphyrin fiber should become

M: 2H   a
Zn(II)   b
Fe(II)   c   **Scheme 4.18**

photoactive if the amphiphilic periphery is doped with an electron acceptor in the structure.

The morphology of the octopus-porphyrin assembly was transformed into a totally different structure by adding alkyl imidazole and/or phospholipid molecules. Aggregates of the octopus-hemin were dispersed with a 20-fold molar excess of 1-dodecyl-2-methylimidazole (DMIm) by vortex mixing in 1 mmolL^{-1} phosphate buffer (pH 7.4), to give spherical unilamellar vesicles with a diameter of ca. 100 nm.

Porphyrinatoiron(III) in the assembly was reduced by the addition of a small excess amount of aqueous ascorbic acid under a nitrogen atmosphere. The visible absorption spectrum of the iron(II) deoxy complex of the octopus-heme vesicle (heme/DMIm, molar ratio: 1/20)(λ_{max}: 436, 535, and 558 m) changed to an O_2-adduct upon exposure to O_2 (λ_{max}: 421 and 545 nm). The spectrum changed reversibly in response to O_2-pressures. The O_2-adduct changed to the corresponding CO-adduct upon bubbling CO gas through the solution (λ_{max}: 422 and 535 nm). The O_2-binding affinity [$P_{1/2}(O_2)$, the O_2-partial pressure at half O_2-binding for the prophyrinatoiron(II)] of the octopusheme vesicle was estimated to be 160 Torr at 25 °C. These results show that the molecular assemblies composed of octopus-heme complexes can mimic the O_2-binding properties of Hb.

4.3.4 Self-Assembled Lipidporphyrin Vesicle and Oxygen Coordination

Thus far, there have been many reports of micelles or phospholipid vesicle embedding porphyrin complexes as hemoprotein models [180−183]. We have also found a phospholipid vesicle embedding porphyrinatoiron(II) derivative in its bilayer, which can form a stable O_2-adduct under physiological conditions and function as a totally synthetic O_2-carrier [184].

Presently, however, few studies have reported the chemical reactions coupled with the aggregate microstructure of the porphyrins in a self-organized amphiphilic porphyrin assembly in aqueous solution. We have recently found that amphiphilic tetraphenylporphyrin derivatives having four dialkylglycerophospho-choline groups on one side of the ring plane (lipidporphyrin; Scheme 4.19) formed a spherical unilamellar vesicle, which can reversibly bind oxygen under physiological conditions [185]. The microstructure of its assembly and the O_2-binding equilibrium and kinetic parameters of the vesicle in the compound in Scheme 4.19 are described.

4.3.4.1 Microstructure of Lipidporphyrin Vesicle

Lipidporphyrin was dispersed in deionized water by sonication ([Scheme 4.19] = 10 ~ 100 μmolL^{-1}) and formed a spherical unilamellar vesicle with a uniform diameter of ca. 100 nm (Fig. 4.26). The particle diameters also agreed with the average sizes (94 ± 19 nm) estimated from a light-scattering experiment. The thickness of the membrane was determined to be 9.5 ± 0.5 nm from cryomicrographs (Fig. 4.26b). Because the thickness of the membrane corresponds to twice the length of the side chains of the compound in Scheme 4.19b; the porphyrin moietes in the outer and inner layers are presumed to be in close contact.

The lipidporphyrin-packing geometry of the bilayer vesicle was estimated from the effective surface area of the lipidporphyrin molecule. The numbers of the lipidporphyrin molecules in the outer and inner layers (n_o, n_i) of the vesicle in the compound in Scheme 4.19b can be calculated from the

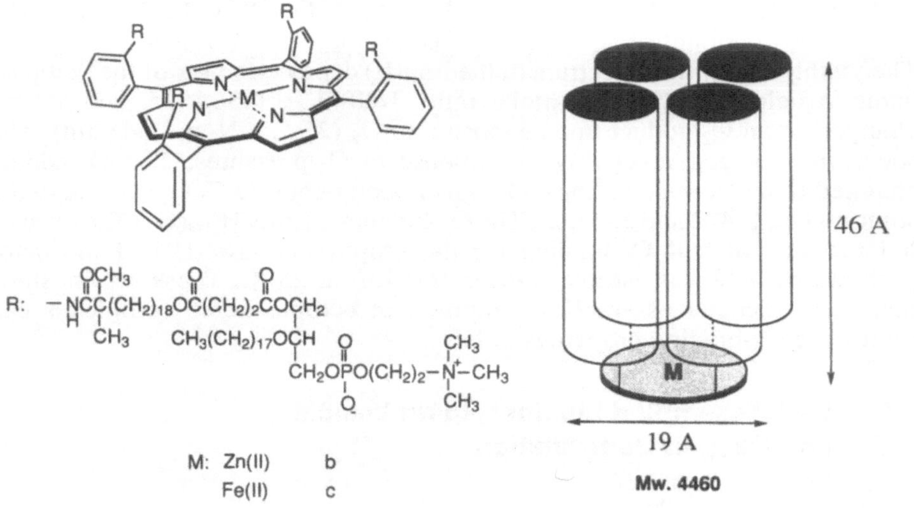

R: $-\overset{\underset{\displaystyle H}{|}}{N}\overset{\overset{\displaystyle O}{\|}}{C}C(CH_2)_{18}O\overset{\overset{\displaystyle O}{\|}}{C}(CH_2)_2\overset{\overset{\displaystyle O}{\|}}{C}OCH_2$

$CH_3(CH_2)_{17}OCH$

$CH_2OPO(CH_2)_2-\overset{\underset{\displaystyle CH_3}{|}}{\overset{\overset{\displaystyle CH_3}{|}}{N}}-CH_3$

M: Zn(II) b
    Fe(II) c

Mw. 4460

46 A

19 A

M

Scheme 4.19

surface area of the vesicle divided by the molecular area of the compound in Scheme 4.19b (287 A^2); $n_o = 9670$ and $n_i = 6160$, respectively. Therefore, the ratio of the numbers of the lipidporphyrinatozinc(II) molecules in the outer and inner layers ($X = n_o/n_i$) is calculated to be 1.57.

4.3.4.2 Porphyrin Arrangement in Lipidporphyrin Vesicle

The visible absorption spectrum of the compound in Scheme 4.19b vesicle showed a Soret band (λ_{max}: 438 nm) with a small shoulder at 405 nm, which was shifted toward the red region relative to that observed in benzene/CH$_3$OH, 1:1 (v/v) (λ_{max}: 425 nm). The Q bands remained essentially unaltered. The red-shifted Soret band can then be attributed to a lateral arrangement of the transition moments of the porphyrin molecules [186–189]. This type of "J-aggregate" was further confirmed based on its photophysical properties. The fluorescence emission intensity for the vesicle of the compound in Scheme 4.19b was slightly red-shifted relative to that of the corresponding monomer in homogenous organic solution, although it was not quenched. Concomitantly, the singlet lifetime for the vesicle of the compound in Scheme 4.19b ($\tau_F = 4.36$ ns) was almost the same as that of the monomer of the compound in Scheme 4.19b in benzene/CH$_3$OH solution.

In contrast, the triplet lifetime for the vesicle of the compound in Scheme 4.19b was extraordinarily short compared with that of the monomer of the compound in Scheme 4.19b in organic solution. The decrease in the triplet lifetime implicates additional nonradiative decay channels from the excited states, due to formation of an oriented multi-porphyrin arrangement.

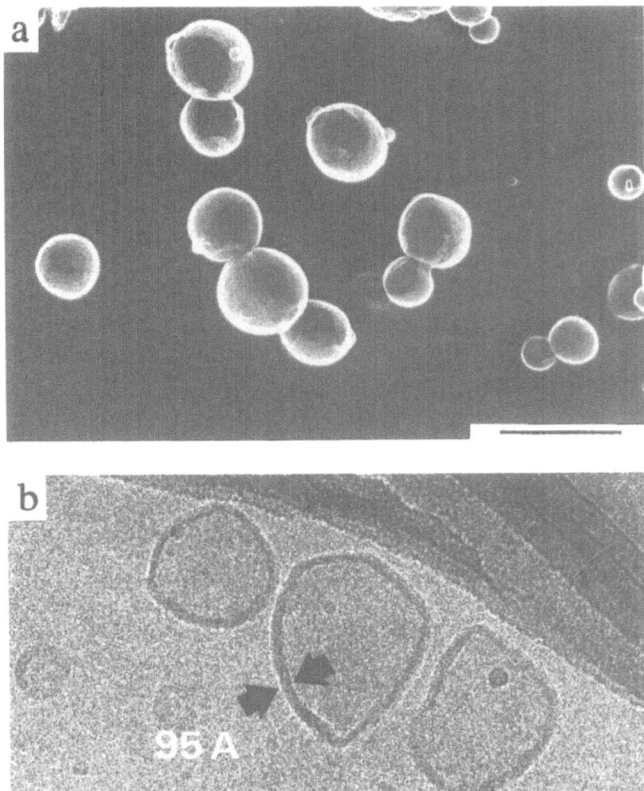

Fig. 4.26. Transmission electron micrographs of freeze-etching replica (**a**) and cryomicrograph (**b**) of lipidprotoporphyrin (Scheme 4.19b) vesicle (bar: **a** 200 nm; **b** 100 nm)

Similar observations have been reported by Schick et al. [186]. They reported a split Soret band absorption for a supported monolayer assembly containing 5,10,15,20-tetrakis[4-(1-oxtyloxy)phenyl]porphinatozinc(II). Barber et al. also reported that 5,10,15,20-tetrakis ($\alpha,\alpha,\alpha,\alpha$-o-hexanamidophenyl)porphyrin J-aggregates in dilute aqueous surfactant solution (i.e., below the cmc) exhibited a red-shifted Soret band absorption ($\lambda_{max} \sim 436$ nm) and strongly reduced triplet lifetimes [188].

The differential scanning calorimetry (DSC) curve of the dispersion of the compound in Scheme 4.19b showed a broadening peak at 56 °C. Interestingly, the Soret band maxima of the vesicle of the compound in Scheme 4.19b showed a temperature dependence in the range of 20 ~ 85 °C. The λ_{max} (438 nm) of the vesible of the compound in Scheme 4.19b was gradually blue-shifted with reduced absorption from 60 °C and reached 434 nm at 80 °C. This characteristic behavior of the λ_{max} was reversibly observed and dependent on temperature changes. This indicates that a broad-phase transition existed near 55 ~ 65 °C in the lipidporphyrin vesicle accompanied by negligible movements of the porphyrin square.

Arrangement of the porphyrin moiety in the vesicle of the compound in Scheme 4.19 b was estimated by exciton coupling calculations [190, 191]. In each monolayer the distance of the transition dipole moments for the edge-to-edge arrangements (r_1) is 19 A, which was estimated based on the close-packing structural model of the four head groups on the lipid-porphyrin. The calculated energy splitting levels for S_x and S_y of the Soret band are shown in Fig. 4.27. From the energy shifts. E_3 (-643 cm^{-1}) and ΔE_3^z (1514 cm^{-1}) r_2 is determined to be 5.9 A, and θ is presumed to be ca. 47°.

4.3.4.3 O₂-Coordination Property of Lipidporphyrin Vesicle

The lipidporphyrin vesicle was obtained in a similar manner in the presence of a small excess molar amount of DMIm as an axial base for the O_2-adduct. The visible absorption spectrum of the iron(II) deoxy complex of the vesicle of the compound in Scheme 4.19c (λ_{max}: 445, 532, and 561 nm) changed to that of its O_2-adduct upon exposure to dioxygen (λ_{max}: 432 and 549 nm). The

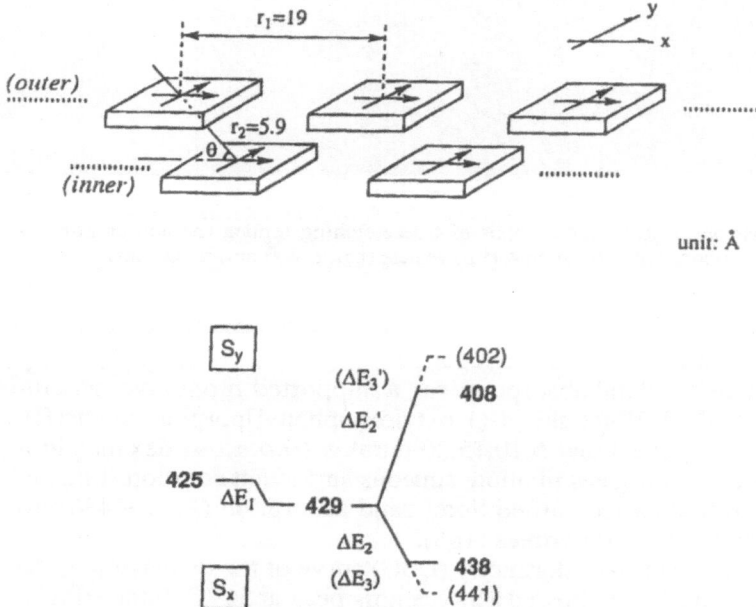

Fig. 4.27. Illustration of geometrical packing of the porphyrin ring in the vesicle of the compound in Scheme 4.19 b

spectrum changed reversibly in response to O_2-pressure. The O_2-adduct changed to the corresponding CO-adduct upon bubbling CO through the solution (λ_{max}: 430 and 532 nm). The O_2-binding affinity [$P_{1/2}(O_2)$] of the DMIm vesicle of the compound in Scheme 4.19c was estimated to be 43 Torr at 37 °C, which is slightly lower than that of a red-cell suspension. The half-life of the oxygeneated species was 50 h under physiological conditions (37 °C, pH 7.4). Most importantly, one spherical vesicle of the compound in Scheme 4.19b (mol. wt. $8.16 \cdot 10^7$ dalton) can bind $1.58 \cdot 10^4$ moles of O_2 molecules, although one particle of Hb (mol. wt. $6.45 \cdot 10^4$ dalton) binds only four moles of O_2.

In order to elucidate the O_2- and CO-binding properties of the vesicle of the compound in Scheme 4.19c, the kinetics of the binding were explored using laser flash photolysis [169, 170, 177]. Kinetic parameters for the O_2- and CO-binding to the lipidporphyrinatoiron(II) vesicle are summarized in Table 4.5.

The $P_{1/2}(O_2)$ of the DMIm vesicle of the compound in Scheme 4.19c is almost the same as that of T-state Hb [192, 193]. However, the association and dissociation rate constants for O_2 of the lipidporphyrinatoiron(II) vesicle are 10^2-fold greater than those of Hb [193–194]. This has been interpreted to indicate that the O_2-binding reaction is not retarded by the diffusion of oxygen in and through the alkyl moieties of the lipidporphyrin bilayer vesicle. Even in the case of the molecules in Scheme 4.19c, which are self-assembled to form a highly-ordered structure, porphyrinatoiron(II) moieties can bind and release oxygen more rapidly than Hb. Thus, this vesicle of the compound in Scheme 4.19c has the ability to act as a totally synthetic O_2-carrier under physiological conditions.

Based on these results, the lipidporphyrinatoiron(II) vesicle, including highly-ordered reactive sites with high density is regarded as a model for superhemoglobin. Furthermore, the lipidporphyrin vesicle would also be useful as a new molecular structure in biochemical reactions such as a regioselective oxidation utilizing the ordered bilayer.

Table 4.5. O_2- and CO-binding parameters of lipidporphyrinatoiron(II) vesicle in phosphate buffer at 25 °C

		$P_{1/2}(O_2)$ (Torr)	$k_{on}(O_2)$ ($M^1 s^{-1}$)	$k_{off}(O_2)$ (s^{-1})
Vesicle (Scheme 4.18c)		7.4	$2.8 \cdot 10^8$	$1.5 \cdot 10^4$
			32	
Hb (T-state) a[a]	7.0–7.4	40	$2.9 \cdot 10^6$	$1.8 \cdot 10^2$
Mb[a]	7.0–7.4	0.37–1	1–$2 \cdot 10^7$	10–30
		$P_{1/2}(CO)$ (Torr)	k_{on} (CO) ($M^1 s^{-1}$)	$k_{off}(CO)$ (s^{-1})
Vesicle (Scheme 4.18c)		7.4	$2.8 \cdot 10^{-2}$	$3.9 \cdot 10^6$	0.14
Hb (T-state) a[a]	7.0–7.4	0.3	$2.2 \cdot 10^5$	0.09
Mb[a]	7.0–7.4	1.4–$2.5 \cdot 10^{-2}$	3–$5 \cdot 10^5$	1.5–$40 \cdot 10^{-3}$

[a] At 20 °C.

4.3.5 Lipidheme Microsphere as Oxygen Carrier

An aqueous suspension of a soybean emulsion stabilized with egg yolk lecithin (lipid microsphere) has been widely used in clinics for parenteral nutrition since the 1960s [195]. The lipid microsphere is infused intravenously as a source of calories in patients who suffer from inflammatory bowel disease, malnutrition, gastrointestinal pain or obstruction, or any other disorder in which the gastrointestinal tract is not functional. Because the average particle diameter of the lipid microsphere is controlled to ca. 0.2 μm, blockage of the blood capillaries is not observed, and a variety of studies suggest that the lipid microsphere is similarly metabolized with chylomicrons [196−198]. Furthermore, new applications of the lipid microsphere have been developed as a carrier for lipophilic drugs and parenteral contrast agents [199, 200].

Recently, we found that on O/W lipid microsphere, which is formulated from triglyceride (e.g., soybean oil) and emulsified with the lipidheme (Scheme 4.20) as a surfactant, gives a red-colored dispersion which is able to reversibly bind oxygen in aqueous medium [201]. The surface of the oil droplet was covered with lipidheme (O_2-binding site), so that the particle size and heme concentration (i.e., O_2-solubility) can be voluntarily controlled. In this section the solution properties, O_2-transporting ability, and performance in a biocirculation system of a new type of finely dispersed lipidheme microspheres are described.

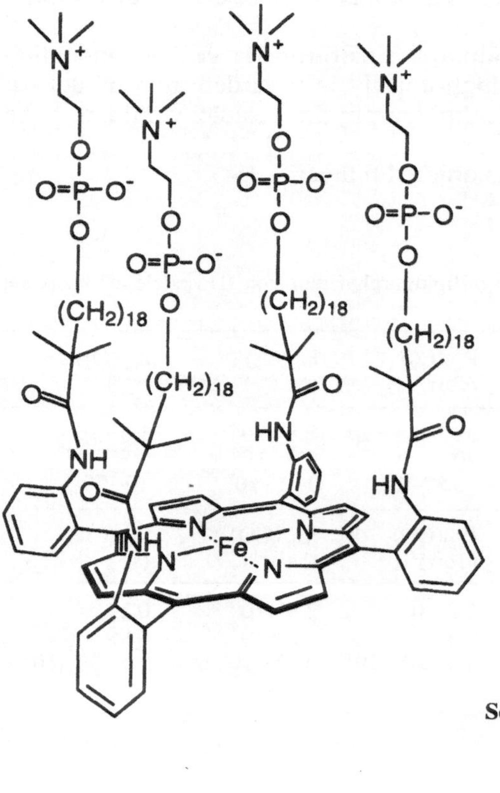

Scheme 4.20

4.3.5.1 Physicochemical Properties of Lipidheme Microsphere

The average particle diameter of the lipidheme microsphere is ca. 0.1 μm based on TEM (Fig. 4.28). The lipidheme microsphere suspension was stable and could be stocked for a few months without precipitation and change in the particle size at $4 \sim 25\,°C$. The physicochemical properties of the lipidheme microsphere are summarized in Table 4.6. Specifically, the suspension was characterized by its low viscosity (1.2 cP, [heme] = $10\,mmolL^{-1}$), which was much lower than that of human blood ($4.5 \sim 5.5$ cP). This is the most important feature of the lipidheme microsphere, and is a great advantage for in vivo administration.

The specific gravity (d: 1.001) was almost the same as that of a saline solution. Osmotic pressure was adjusted to the physiological value (ca. 300 Torr) by adding glycerine or NaCl. The colloid osmotic pressure of the lipidheme-microsphere suspension ([heme]: $10\,mmolL^{-3}$) was <2 Torr. It was adjusted to that of human blood (25 Torr) by adding a water-soluble polymer, e.g., dextran (Dex: mol. wt. 40000) or human albumin (Alb). The optimal added amount of these water-soluble polymers was Dex: 2 wt.%, Alb: 5 wt.%. During the operation an increase in viscosity was not observed. Blood compati-

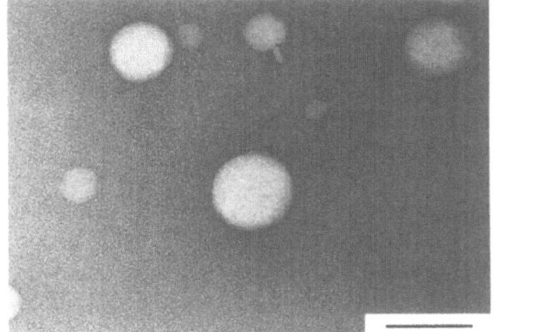
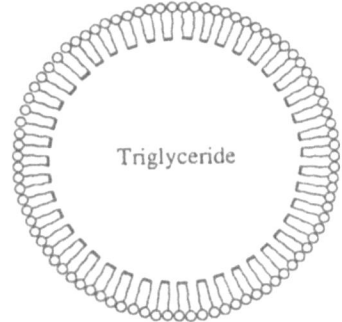

Fig. 4.28. Transmission electron micrographs of lipidheme microsphere (bar: 100 nm)

Table 4.6. Solution properties of the lipidheme microsphere suspension

	d	Viscosity[a] (cP)	OP (mOsm/kg)	COP (Torr)	O_2-solubility (mlO_2/100 ml)
LH-M [LH]: 5[b]	0.998	1.1	280 ~ 300	~ 2	14
10[b]	1.001	1.2	280 ~ 300	~ 2	25
15[b]	1.003	1.5	280 ~ 300	~ 2	36
Human blood	1.055 ~ 1.065	4.5 ~ 5.0	280 ~ 290	25	23
Serum	1.027	1.33	285 ± 11	~ 2	2
Saline	0.999	0.75	285	0	2
Intralipid	0.981	1.5	285	2	2

[a] At 37 °C. OP osmotic pressure; COP colloid osmotic pressure.
[b] $mmolL^{-1}$.

bility of the lipidheme microsphere suspension was preliminarily estimated by viscosity measurements of the mixed solution. Even for the lipidheme microsphere and human blood mixed at 1/1/ (v/v), aggregation and/or precipitation were not observed, and its apparent viscosity was maintained at 2.5 cP (at share rate: $230\,s^{-1}$). This result indicates that the lipidheme microsphere has a high compatibility with blood.

4.3.5.2 O₂-Solubility of Lipidheme Microsphere

The surface of the lipidheme microsphere was covered with lipidheme as the O_2-binding sites, so that the O_2-solubility of this solution could be voluntarily controlled by prescription or by the diameter of the particle. The O_2-solubility of the human blood suspension was 23 mL/(100 mL medium) because of its heme concentration of $9.2\,mmolL^{-1}$; however, more oxygen was dissolved in the lipidheme microsphere suspension depending on its heme concentration. The measured volume of the dissolved oxygen in the lipidheme microsphere suspension ([heme] = $10\,mmolL^{-1}$) actually corresponded to the calculated value (25 mL/(100 mL medium)). This result also indicated that lipidhemes bound equivalent moles of oxygen, i.e., the lipidheme-microsphere suspension ([heme]: $15\,mmolL^{-1}$) is able to uptake O_2 up to 35 mL/100 mL medium, which is much higher than that of human blood. Therefore, the small size of particles, which is approximately 80 times smaller in diameter than that of red blood cells, and low viscosity of the lipidheme microsphere facilitate circulation and permit a better exploitation of the collateral microcirculation. Administration of the lipidheme microsphere suspension would now be useful for desirable therapy in such situations as hemorrhagic shock associated with myocardial or cerebral thrombosis.

The oxygen equilibrium curve of the lipidheme microsphere is shown in Fig. 4.29. The $P_{1/2}(O_2)$ was adjusted to ca. 40 Torr, slightly lower than that of human blood. This suggested that efficient oxygen delivery has been achieved from the synthetic heme to hemoglobin in the mixed solution. The Hill coefficient (n) of the lipidheme microsphere was 1.0; however, the O_2-transporting efficiency from the lungs ($P(O_2)$ = ca. 110 Torr) to Mb in muscle tissues ($P(O_2)$ = ca. 40 Torr) was almost the same as that of Hb. The values of $k_{on}(O_2)$ and $k_{off}(O_2)$ were higher than those of Hb, and as a result, O_2-molecule associates and dissociates to the synthetic heme more rapidly when compared with Hb, even in the mixed system.

4.3.5.3 Biocirculation System as O₂-Carrier

The in vivo O_2-transporting ability of the lipidheme microsphere dispersion as a red cell substitute has been studied by exchange transfusion in hemorrhagic shocked dogs. The sample for the in vivo test ([heme] = $5\,mmolL^{-1}$) was filtered several times with a pyrogen-free cellulose acetate membrane. Six beagles (8 kg) were anesthetized and paralyzed. A cuffed endotracheal tube was inserted, and the beagles were connected to a Harvard respirator and mechanically ventilated under ambient air. A Swan-Ganz catheter was inserted into the pulmonary artery through the jugular vein, and an arterial line was

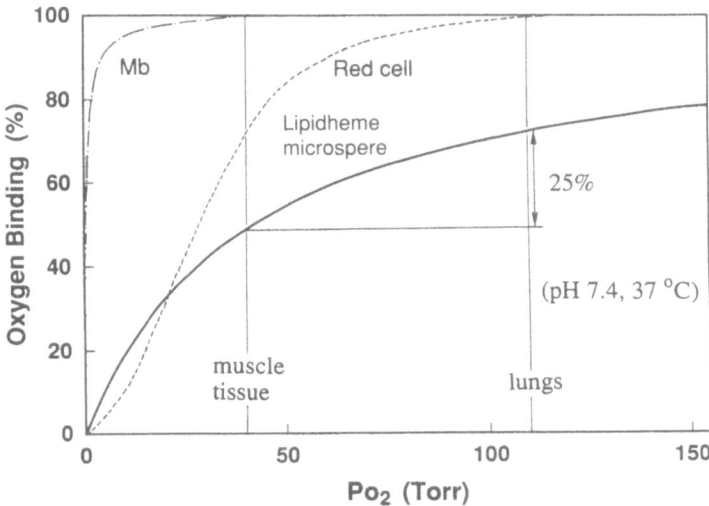

Fig. 4.29. Oxygen-binding equilibrium curve of lipidheme microsphere

connected to the femoral artery. Thirty mL kg^{-1} of blood was then shed via the arterial line. Next, 30 min after the shedding, 30 ml kg^{-1} of the lipidheme microsphere solution (125 ± 23 nm) was intravenously injected (total 240 mL).

In the mixed system of the totally synthetic artificial red cell and blood suspension, hemolysis, blood coagulation, and platelet aggregation were not observed. The intravascular persistence of the lipidheme in the bloodstream is shown in Fig. 4.30. The apparent half-life (50% disappearance time of the lipidheme; $\tau_{1/2}$) of the lipidheme was estimated to be ca. 12 h, which satisfies the minimum term of a red cell substitute. The average particle diameter of the lipidheme microsphere did not change up to 5 h after injection, indicating that the lipidheme microsphere was maintained its stable structure in the bloodstream.

The visible absorption spectrum of the supernatant of the withdrawn blood showed a change in the deoxy and oxy forms of the lipidheme microsphere upon introduction of bubbling nitrogen and oxygen, respectively. The indicated that the lipidheme microsphere could deliver oxygen to the bloodstream.

Then, 30 min after shedding, the cardiac output (Q) decreased to 0.5–0.6 L min^{-1}, which corresponded to 30–34% of the control level. The decreased Q immediately increased upon injection of the lipidheme-microsphere suspension. A recovery ratio was 67% of the control value. This ratio was almost the same as that of a saline-injected system (ca. 70% of the control level). The pH and Pco_2 were not influenced by the exchange transfusion with the lipidheme microsphere.

Oxygen consumptions (Vo_2) of the lipidheme and Hb were calculated using Q, the lipidheme or Hb concentration, and arterial and venous O_2-saturation (Sao_2 and Svo_2):

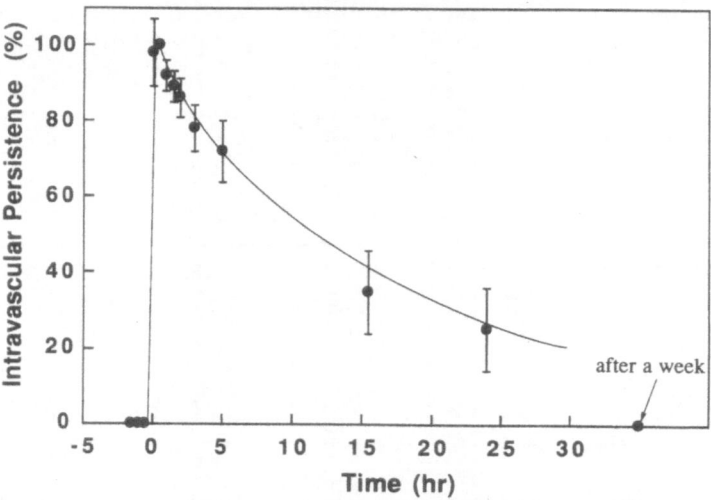

Fig. 4.30. Circulation lifetime of lipidheme microsphere in the blood stream

$$V_{O_2}(\text{lipidheme}) = Q \cdot [\text{lipidheme}] \cdot (S_{aO_2} - S_{vO_2}) \qquad (1)$$
$$V_{O_2}(\text{Hb}) = Q \cdot [\text{Hb}] \cdot (S_{aO_2} - S_{vO_2}) \qquad (2)$$

The decreased total O_2 consumption immediately recovered after injection of the lipidheme microsphere dispersion. Furthermore, the V_{O_2} ratio by the lipidheme microsphere in the total amount of O_2-consumption was ca. 15%, which agreed with the concentration ratio of lipidheme. These results suggested that the O_2-consumption ratio of the totally synthetic system could be regulated by control of the lipidheme concentration or the exchange volume.

Acknowledgement. This work was partially supported by a Grant-in-Aid for Scientific Research (nos. 05403028, 05236103, and 053666) from the Ministry of Education, Science, and Culture, Japan.

4.4 Catalysis in Macromolecular Metal Complexes

F. Ciardelli

4.4.1 General Concepts

Whereas many important applications are today known for transition metal complexes attached to macromolecular ligands as exhaustively described in the various chapters of the present book, the application as heterogenized catalysts is certainly one of the most attractive, and has been largely investigated [202–204]. Clearly, many aspects of catalysis involve polymer-bond transition metal complexes; thus, mention of these properties is made along the whole book, even in chapters primarily devoted to synthesis or characterization (e. g., Chap. 2). This is particularly evident for polymer-metal complexes in living systems (Chap. 3), in mediated electronic processes (other section of Chap. 4), and in photoinduced processes (Chap. 5).

In this context the present section focuses on systems where catalysis was the actual objective of the preparation of macromolecular metal complexes (MMC). Also, reference is made to systems used in traditional catalytic processes mainly employing olefinic substrates. Thus, the catalytic processes discussed generally concern isomerization, hydrogenation, carbonylation, oligomerization, and polymerization reactions.

Both cross-linked and linear organic and inorganic polymers have been largely used in the past few decades for heterogenizing transition metal catalysts (Sect. 2.2, 2.3 and 2.4). These supports can be functionalized to give ligands for attaching transiton metal complexes through coordination bonds as in the corresponding soluble monomeric analogs.

The main goal of this approach was the preparation of catalytic systems displaying the good activity, selectivity and reproducibility typical of homogeneous catalysts, combined with the easy separability and recovery characteristics of heterogeneous catalysts.

In case the structure of metal centers coordinated to the polymer is the same as in the monomer, the large steric hindrance by the polymer matrix (Sect. 2.1) reduces the catalytic activity and is only partially attenuated by the swelling or dissolving of the systems with a proper liquid medium. This aspect suggests that the macromolecular ligand, in contrast to a traditional inert support, can modify catalytic behavior by physically interacting with solvent media and thus modulating the reaction rate. In addition, the macromolecular matrix can affect the selectivity and stereoselectivity, both from steric and geometrical origin. The former is observed when the macromolecular ligand produces a specific steric environment around the metal center in the active sites, and can take place either in solution or in heterogeneous systems. Some typical examples are known, but more work should be done in this context. This concept is typically shown by the preparation of catalysts for asymmetric synthesis either by using optically active polymer ligands or by fixing optically active metal complexes on polymer matrices [205, 206]. Moreover, shape selectivity can be achieved in the heterogeneous phase with metal complexes supported on cross-linked polymers, where the dimension of pores has been controlled by modulating the degree and type of cross-linking.

The rigidity of the macromolecular ligand, due either to its molecular structure or to proper design of the network structure, can provide a useful method to stabilize reactive intermediates, thus allowing their isolation and identification. Indeed, fixation of isolated metal species along the polymer backbone, whose rigidity indered bimolecular interactions between active centers, results generally in longer-living species.

This type of heterogenized system can then provide information on the mechanism of the same reaction carried out with the corresponding homogeneous catalysts, giving labile nonisolable intermediates. Also, the improved stability of active species can be used to prepare catalytic systems capable of maintaining their activity during recycling. Furthermore, coordinatively unsaturated polymer-bound metal species have been obtained in several cases resulting in improved catalytic behavior of heterogenized complexes.

The formation of finely dispersed metal species, isolated atoms, and clusters within a polymer matrix (Sect. 2.4) has been observed in relation to the very high local concentration of functional groups supplied by the macromolecular ligand, which can lead to more extensive displacement of the low molecular weight ligands present in the metal precursor. The resulting "naked" metal atoms can then be coordinatively embedded and stabilized by the polymer as isolated metal atoms or aggregates depending on the chemical species involved [207].

The polymer effect on catalysis by macromolecules–transition metal complexes was discussed in a previous paper [208] pointing out that the macromolecular ligand can affect the structure of the metal complex giving rise to very special situations with only few cases where identical structures exist between monomeric homogeneous and polymeric-supported species.

The effect on catalytic properties is examined in three distinct areas taking into account the previously mentioned concepts. In the first area examples are considered where the transition metal atom maintains in the macromolecular complex with an organic polymer a mononuclear structure either if the later is substantially the same structure as in the monomeric analog, or if the presence along the organic polymer macromolecule backbone of many binding groups gives a different type of structure with regard to that with low molecular weight ligands containing one metal-binding site only.

These aspects are reconsidered by shortly summarizing well-established results and examining in more detail recent data and their fitting in the above frame.

In the second area attention is focused on the wrapping of macromolecules not only around transition metal atoms, but also around metal clusters. Catalysis can be regarded as one of the possible applications of these systems, which can provide a catalytic behavior similar to metal atoms [209].

Finally, a third area deals with transition metal catalysts attached to inorganic macromolecular systems in a broad sence and thus including also silica, alumina [210], and zeolites [211].

It is useful also to mention that a very broad description about the specific characteristics of catalysis by polymer-immobilized complexes was recently published by Pomogailo [212]. Some reference to this last review is made also taking into accunt the very important role of the peculiar structural features of MMC as discussed in Sect. 2.1.

4.4.2 Organic Polymer-Supported Catalysts Based on Mononuclear Transition Metal Complexes

4.4.2.1 Same Structure as in Monomeric Analog

When the structure of the polymer-metal complex is well defined and resembles that of a monomeric analog, a lower catalytic activity is generally observed because of the heterogeneous character, which reduces the number of active centers available to the substrate. Moreover, steric effects become more important, providing an additional decrease in the reaction rate, particularly with bulky reagents. The observation has been made on several occasions, and a very typical example is offered by $Ru(C_3H_5)X(CO)_2(PPDPS)$, where PPDPS indicates a phosphenated unit of linear polystyrene samples with different molecular weights and tacticity. The complexes thus obtained were used for the isomerization of 1-butene to a mixture of Z- and E-2-butene in a liquid medium (toluene) capable of dissolving the monomeric complex with triphenyl phosphine, but onyl able to swell the various polymeric catalysts. It was observed that the reaction rate decreased for various phosphenated ligands in the order PPh_3 > low molecular weight atactic PPDPS > high molecular weight atatctic PPDPS > high molecular weight isotactic PPDPS (where PPDPS = poly-p-diphenylphosphinostyrene). Such results clearly indicate that the better swelling of the polymeric ligand by the liquid reaction medium increases the portion of available reaction sites, thus helping the polymer-attached catalyst to approach the activity of the hemogeneous system where all Ru atoms act as catalytic sites [213].

The above situation is, of course, observed when the specific polymer ligand/metal complex reaction gives the same metal environoemnt as the monomeric system, with the polymer acting as a chemically inert support. Indeed, other factors due to the macromolecular structure can be involved, and the catalytic process is consequently subject to several polymer effects [208].

Polymer-attached mononuclear Rhodium complexes can act as catalysts for several reaction such as isomerization [214] hydrogenation [215−217], synthesis of pyridine [216], and hydroformylation and decomposition of hydrogen peroxide [217].

The anchoring of $RhCl(PPH_3)_3$ to a polymeric phosphine such as diphenylphosphinated styrene/divinylbenzene copolymers with different cross-linking degree (1−4%) improved selectivity in the hydrogenation of linear α-olefins with regard to cycic olefins [218], thus showing the occurrence of shape selectivity arising from the porous nature of the polymer support.

Also, in polycarboxylatotriphenylphosphine rhodium complexes [219], even if the local metal complex structure resulted very similar to the monomeric analog, a larger number was observed of coordinatively unsaturated metal species that were longer-living due to the rigidity of the polymer ligand, thus giving an improvement of the catalytic activity toward the isomerization and hydrogenation of olefins. These polymeric systems had the same sensitivity as the homogeneous analogs to the partial replacement of PPH_3 by CO ligand, thus confirming the same chemical mechanism [219].

Cyclopentadienylrhodium complexes attached to various resins showed activity in the synthesis of pyridines from alkynes and nitriles dependent on the nature of the ancillary ligand and the cross-link density of the polymer. However, activity and chemioselectivity were lower than for the homogeneous reactions catalyzed by the monomeric analogs [216].

Rhodium (I) complexes anchored to divinylbenzene cross-linked phosphinated polystyrene were revisited as catalyst for 1-heptene hydroformylation. Inactivation of the catalyst after several recycles was explained on the basis of rhodium losses and some reorganization of the phosphine ligands with subsequent oxidation of Rh(I) species [220].

Additional physical effects arise from the dependence of chain mobility of the polymer support on temperature [221]. This has been observed for a rhodium complex anchored to three types of phosphineated polystyrenes that were cross-linked with divinylbenzene or ethylene dimethacrylate. Glass transitions were separately observed for the flexible polymer supports and their polymer-anchored rhodium complex catalysts. Discontinuous drops in the Arrhenius plot of the ethylene hydrogenation activity were observed for these flexible polymer catalysts, whereas for the rhodium complex anchored to a highly cross-linked rigid polymer, neither a glass transition nor such a discontinuity in the catalytic activity could be detected. This reversible activity drop is discussed in terms of changes in the stability and the coordination state of the ethylene complex influenced by the micro-Brownian motions of the polymer network.

The "negative" physial effect of the resin on catalytic activity has also been observed in the butene isomerization by triosmium clusters anchored to resin carrying primary alcohol groups [222]. The structure of coordinated metal species was shown by IR analysis to be substantially identical to the corresponding monomeric analog, the latter being, however, much more active.

In addition to the previously mentioned microscopic effects connected with the physical behaviour of the support, other effects may arise at the molecular level particularly when the polymer contains funtional groups not involved in the metal binding and capable of exerting a synergistic effect in the catalytic process, for instance by specific interactions with the substrate, thus providing a high local concentration of reagents or by controlling deactivation reactions. In this connection some examples still deserve consideration as the three kinds of supported catalysts have been prepared by reacting $Ru(O_2CCF_3)(CO)(PPh_3)_3$ with polystyrene–divinylbenzene resins containing diphenylphosphine, carboxylic, or both functional groups, respectively [223]. The IR analysis of characteristic vibrational bands of carboxylate and carbonyl groups indicated a structure of the ruthenium complex after fixation similar to that of the starting complex. These systems have been used as catalysts for the dehydrogenation reaction of alcohols, which is known to be activated by the presence of free carboxylic acids [224]. The resin-attached complexes generally show a reduction of activity when compared with the homogeneous catalyst as expected by considering diffusion limitations. In addition, effects of the structure of the polymeric ligands are also observed, the phosphine-containing supports giving higher activity than those containing only carboxylic groups.

Ruthenium (II) complexes supported on polycarboxylate matrices were used as catalysts for the hydrogenation of aldehydes with alcohols as the hydrogen source. In this case the excess of free carboxylic groups present in heterogeneous systems obtained by reacting $RuH_2(PPh_3)_4$ with preformed carboxylic acid resins or atactic poly(acrylic acid) promotes side reactions involving the aldehyde substrate. This drawback can be only partially overcome by increaesing the metal loading of the heterogeneous catalysts. Catalytic activities and selectivities comparable to those of homogeneous carboxylate ruthenium complexes can be achieved only by following a completely different route for the preparation of metal-polymer systems based on the co- or terpolymerization of bis(acrylato)bis(triphenylphosphine)Ru(II) monomers with methyl methacrylate and 1,4-butanediol dimethacrylate [225].

A similar structure and catalytic behavior as the monomer analog can generally be achieved with chelate ligands attached to the polymeric support [208]. This idea has been pursued more recently for polymer-supported ruthenium(trimethylendiamine). The polymer ligands were prepared starting with 3 and 10 % divinylbenzene–styrene copolymer porous beads by sequential attachment of the bifunctional ligand followed by tratment of the metal salt with the functionalized support [226].

A similar approach has been followed for palladium (II) complexes using the same functionalized polymeric supports. The hydrogenation of cyclohexene was taken as the model reaction to test the catalyst. The influence of temperature, concentration, substrate, and quantitiy of the solvent on the catalytic behavior was reported [227]. Experiments at different temperatures allowed the evaluation of the activation energy, which resulted lower for the polymeric catalyst with regard to the monomeric model.

Also, in case of heterogenized nickel catalysts, the chemical approach based on the anchoring to polystyrene resins through the phenyl groups of triphenylphosphine ligands was replaced by the anchoring through a carbon atom of the chelate ring formed by a bidentate P/O ligand [228] in order to end up with better-defined, less sterically constrained and congested metal centers. The anchoring to polystyrene was carried out through a phenyl ring of the chelate ligand or a polymethylene spacing group between the resin and the chelate ligand (Scheme 4.21).

Microporous and macroporous resins containing 0.01–0.2 mmol of chelating groups per gram were used with Ni(COD)$_2$-added PPh$_3$ in toluene. The supported nickel catalysts gave high molecular weight polyethylene under conditions in which the model homogeneous complex gives oligomers. The macromolecular texture of the resins strongly affects the activity of the anchored catalysts, thus suggesting that physical effects are responsible for the conversion of the homogeneous oligomerization catalyst into a heterogeneous polymerization catalyst. This consideration and the synthetic procedure suggest that in the macromolecular matrix the metal species have the same structure as the monomeric analogs [229].

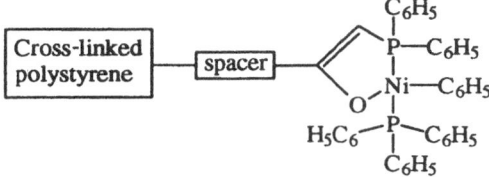

Scheme 4.21. Ni-complexes anchored to the polymer matrix through the =CH– group of the ylidic ligand [229]

In any case it is interesting to emphasize that the greatly different catalytic behavior is an evident indication that the transition metal complex is bound to the polymeric matrix. This bond has to be strong enough to be unaffected during the catalyzed reaction process.

Nickel (0) phosphine complexes immobilized on phosphenated poly(styrene/divinylbenzene) support show a different behavior. Indeed, the complex prepared by phosphine exchange between tetrakis(trisphenylphosphine)nickel and diphenylphosphenated cross-linked polystyrene gives propadiene oligomerization to the same oligomer mixture as in homogeneus reaction. Evidence against catalysis by free Ni(0) in solution in the absence of free triphenylphosphine was obtained by an oligomer selectivity pointing to the coordination of phosphine ligand during the reaction [230].

In an extended and recent study on polymer-supported Wacker-type catalysts, Sherrington and Tang [231, 232] have prepared several different catalysts employing ligands with polybenzoimidazole backbones, polystyrene skeleton, and polyacrylonitrile backbone (Scheme 4.22). As summarized by the authors these complexes with $CuCl_2$ cocatalysts oxdize dec-1-ene primarily to the methylketone under normal Wacker oxidation conditions and in some instances are more active than the $(CH_3CN)_2PdCl_2$ monomeric model. The most active species are obtained with a highly rigid cyanomethylated polybenzoimidazole. Some of these metal centers may be coordinatively unsaturated, however, the catalyst displays remarkable thermo-oxidative stability up to ~ 400 °C. Moreover, the polymer-supported species remain very active up to high temperature (~ 120 °C) and require addition of no hydrochloric acid to avoid irreversible precipitation of Pd(O) species in a complete contrast to homogeneous $PdCl_2$. Pd(O) complexes immobilized on the polymer seem to be "site-isolated" and unable to aggregate, thus allowing easy reoxidation.

The polymer-supported species can be recycled several times with an initial fall in activity after three cycles, and then remains essentially constant.

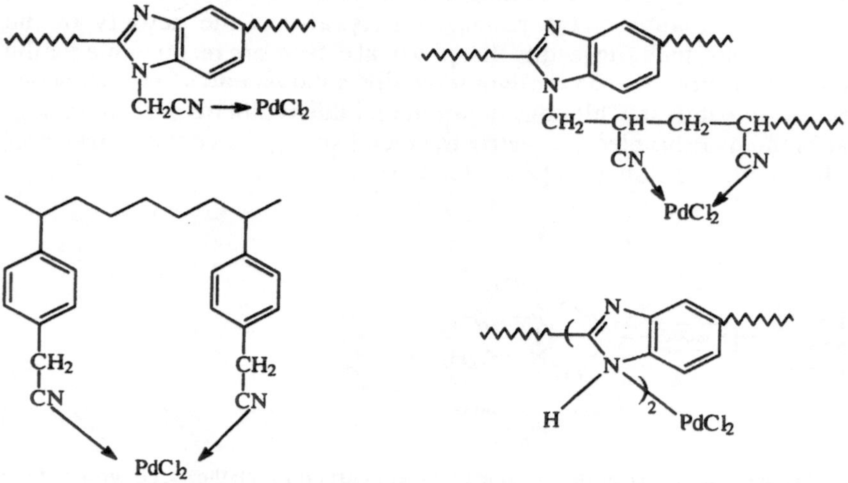

Scheme 4.22. Polymer-supported $PdCl_2$ Wacker Catalysts [231]

Pd leaching also occurs in the first reaction, but is rapidly arrested, and after six cycles Pd loss is only ~ 1 ppm per cycle.

No additional Cu(II) needs to be added following the use of $CuCl_2$ in the first cycle, and sufficient cocatalyst appears to be carried through with the isolated polymer-supproted Pd(II) species. Alk-1-enes isomerization to the more thermodynamically stable internal alkenes is very much faster than oxidation, but the major product in all cases is the 2-one. The latter almost certainly arises from rapid oxidation of a small stationary concentration of alk-1-ene, with shift of the alkene equilibria maintaining the latter. Direct oxidation of the higher alkenes to higher ketones occurs more slowly, but this is significant. These data show that the supportation of $PdCl_2$ on polymeric ligands provides several advantages, even if the structure of the polymer-supported metal complex resembles that of the homogeneous Pd(II) complex [231,232].

New polymer-supported rhodium hydroformylation catalysts have been obtained by van Leeuwen et al. [233] with ligands prepared by attachment of phosphites to styrene copolymers. The influence of chain loading on the activity and complex formation of three types of copolymer-bound rhodium hydroformylation catalysts in comparison with their low molecular weight analogs has been studied. The polystyrene-bound system with the most bulky phosphite shows catalytic activity identical to that of the low molecular weight analog. Only one phosphite is coordinated to the rhodium complex in its active form, whereas the equilibrium between this complex and an inactive complex without phosphite ligands prohibits its use in continuous-flow reactors.

When a perfectly random copolymer of styrene and less bulky 3,3',5,5'-tetraterbutylbiphenyl-2,2'-2,2'-diyl p-vinylphenylphosphite is used as polymer support, the chain loading has a large influence on the complex formation of the catalyst. Thus, whereas high chain loading gives moderately active bis-phosphite complexes, low chain loading gives active, easily accessible mono-phosphite complexes. The activity of the copolymer-bound catalyst toward the hydroformylation of cyclooocetene is as high as that of the low molecular weight analog with styrene: however, this polymer-supported system yields a slower catalyst than the low molecular weight analog.

Silica-grafted, polymer-bound, phosphite-modified rhodium complexes can be used in continuous-flow reactors yielding constant conversion over a period of at least 10 days during the hydroformylation of styrene at moderate pressure (P, $CO/H_2 = 3$ Mpa) and temperature (T = 100 °C). Still, in the hydroformylation of olefins, Pan and Zong [234] reported selectivity to linear aldehydes up to 95 % by catalysts containing Pt–Sn complexes with a number of linear and cross-linked polymer ligands containing P, N, or S as coordiantive atoms.

A considerable amount of work has been performed also with some polymer-metal complexes used as catalyst in oxidation processes. Because these systems are treated in more details in other chapters, they are only shortly mentioned here for the sake of completeness. Indeed, these systems are more relevant in electron transfer processes (Sec. 4.3 and 4.5) and living systems (Chap. 3).

Rapid and reversible binding of molecular oxygen was described for polymer-coordianted cobalt–Shiff base complexes [235] as well as for poly(1-

vinyl-2-pyrrolidone)- and dextran-bound protohoeme mono(-N-imidazol-derivatives [236]. These concepts were recently developed to a supported catalyst with a better turnover than any of the homogeneous systems described up to now. Such a catalyst consisting of a poly(pyrrole manganese porphyrin) film electrode was used in electro-assisted oxidation of *cis*-cyclooctene and 2,6-*di*-*t*-butylphenol with molecular oxygen [237]. In the same line are the catalysts derived from macromolecular systems in which metal ions, Mn(III), Cr(III), and Fe(III), are anchored to a polymeric Schiff base backbone. These systems are active toward epoxidation of alkenes, and their activity is enhanced by addition of aromatic amines. These last bases coordinate to the metal at an axial position and facilitate formation of the metal oxo complex [238].

Further examples were also added to the previously widely explored [204, 239] area of macromolecular copper complexes used in oxidation processes. Indeed, improvement of catalytic properties and reusability were reported in oxidation reactions catalyzed by macromolecular copper systems where the macromolecular ligand was designed in order to facilitate the fundamental steps of the catalytic process.

These examples concern the oxidation of hydroquinone in the presence of poly(4-vinylpyridine-co-N-vinylpyrrolidone) [240], and oxidation of benzyl alcohol with Cu(II) mediated by polystyrene with nitroxyl radicals covalently attached to the side chains [241].

Several copper–amine–cellulose complexes were used for the controlled and selective autooxidation of a variety of organic substrates [242].

Various polymer-bonded transition metal complexes for oxidative and electron transfer processes have been studied by Tsuchida [243]. However, these examples are reported in detail in the other section of Chap. 4.

To conclude this area of discussion, mention is made of recent study concerning the catalytic activity in the oxidation of sulfoxides to sulfones by *t*-butyl hydroperoxide exerted by vanadyl complexes anchored to polystyrene resins with polidentate ligands [244]. The latter were obtained by attaching salicylaldoxime, 3-2(thenoyl-1,1,1-trifluoroacetone or *o*-phenylene-bis-salicylimine to cross-linked polystyrene. The polymeric complexes show comparable or lower activity with regard to the homogeneous monomeric analogs, but can be reused at least three times with practically the same activity [244].

4.4.2.2 Different Structure from Monomeric Analogs

As previously mentioned, the use of macromolecular ligands provides, with regard to monomeric analogs, a more complex situation that usually affects catalytic behavior. Indeed, the facilitated formation of coordiantively unsaturated species (typical of heterogeneous catalysts), steric effects arising from the macromolecular nature of the ligand, stabilization of reaction intermediates, and formation of unexpected metal environments can give rise to higher reactivity, different selectivity, longer life of active species and new catalytic reactions.

Coordinatively unsaturated metal species were proposed in 1972 to explain the higher activity observed in some cases for transition metal complexes attached to phosphenated polystyrene resins [245]. Similar explanations have been put forward for complexes obtained from $RuH_2(CO)(PPh3)_3$ and polymer with carboxylic groups. Indeed, these complexes show increasing catalytic activity by decreasing the P/Ru ratio to below 2, which is expected on the basis of the stoichiometry of monomeric models [219].

One or the other of these polymer effects have been observed in the example described before, and this makes if difficult to really estabish if the active species in the polymer-supported catalyst have the same nature as in monomeric analogs.

A typical case deserving mention concerns with the imprinting of active-site structure thrugh the design of the polymer support. Efendiev [246] had described in detail a new approach to preparation, regulation of properties, and formation of three-dimensional structure of complexing polymer sorbents and metal-polymer complex catalysts where specially designed structure of active centers has been employed. The approach is based on the use of "memory" of polymer compositon and consists of rearrangement of macromolecules of initial complexing polymer or metal-polymer complex catalyst to sorbing metal or hydrocarbon substrate followed by fixation of optimum conformation for particular substrate by intermolecular cross-linking. A number of cross-linked complexing polymer sorbents and metal-polymer complex catalysts containing phosphorylic, carboxylic, pyridine, amine, and imine functional groups have been prepared. It has been shown that the rearrangement leads to essential improvement of basic sorption and catalytic properties of the cross-linked complexing polymer. Indeed enhanced sorption capacity, rate of uptake, selectivity, and catalytic activity were observed for the reactions of oxidation of toluene to carboxylic acid or anhydride, and of ethylbenzene to hydroperoxide.

From the point of view of the catalysis, one can say that generally the MMC show a different behaviour connected to the previously discussed "physical effects" and in some cases to different structural features at molecular levels. The latter can particularly be responsible not only for varied catalytic activity and selectivity, but also for a different catalytic process. In this last connection clusters and inorganic supports offer more evident examples as discussed in the next section.

4.4.3 Macromolecular Systems Involving Transition Metal Clusters

Metal atoms deposited into polymers can be entrained in the form of organometallic complexes, clusters, and colloids [247] (see also Sect. 2.4). In these systems no ancillary low molecular weight ligands are present around metal species, which therefore can interact only with the macromolecule and/or other metal atoms. Applications of these macromolecule/transition metal systems are anticipated in the areas of redox processes, bioseparations, magnetic and magnetiooptic processes, nonlinear optics, and catalysis [247].

Consistently with the topic of this presentation we discuss here some examples of catalytic behavior of transiton metal clusters wrapped into macromolecules.

The solid product derived from the reaction of Ruthenium(0)(cyclooctadiene) (cyclooctatriene) [Ru(COD)(COT)] with polystyrene under hydrogen resulted as consisting only of ruthenium and polystyrene (Scheme 4.23). Furthermore, it showed unique catalytic properties for the hydrogenation of aromatics and other functional groups [248].

The structural analysis of this system allowed determinations that several ruthenium atoms are each bonded to a single phenyl ring of the polystyrene side chains (Scheme 4.24). Moreover, other ruthenium atoms could be detected in the second shell of the coordination sphere with Ru–Ru distance close to that observed in small, well-defined ruthenium clusters. The combination of these data with additional instrumental and chemical evidence led to the conclusion that the polystyrene/ruthenium catalyst consisted of polystyrene chains bridged by small Ru-clusters (Scheme 4.23).

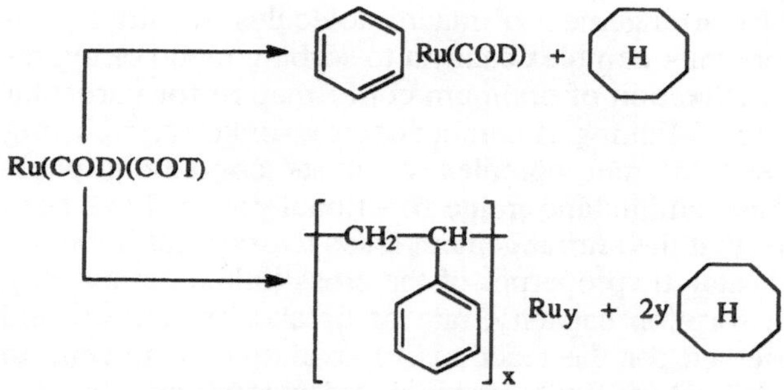

Scheme 4.23. Different reactions of Ru(COD)(COT) with benzene (upper reaction) or polystyrene (lower reaction) [248]

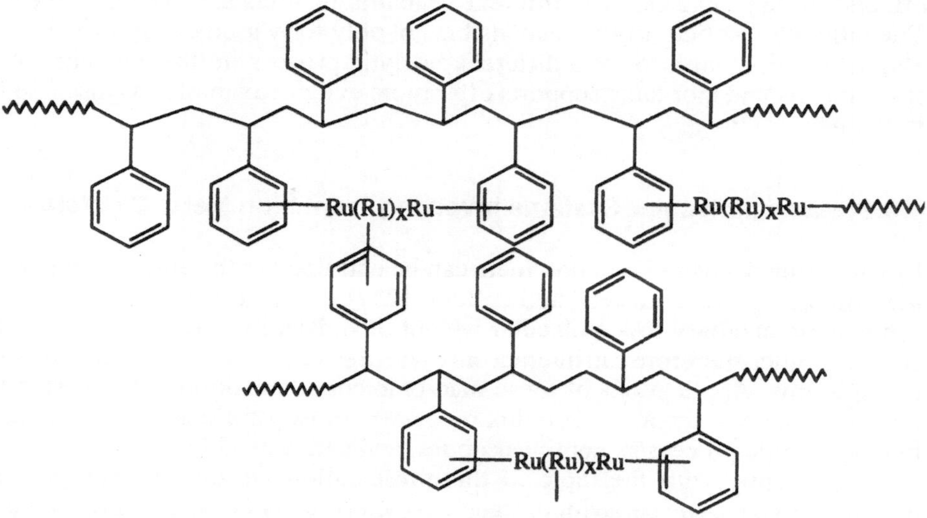

Scheme 4.24. Proposed structure of polystyrene-wrapped Ru clusters [249]

One the basis of these results is has been shown that the reaction of a transition metal complex bearing easily displaced low molecular weight ligands, with the ligand containing at least two aromatic rings in the proper relative positions, such as the polymer of an aromatic vinyl monomer (styrene or 1-vinyl-naphthalene), can provide finely dispersed metal particles or small clusters complexed to the polymer matrix, thus stabilizing the substantially "naked" metal atom clusters [250]. This process is shown in Scheme 4.25, where the formation of polymer-ligand-attached clusters in described as compared with the cluster formation without the polymer.

Also, the dependence of catalytic properties on the nature of the aromatic polymer ligand, metal loading, and substrate, as well as the reaction stereochemistry, can be better explained by taking into account substrate activation by coordination and steric effects at the molecular level, as observed for homogeneous Ru and Rh complexes active for arene hydrogenation. It is of interest to note that these polystyrene or poly-1-vinylnaphtalene-supported ruthenium clusters show extraordinary catalytic activity in hydrogenating aromatic compounds and other functional groups (-C=O, $-NO_2$, -CN, and $-NO_2$) in addition to the olefinic double bond, which is the only active substrate for the corresponding homogeneous catalyst $Ru(C_6H_6)(COD)$ (Scheme 4.26) [248, 249]. Moreover, the metal species remain bonded to the aromatic polymer during the entire reaction time, and no loss of activity is observed upon reuse. This stability is connected to the lack of hydrogenation of the aromatic side chains of the macromolecular ligand. Indeed, even linear polystyrene is not hydrogenated under the reaction conditions, thus indicating the substantially heterogeneous character of the catalyst [251].

Palladium cluster catalysts supported on chelate resin containing iminodiacetic acid moieties have been described by Toshima et al. [252]. Increase of surface area of the chelate resin can be obtained by washing with an organic solvent miscible with water and by complexing with multivalent cations. The supporting process of the palladium cluster on chelate-resin metal complexes is made by two methods in which the order is reversed between "complexing of metal ions" and "supporting of palladium clusters." The final supported Pd clusters catalyze the hydrogenation of olefinic double bonds, the catalytic activity being markedly affected by the metal ions used in the complexation and the preparation method.

The catalytic synergism between the two metallic centers observed in the system with monomeric ligands points to the future interest concerning MMC containing heterobimetallic sites [253]. The synthesis of the latter can be approached by sequential or contemporary evaporation of two metals on the macromolecular ligand.

In our opinion, polymer-wrapped clusters can overcome some of the problems encountred with mononuclear MMC, and can provide new composite materials with appealing catalytic features, even if clearly the role of ligands and complex structure can be partially lost.

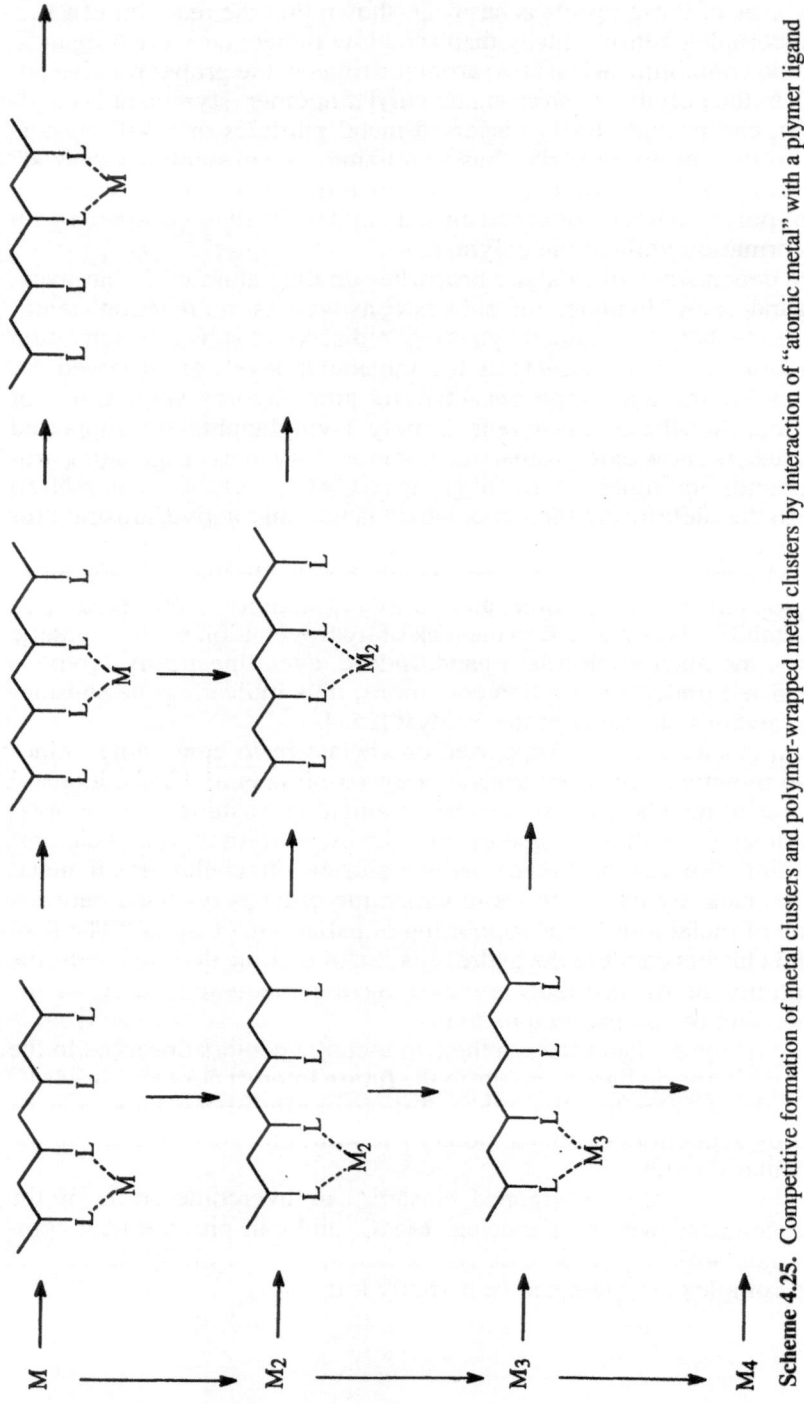

Scheme 4.25. Competitive formation of metal clusters and polymer-wrapped metal clusters by interaction of "atomic metal" with a plymer ligand

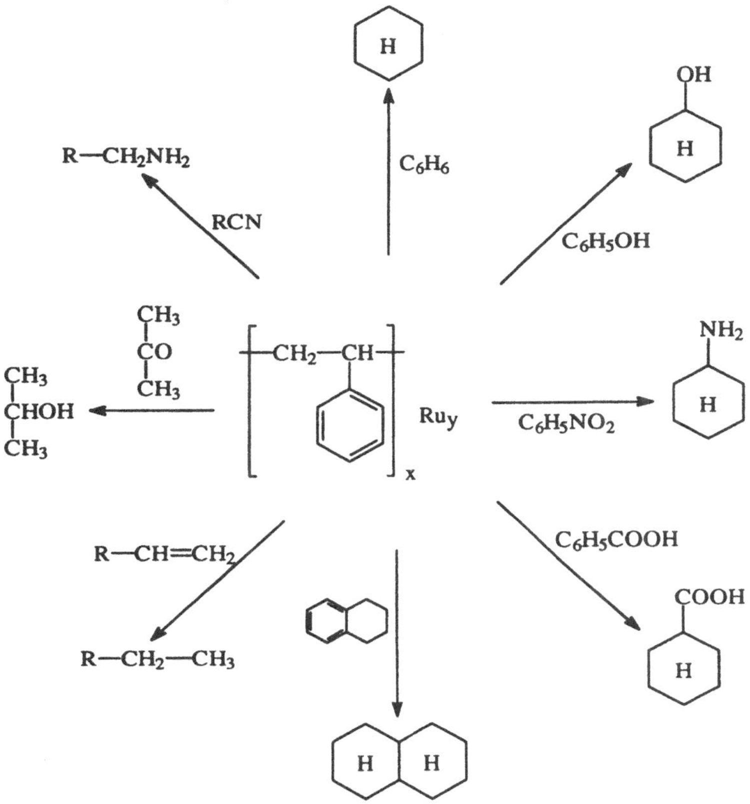

Scheme 4.26. Versatile catalytic hydrogenation by polystyrene-wrapped Ru Clusters [248, 249]

4.4.4 Transition Metal Complexes on Inorganic Supports

Even if inorganic supports were used well before organic polymers for preparing heterogeneous catalysts, the possibility of anchoring to an inorganic matrix a transition metal complex is a relatively recent area of research. Indeed, the structure of several inorganic materials resembles in some way that of a regular three-dimensional polymer network. Thus, the surface of alumina, silica, and aluminosilicates are characterized by the presence of a chain of atoms connected by substantially covalent bonds providing a structure similar to that of organic macromolecules. These surfaces contain functional groups, such as Lewis acid, hydroxyl, or hydroxylate groups, which can be used for more or less controlled binding of transition metal complexes [254, 255]. Otherwise, these groups can be chemically modified and converted into different groups or ligands that can provide a more definite binding of the metal species.

The low concentration of functional groups (e. g, isolated hydroxyl groups) normally gives the formation of mononuclear species of the kind discussed in Sect. 4.4.2.1. On the other side, the binding to the surface of multidentate ligand moieties provides an even more controlled route to structurally defined metal complexes attached to inorganic polymeric systems.

Although many examples are reported in the literature using the inorganic supports, only a limited part is of interest for the present book, because in many cases not enough attention was placed on the concept of the matrix metal complex structure and its relevance to catalytic reactions. Thus, several systems have been described concerning Pd, Pt, Ni, and Rh supported on aluminum ortophosphate as interesting hydrogenation catalysts, but the structure of the complex has not been studied in detail [256].

A more systematic study concerns the use of pumice as support of metal catalysts. In these systems pumice behaves as an alumino-silicate naturally doped with alkali-metal ions, and allows high catalytic activity also at very low metal loading [257]. The preparation of metal catalysts is carried out starting with diallyl metal complexes and successive reduction (Scheme 4.27) [258] with dihydrogen of the anchored metallorganic substrate where M was palladium, platinum, or both.

The obtained palladium, platinum, and Pd–Pt alloys supported on pumice have been used in the selective hydrogenation of alkadienes and alkynes in the liquid phase. The dispersion of the metallic phase, which is also a function of metal loading, can be controlled by the reaction temperature during the attachment of the metal derivative and the successive reduction with dihydrogen. Aanlysis by XPS and Auger spectroscopy shows a negative shift of the representative peaks of both Pd and Pt with regard to the corresponding peaks of the pure metal. Whereas this indicates an electron density transfer from the support to the metal, it also demonstrates the active role of the support in the production of high-activity catalyst. This catalytic behavior of the pumice-supported metal catalysts is attributed to the presence, on the pumice surface, of alkali-metal ions, which, by suppling electron density to the metal (very likely through electron-donating metal-alkali ion oxygen composites), counterbalances the loss of metallic character usually found in supported metal catalysts at high metal dispersion [257].

The use of inorganic supports, such as alumina, silica, and zeolites, has been extended to supporting Ti, Zr, and Ni, and these systems have been used as heterogenized catalysts active in the polymerization and oligomerization of monoalkenes.

A real corner stone of this topic has been placed by Marks with his recent review concerning metalhydride alkyls bound to inorganic surfaces [259]. Both partially dehydroxylated alumina (PAD) and fully dehydroxylated

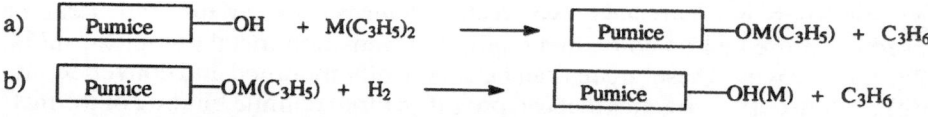

Scheme 4.27. Preparation of pumice-supported metal catalyst (M = Pd or Pt or Pd/Pt) [257]

alumina (DA) (Scheme 4.28) were used as inorganic macromolecular ligands. Interactions of these ligands with $Cp_2Zr(CH_3)_2$ is accompanied by CH_4 evolution in compartively lower amounts with regard to the analogous Ti complex, probably because of the lower polar character of the Zr–alkyl bond [260].

Reduction of $Cp_2Zr(^{13}CH_3)_2$/PDA to Zr(III) resulted to be insignificant. The CPMAS ^{13}C-NMR spectra $Cp_2Zr(^{13}CH_3)_2$/PDA indicated the presence of both μ-oxo (A) and cationic (B) species (Scheme 4.29). The formation of the cationic species (zirconocene alkyl cations) explained the activity observed with these systems as hererogenized catalysts for ethylene polymerization [259].

Collins et al. [261], supported ethylenebis(η-5-indenyl) zirconium dichloride (Et[Ind]$_2$ZrCl$_2$) and ethylenebis(η-5-tetrahydroindenyl)-zirconium dichloride (Et[IndH$_4$]$_2$ZrCl$_2$) on fully hydroxylated, partially dehydroxylated, and dehydroxylated SiO$_2$ and Al$_2$O$_3$. Also, PDA and PDS pretreated with AlMe$_3$ prior to the addition of the metallocene complex were investigated, as this reaction converts the free-OH groups on the solid support into dialkylaluminato species (Scheme 4.30)

The amount of metallocene that can be attached to the surface of these supports is much lower than expected on the basis of simple physical absorption. This seems to indicate that attachment takes place at specific sites of the surface. In any case the amount of hydroxyl groups seems to play a determining role as shown by structural analysis and catalytic activity of olefin polymerization of the resulting heterogenized complexes.

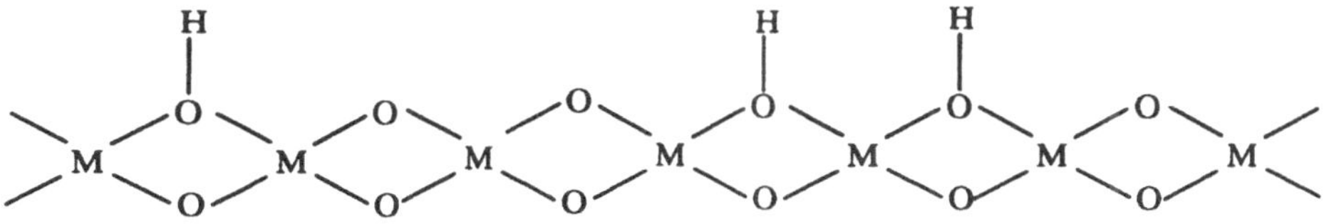

Partially dehydroxylated alumina (PDA, M = Al) or Silica (PSA, M = Si)

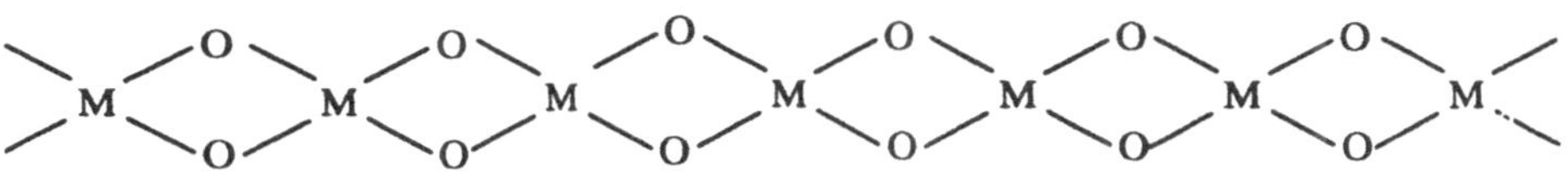

Scheme 4.28. Fully dehydroxylated alumina (DA, M = Al) or silica (SA, M = Si)

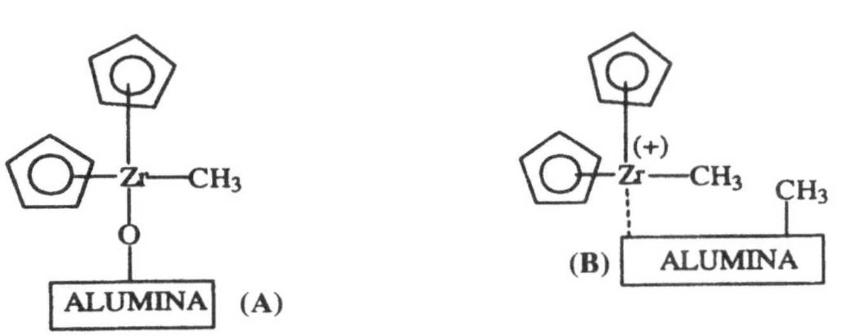

Scheme 4.29. μ-Oxo (A) and cationic (B) zirconocene species attached to alumina

Indeed, the best catalytic activity was obtained in all cases by activation with methylalumoxane (MAO) of the zirconocene supported on partially dehydroxylated silica or alumina pretreated with $AlMe_3$. When the surface concentration of hydroxyl groups is large, the main reaction seems to be the removal of the metallocene ligand, which gives $ZrCl_2$ μ-oxo species attached to the surface. It is reasonable to suppose that when the support has been pretreated with $AlMe_3$, the formation of $-OAlMe_2$ groups on the surface allows the fixation of the unaltered metallocene. This conclusion is in agreement with the observation that an isotactic polymer is produced and its microstructure is very similar to that obtained with the corresponding homogeneous catalyst. Also, molecular weight distribution is much narrower than that normally obtained with a heterogeneous catalyst.

In substantial agreement with these data, Soga and Kaminaka obtained highly active catalysts by supporting zirconocenes on SiO_2 pretreated with a small amount of MAO [262]. These authors propose the formation of cationic species on the inorganic surface in a very analogous way as proposed in solution (Scheme 4.31).

Scheme 4.31 was derived from that reported by Chien and He [263], who, apart from the different nature of the zirconocene complex, proposed the structure shown in Scheme 4.32. In both cases one cannot exclude that the primary reaction of the surface hydroxyl group occurs with the free $AlMe_3$ present in the MAO. In this case the inorganic support surface would be modified as suggested by Collins et al. [261].

The data herein are representative of a large number of papers dealing with supportability of metallocenes of transition elements of the IV group with the aim of obtaining heterogeneous catalysts with active sites of one type only.

As shown here the difficulty in determining the molecular structure of these complexes is due mainly to their insolubility and the limited concentration of metal complex that involves only the inorganic surface. Also, the macromolecular nature of the matrix is certainly responsible for the complexity of

Scheme 4.30. Reaction of surface hydroxyl groups with aluminiumtrimethyl

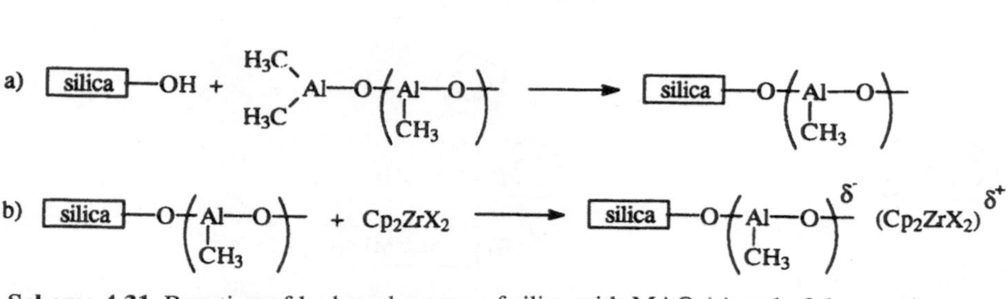

.**Scheme 4.31** Reaction of hydroxyl groups of silica with MAO (a) and of the reaction product with zirconocene (b) [262]

the structure with the possibility that the transition metal complex with different number of functional groups on the surface. Clearly, the use of ligands attached to the surface through a spacer, would help in obtaining more-defined complexes between the inorganic matrix and metal derivative. In this case, however, the matrix would act simply as a heterogenizing support, thus loosening its possible chemical effect on the catalytic process and on the properties of the final material.

Taking into account these considerations, Ciardelli et al. [264] performed a study concerning the supportability of soluble titanium and zirconium complexes on amorphous silica and crystalline YH-zeolite. In order to attempt coordination to the solid matrix without loosening the original structure of the complex, a different approach was originally followed. Thus, a titanium complex with a bidentate ligand (dichlorotitanium pirocathecolate) [PcTiCl$_2$] was prepared [265], and this successively attached to silica or zeolite. The presence of vicinal hydroxyl groups on the surface of the support gave, however, irregular binding, due to the formation of one or two bonds between the titanium complex and the inorganic surface (Scheme 4.33).

Additional heterogeneity was observed in some other cases, due to the competition of other ligands with chlorine. For this reason ^{13}CPMAS NMR was used in the hope of going deeper in characterizing the structure of the final complex. Zeolite was selected as support, because it was considered more informative with the possibility of following the evolution of Al and Si resonance after the various chemical-surface reactions in which the support was involved.

The dealuminated zeolite contains both Si–OH groups and extrareticular aluminum. The latter can be removed by exhaustive extraction at 50 °C with a solution of acetylacetone in ethanol. After this treatment the ^{27}Al-NMR

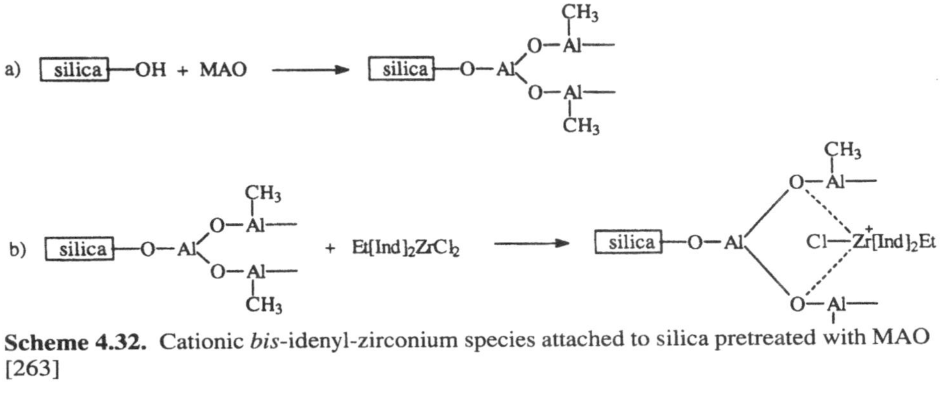

Scheme 4.32. Cationic *bis*-idenyl-zirconium species attached to silica pretreated with MAO [263]

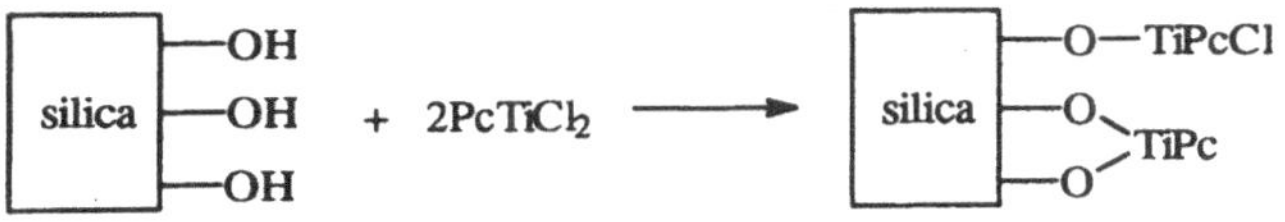

Scheme 4.33. Mono- and Double-bonding of dichlorotitanium pirocathecolate to hydroxyl groups of silica [265]

(MAS) spectrum shows a single resonance at 57.45 ppm of the tetrahedral Al [266], whereas resonances at 0 and 30–50 ppm of the extrareticular Al are completely lacking [267]. Treatment of this support with $Cp_2Zr(CH_3)_2$ is accompanied by CH_4 evolution associated with the Zr–C bond cleavage by Si–OH groups. In order to control the last reaction the zeolite was primarily treated with $Al(CH_3)_3$. The resulting support shows in the ^{27}Al-NMR (MAS) spectrum three resonances at 60.4, 33.1, and 2.2 ppm suggesting that two different species of aluminum, associated with absorbed $Al(CH_3)_3$ and the reaction product of $Al(CH_3)_3$ with Si–OH groups, are now present. The addition of $Cp_2Zr(CH_3)_2$ to the modified zeolite does not produce any CH_4 evolution, and the ^{27}Al-NMR (MAS) spectrum shows the same resonance as before the zirconocene addition, the resulting relative intensities only moderately changes. In the ^{29}Si-NMR (MAS) spectrum only the resonance of the Si(OAl) species at −107 ppm can be observed indicating that no changes are detectable for Si after addition of the zirconocene to the zeolite pretreated with $Al(CH_3)_3$. Certainly more information could be obtained by using ^{13}C-enriched $Cp_2Zr(CH_3)_2$ [268], which would allow completion of the analysis by NMR. However, presently it is reasonable to conlcude that fixation of $Cp_2Zr(CH_3)_2$ on the inorganic support occurs without formation of strong chemical bonds, thus allowing prediction that the structure of the metal complex is maintained. Indeed, when activated with MAO (Al/Zr = 3000) the HY-zeolite-supported $Cp_2Zr(CH_3)_2$ gives polyethylene with the same productivity (590 kg PE/mol Zr.h) as the complex in solution.

Parallel to the previiusly described work concerning Ni complexes attached to polystyrene resins, structuraly defined systems were also prepared with silica and alumina as inorganic supports. As for IV group metals, even for nickel P/O chelate it was observed that activity and selectivity rose when the inorganic support was pretrated with $Al(CH_3)_3$ [269]. Following this research another type of highly active catalyst was prepared attached by specific bonds at the modified silica surface [270]. The preparation route (Scheme 4.34) gave a product having 0.78 Ni/Al ratio while one mole of methane per g-atom of Ni was evolved.

This suggests that the predominant structure of the complex is that depicted in Scheme 4.34, as also confirmed by IR data. Systems containing two transition metals in each complex (Scheme 4.35) were also obtained by replacing $Al(CH_3)_3$ with tetrabenzyltitanium during the pretreatment of the surface.

Silica-bond well-defined nickel complexes can also be prepared by supporting a phosphine on silica. Thus, modification of the inorganic surface with $(C_6H_5)_2P(CH_2)_2$-Si(OEt)$_3$ gave a supported diphenylalkylphosphine, which, by reaction with equimolar amounts of bis(cyclooctadiene)nickel and triphenylphosponium 2-oxo-2-phenylethylide, gave a heterogenized nickel catalyst where the Ni atom is bound to the support through a phosphine ligand spaced by two methylene groups. However, the preparation approach described in Schemes, 4.34 and 4.35 seems to be preferred, because the bonding by the donating P-bond might not hinder substantial metal leaching [271].

The possible and predictable formation of different organometallic species during the supporting reaction on an inorganic matrix has been clearly

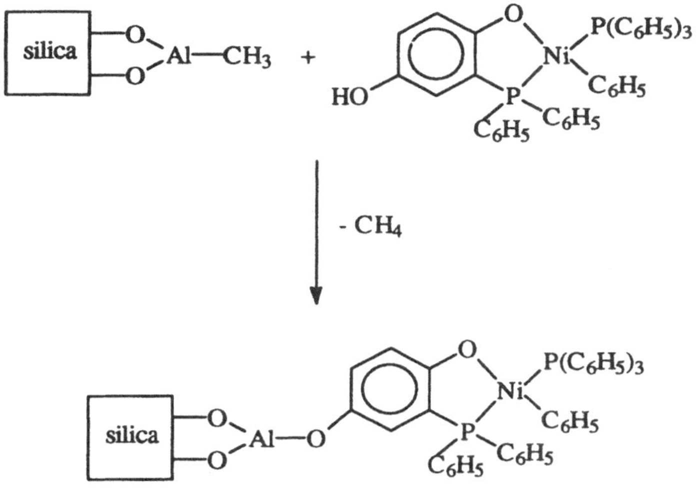

Scheme 4.34. Silica-supported Ni catalyst [270]

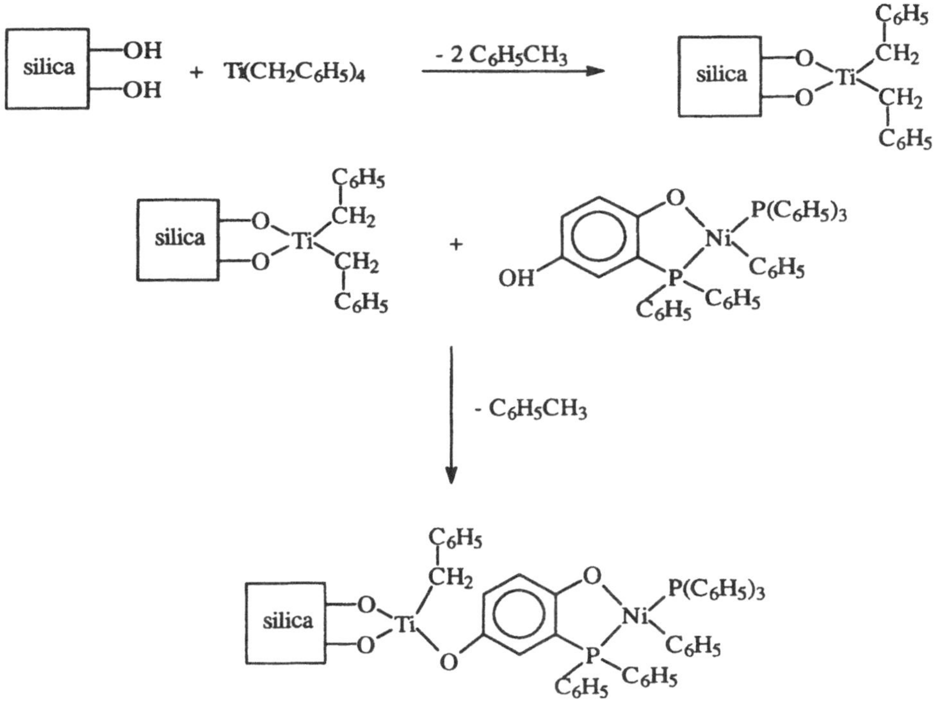

Scheme 4.35. Preparation of silica-supported Ti/Ni catalyst [271]

evidenced in the case of alumina/Cp*TiCl$_3$ (Cp* = pentamethylcyclopenta-dienyl) or alumina/CpTiCl$_3$ systems [272]. The resulting heterogeneous complex when activated with Al(i.C$_4$H$_9$)$_3$ gave both syndiotactic and isotactic polystyrene. This result was attributed to the presence in the alumina of a considerable amount of Lewis acid sites in addition to the surface hydroxyl groups, the reaction with CpTiCl$_3$ giving cationic and μ-oxo species respectively. The former would be responsible for the syndiotactic polymerization, and the latter for the isotactic polymerization.

This result emphasizes the need to use "purified" inorganic supports where only one binding site for the organometallic complex is present.

4.4.5 Conclusion

In this section an attempt was made to describe the present situation and scientific significance of the reseach concerning the catalysts prepared by supporting transition metal derivatives on either organic or inorganic macromolecular compounds. In order to reach this objective, reference was first made to previous examples relevant to establishing some general characteristics of macromolecule–metal catalyst and to show the different roles played by the polymeric supports. These general concepts were successively used to discuss recent examples.

It is our feeling that on these bases it is now possible to select the proper system, macromolecular compound and metal derivative, for a specific pur-

Table 4.7. Examples of chemical reactins catalyzed by mononuclear macromolecule–metal complexes

Metal	Macromolecular matrix	Catalytic reaction	Reference
Ru, Rh	Phosphenated polystyrene	Isomerization of olefins	213, 214
Rh	Phosphenated polystyrene	Hydrogenation of olefins	215–218
Rh	Phosphenated polystyrene	Synthesis of pyridines	216
Rh	Phosphenated polystyrene	Hydroformylation of olefins	217
Rh	Phosphite on styrene copolymers	Hydroformylation of olefins	232
Pt–Sn	Polymer ligands containing P, N, or S	Hydroformylation of olefins	233
Ru–Co	Phosphine bound to silica	Hydroformylation of olefins	233
Ni	Polystyrene with P/O ligand	Polymerization of ethylene	229
Pd	Nitrile and benzoimidazole bound to various polymers	Oxidation of olefins	231
Cu	4-vinylpyridine polymers	Oxdation of phenols	239–241: Sect. 4.5.2
Mn, Fe	Poly(pyrroleporphyrin)	Oxdation of olefins and phenols	234
	Polystyrene with N/O ligand	Oxidation of sulfoxides	244
	Various ligands	Oxidating polymerization of disulfides	Sect. 4.5.3
Ti, Zr	Zeolite	Polymerization of olefins	264
Ni	Phosphenated silica	Polymerization of ethylene	269–271
Co	Silica with bound phthalocquanines	Oxidation of thiols	273

pose related to a certain chemical reaction and process conditions. Indeed, whereas selection of the metal is first of all dependent on the reaction of interest (Table 4.7), the selection of the polymer can be made depending on external parameters such as solvent, temperature, reagents, products, and reactor.

In principle the number of catalysts one can obtain is illimited, due to the number of metal complexes and different supports that can be used. Moreover, in several cases new systems can be obtained, due to the very particular situation provided by the possible high local concentration of ligands as when metal atoms are produced in the presence of the polymeric ligand. Some examples shown in the table, as well as those described in Chap. 3 for living systems and in Sect. 4.1, 4.3, 4.5, and Chap. 5, indicate that future trends of catalysis by MMC will be increasingly in processes involving redox reaction and electron transfer, which are made possible or at least facilitated by the presence of the macromolecular matrix.

4.5 Recent Advances in Oxidative Polymerization Through Multielectron Transfer

K. Yamamoto and E. Tsuchida

4.5.1 Introduction

Electron transfer plays a significant role in many chemical reactions. Molecular conversions based on efficient multielectron transfer facilitated by polynuclear complexes is a topic of considerable recent research. It is particularly important in chemical syntheses because of the possible reduction in the amount of energy that must be supplied.

Multielectron transfer in the photosynthetic reaction center as mediated by a polynuclear manganese–protein complex has recently been elucidated. However, human-made chemial systems for accomplishing similar molecular conversions remain to be realized. The need for such systems is great. For example, if a new chemical system for the formation of polysaccharides could be devised, it would be of relevant chemical and practical significance.

4.5.2 Multielectron Transfer of Metal Complexes and Molecular Conversions

Detailed observations of nature have revealed the existence of systems based on efficient multielectron processes. Studies on the individual multielectron process of biological systems [e.g., the role of oxygen reduction (four-electron process) by multinuclear copper protein in the well-known process of melanin formation in animals or the oxidation of H_2O (four-electron, four-proton process) accompanied by O_2 evolution by a multielectron system based on the polynuclear manganese protein of plants] is in progress. However, the whole photosynthetic system for the formation of sugar has not yet been accomplished.

During the oxidation process of H_2O, the apparent one-step four-electron transfer is mediated by the di- to tetra-nuclear manganese complex [274, 275]. The oxidation potential of H_2O in this system is 1.23 V (vs NHE), which is much less positive than that for the two-electron process (1.77 V) or one-electron process (2.74 V). Such cathodic shift of the oxidation potential is based on the characteristic system of plants where the active species produced in the elemental steps is captured and readily subject to the next process, which results in the efficient O_2 formation by the oxidation of H_2O.

Recently, the corelationship between the dynamics of electron transfer and the structure of redox-active molecules has attracted much oxidation with the goal or aritficially developing such systems based on the control of the elemental processes. This chapter describes the recent approaches to developing a molecular conversion system via the multielectron transfer process.

4.5.2.1 Multielectron Transfer and Polynuclear Complex

The polynuclear complex, which shows a one-step multielectron transfer, is very rare. The one-step electron transfer is not accomplished by merely assembling the metal complexes. The polynuclear complex (Scheme 4.36) is composed of a ferrocenyl groups ($E = 0.55$ V vs nHE) bridged by an acetylene. The cyclic voltammogram of (Scheme 4.36) shows a two-step wave (0.60 V and 0.95 V, Fig. 4.31a). The one-electron oxidation of (Scheme 4.36) [Fe^{II}-Fe^{II}-Fe- = Fe^{II}-Fe^{III}] results in the electrostatic repulsion based on the increase in the charge ($+2$ to $+3$), which shifts the next oxidation potential onto the more positive side [276]. The one-step two-electron transfer is not achieved even by bridging the complex with a conjugated bond.

In the other hand, the iron carbonyl polynuclear complex (Scheme 4.37) shows a two-electron transfer at almost the same potentials [$E_{2/2} - E_{2/3} = \Delta E$ < 0.0356 V (297K)] as shown in Fig. 4.31 b [275]. There are few polynuclear complexes that display such a one-step multielectron transfer [e.g., copper triketonate complex (Scheme 4.38), ruthenium complex (Scheme 4.39), and some metallocene complexes) [277–279]. Such complexes are restricted, because the condition that is essential for the multielectron transfer is still not clear.

The two-electron transfer of the dicopper triketonate complex is also accompanied by a change in the coordination structure. An alkali-metal ion is essential for the one-step two-electron transfer in this case (Fig. 4.32; Cu^I-Cu^I-$Li^+ + 2e = Cu^{II}$-$Cu^{II} + Li^+$) [280]. The alkali-metal ion induces the warp in the structure of the complex, which results in the coordination change from planar to tetrahedral structure. The potential for the two-electron oxidation of Cu^I–Cu^I shifts to the more positive side than in the absence of alkali metal, which enables the two-electron transfer (Fig. 4.33).

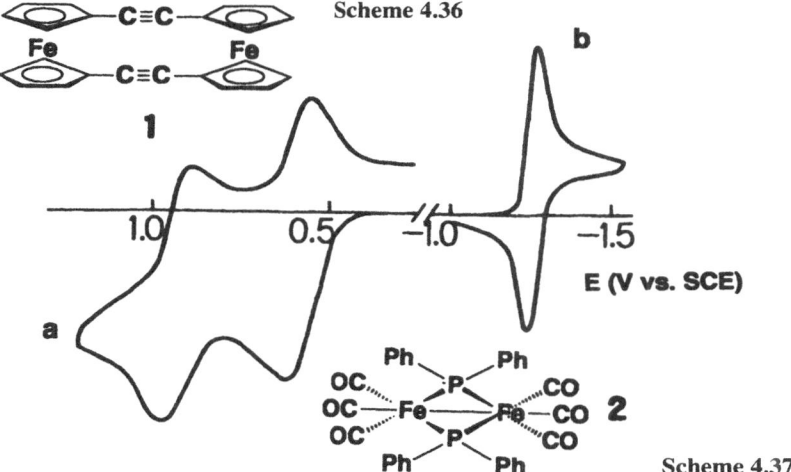

Scheme 4.36

1

2

Scheme 4.37

E (V vs. SCE)

Fig. 4.31. Cyclic voltammograms **a, b** of dinuclear ferroceny complexes (**1, 2**)

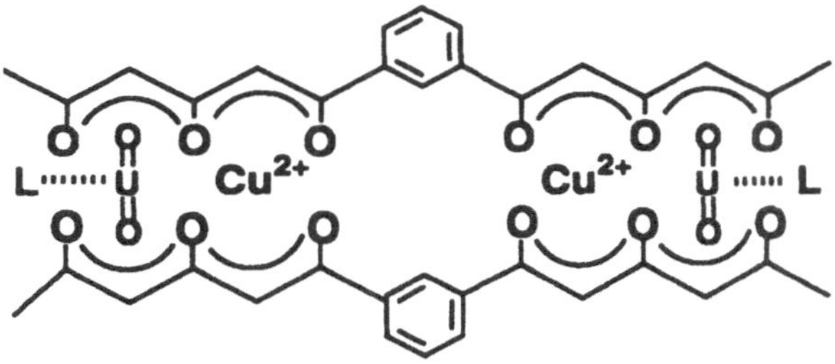

Scheme 4.38

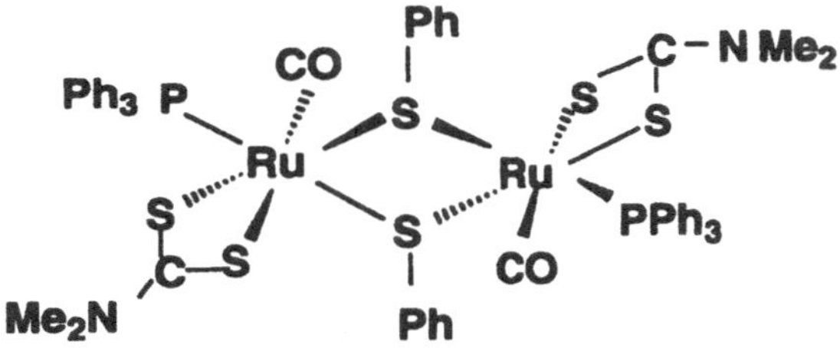

Scheme 4.39

As another example, it has been recently reported [281] that a diruthenium(II) complex involving the tribridged cores (1-MeIm = 1-methylimidazole), $[\{1\text{-MeIm}\}_3\text{Ru}\}_2(\mu\text{-O})(\mu\text{-O}_2\text{CMe})_2](\text{ClO}_4)_2$ (Scheme 4.40) provided the electrochemical evidence for a quasireversible, two-electron reduction process (Fig. 4.34). The redox properties of the complex were studied by cyclic voltammetry. The complex in acetonitrile containing $0.1\,\text{molL}^{-1}$ tetrabutylammonium perchlorate showed three responses near $+1.5$, $+0.4$, and -1.1 V vs SCE. The redox couple appearing at 0.36 V corresponded to the reversible, one-electron oxidation of the diruthenium(III) unit as evidenced from the ΔE_p value of 60 mV, the i_{pc}/i_{pa} ratio of unit with a scan rate of $10-100\,\text{mV/s}$, and an n value of 1.0 from coulometric oxidation where i_{pc} and i_{pa} were the anodic and cathodic currents, respectively. The quasireversible voltammetric response at 1.52 V ($\Delta E_p = 70-80\,\text{mV}$) was assignable to the $\text{Ru}^{III}\text{Ru}^{IV} = [\text{Ru}^V]_2 + e^-$ couple. The significant aspect of the redox behavior is a one-step two-electron transfer process occurring at -1.07 V vs SCE. The peak-to-peak separation (ΔE_p) of $30-50\,\text{mV}$ at different scan rates ($5-100\,\text{mV/s}$), the i_{pc}/i_{pa} ratio of unity at various scan rates, and an n value of 2.09 from the coulometric reduction at -1.3 V suggest the quasireversible nature of the $[\text{Ru}^{III}]_2 + 2e = [\text{Ru}^{II}]_2$ couple. A similar complex, $[\{(\text{py})_3\text{Ru}\}_2$

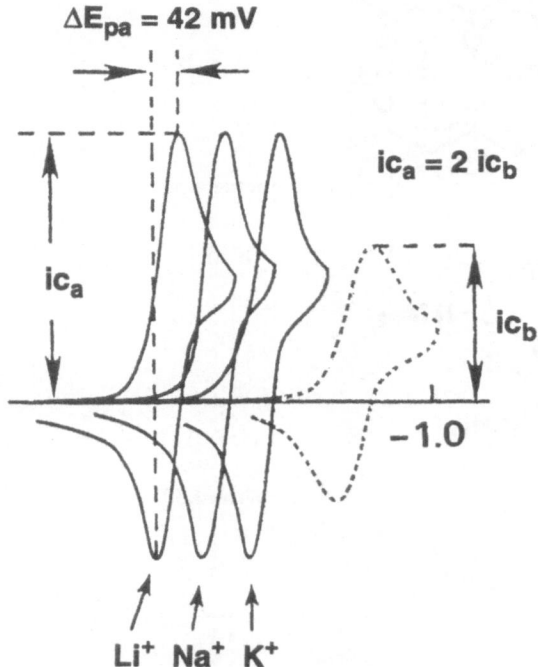

Fig. 4.32. Cyclic voltammograms of copper triketone complex (Scheme 4.38). Effect of alkaline ions on multielectron transfer process

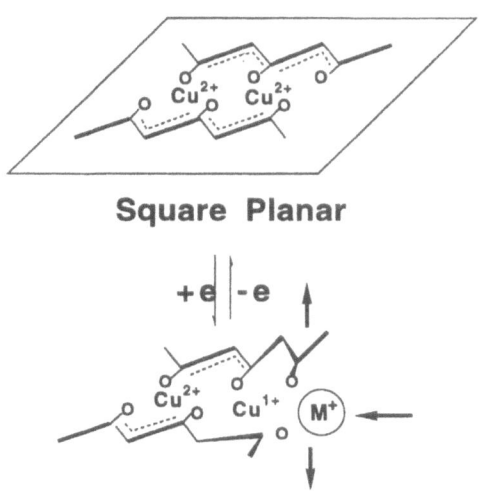

Fig. 4.33. Mulitelectron transfer mechanism of copper triketone (Scheme 4.38)

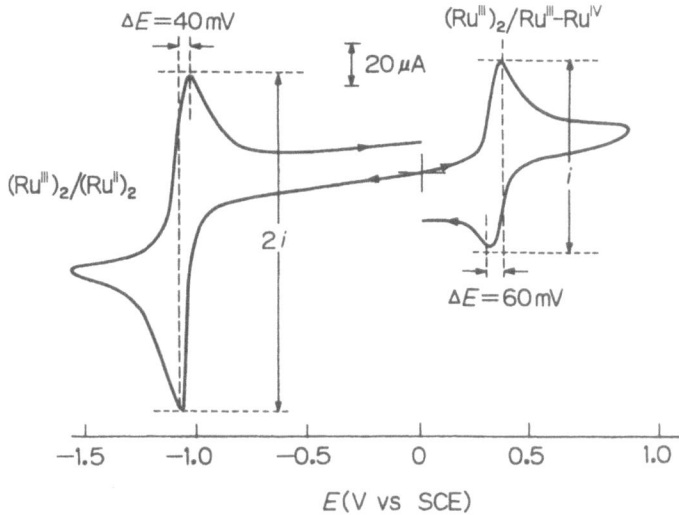

Fig. 4.34. Cyclic voltammogram of compound (Scheme 4.39)

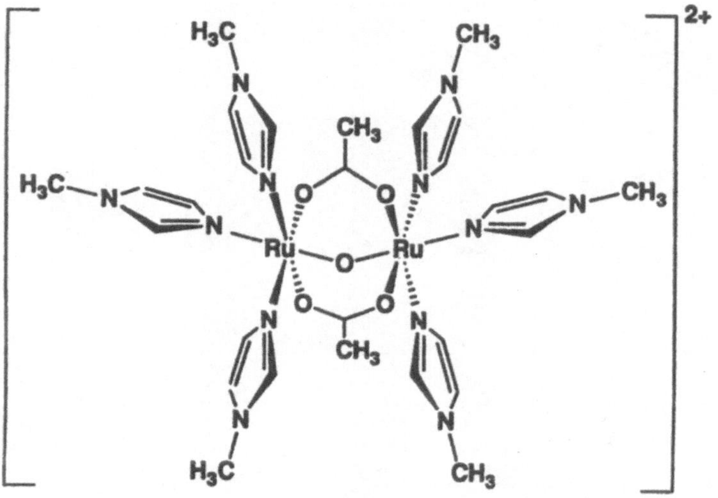

Scheme 4.40

$(\mu\text{-O})(\mu\text{-O}_2\text{CMe})_2](\text{PF}_6)_2$, where ruthenium is coordinated by pyridine (Py) in place of 1-methylimidazole exhibited two one-electron reductions at -0.85 and -1.4 V [282]. The mechanism presumed to be responsible for the one-step two-electron transfer of the 1-methylimidazole coordinated complex $[\{(1\text{-MeIm}]_3\text{Ru}\}_2(\mu\text{-O})(\mu\text{-O}_2\text{CMe})_2](\text{ClO}_4)_2$ is the ligand (1-MeIm) dissociation after the one-electron transfer $([\text{Ru}^{II}]_2\text{Im} - \text{e-} = \text{Ru}^{II}\text{Ru}^{III}\text{IM}: \text{E1};$ $\text{Ru}^{I\text{-}I}\text{Ru}^{III}\text{IM} - \text{e} = [\text{Ru}^{III}]_2 + \text{Im}: \text{E2})$ by which the potential gap for each step $(\Delta E = \text{E1}-\text{E2}$ becomes small. At this point polynuclear complexes for multielectron transfer are rare, because the structure or other factors required for multielectron transfer have not been determined. In addition, only a few examples have been reported for the application of the multielectron reduction of substrates such as O_2 or CO_2.

4.5.2.2 Reduction of Oxygen Through Multielectron Transfer

The thermodynamically allowed potential for the two-electron reduction of O_2 in acidic media $(\text{O}_2 + 2\text{e-} + 2\text{H}^+ = \text{H}_2\text{O}_2)$ is located at 0.68 V (vs NHE), whereas that for the four-electron reduction $(\text{O}_2 + 4\text{e} + 4\text{H}^+ = 2\text{H}_2\text{O})$ is at 1.23 V (vs NHE). The direct electron transfer from the electrode to O_2 in the electrolyte solution is usually kinetically very slow at moderate potentials, and electron mediators are frequently employed to enhance the rate of electron transfer. A strategy to achieve rapid multielectron transfers (while avoiding undesired intermediates) that has been tested with a number of substrates such as O_2, NO, and CO_2 involves the synthesis of catalysts containing multiple centers that can serve as electron donors or acceptors. Thus far, most of the mediators that have been used as a catalyst for the four-electron electroreduction of O_2 near the thermodynamic potential are polynuclear complexes such as dimeric cofacial cobalt porphyrin [283, 284], some related derivatives [285, 286], and a dimeric iridium porphyrin [287].

In recent studies a method for the coordination of four $Ru(NH_3)_5^{2+}$ groups to (5,10,15,20-tetrakis(4-pyridyl)porphyrinato)cobalt(II) (CoP(py)$_4$ (Scheme 4.41) incorporated into Nafion coatings on electrode surfaces or adsorbed directly on the surface of graphite electrodes was described [288–290]. The resulting tetraruthenated complex, $(CoP(pyRu(NH_3)_5)_5^{8+}$, is an effective catalyst for the four-electron reduction of O_2 to H_2O [291].

The remarkable features of the catalysis using the ruthenated cobalt porphyrin include the dependence of the catalytic activity on the number of coordianted ruthenium centers attached to the porphyrin ring [292]. The triruthenated complex $(CoPPh(pyRu(NH_3)_5)_3)^{6+})$ also catalyzes the four-electron reduction of O_2 to H_2O, whereas the mono$(CoPPH_3(pyRu(NH_3)_5)^{2+})$ and diruthenated $(CoPPh_2(pyRu(NH_3)_5)$ complexes catalyze only the two-electron reduction to H_2O_2. The nonruthenated precursor $(CoP(py)_4)$ and (5,10,15,20-tetraphenylporphyrinato)cobalt(II) (CoPPh$_4$) also catalyze only the two-electron reduction. Evidence was presented that the four-electron reduction of O_2 occurred only when the reducing electrons were passed to the O_2 via the coordianted $Ru(NH_3)_5^{2+}$ centers. In addition, a more selective four-electron reduction of O_2 (85 %) was observed in the catalysis of the tetraruthenated complex as compared with the trisubstituted complex (54 %). Although the mechanism by which the complexes operate in catalyzing the electroreduction of O_2 has not been determined, it is clear that complexes having multiple centers whose redox couples can be rapidly cycled are essential for the multielectron reduction of O_2.

The four-electron transfer of the tetrarutheanted complex (Scheme 4.41) on the electrode was observed by a single redox couple $(CoP(pyRu(NH_3)_5)_4^{2+}$

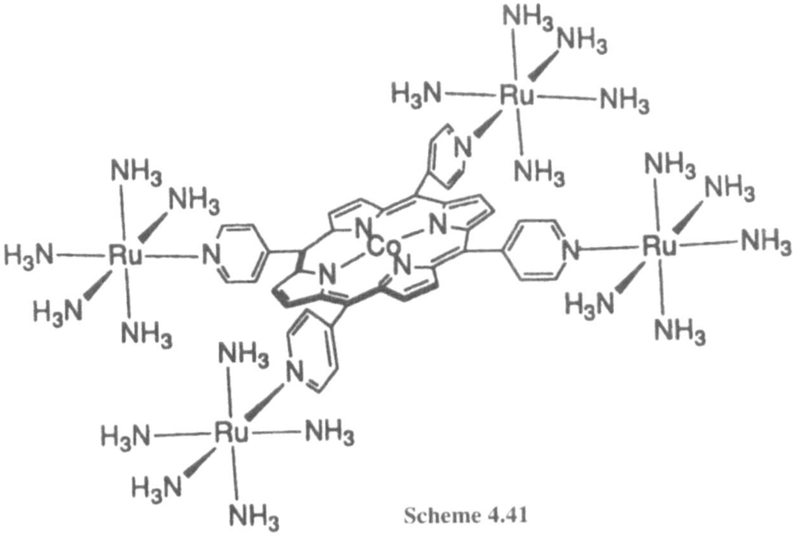

Scheme 4.41

+ 4e = CoP(pyRu(NH$_3$)$_5$)$_4^{8+}$ in the cyclic voltammograms, which provided a definite example that the one-step multielectron transfer is effective for the reduction of O$_2$.

It can be presumed that the four-electron reduction of O$_2$ is facilitated by a catalysis involving multielectron transfer processes that are accomplished during the short period when the substrate is captured on the electrode surface. Electron mediators, which undergo a one-step multielectron transfer, often serve as effective catalysts. However, only few complexes exhibit a one-step multielectron transfer, and the mechanism has not been elucidated. The structural changes accompanied by an electron transfer presumably affect the redox potential by which the one-step multielectron transfer is believed to occur when the potential gap between each step is negligibly small and assumed to be identical.

An apparent one-step four-electron transfer of the a-pyrrolidone-bridged tetranuclear platinum(III) complex ([Pt$^{III}_4$(NH$_3$)$_8$(C$_4$H$_6$NO)$_4$]$^{8+}$ + 4e− = 2[Pt$^{II}_2$(NH$_3$)$_4$(C$_4$H$_6$N)O)$_2$]$^{2+}$, 0.5 V vs SCE) has been reported [293, 294]. A similar octanuclear platinum complex (Scheme 4.42) bridged by acetamide exhibited a redox wave near 0.45 V vs SCE (Fig. 4.35). The electroreduction of O$_2$ in acidic media proceeded at 0.5 V by using these polynuclear complexes as electron mediators on the electrodes.

4.5.3 Oxidative Polymerization of Phenols

Facile oxidative polymerization utilizing the abundant and cheap oxidant, oxygen, to make aromatic polymers, such as high-performance engineering plastics and electroconductive polymers, provides a desirable and clean process for upgrading the value of a material, e.g., the commercial production of poly(2,6-dimethyl-1,4-phenylene oxide) through oxygen oxidative polymerization of 2,6-dimethylphenol with a copper amine catalyst [288, 289].

Recently, some papers have reported that multielectron transfer processes are applicable to the molecular conversion by merely changing the metal species, where polynuclear complexes play an important role as a multielectron transfer mediator.

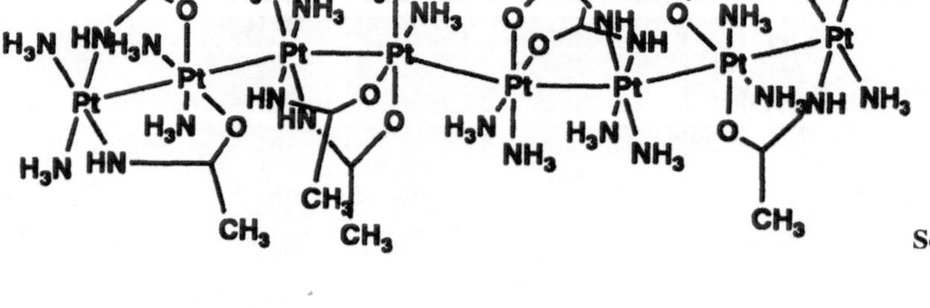

Scheme 4.42

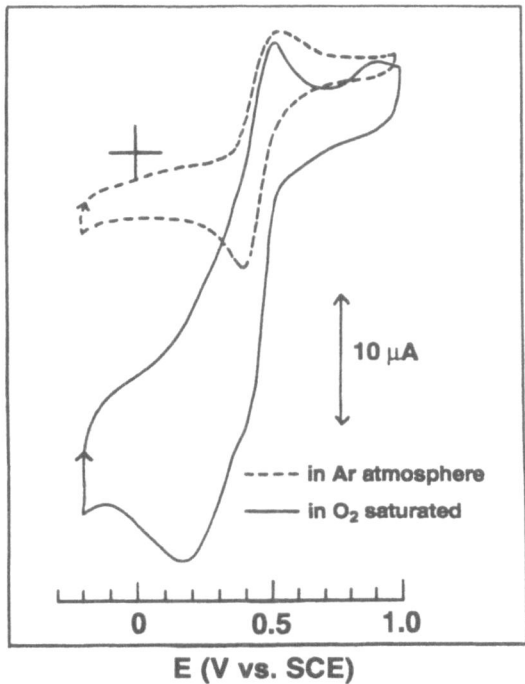

Fig. 4.35. Four-electron reduction of oxygen mediated by compound (Scheme 4.42)

10 μA

- - - - in Ar atmosphere
——— in O$_2$ saturated

0   0.5   1.0

E (V vs. SCE)

4.5.3.1 Oxidative Polymerization

Organic compounds having labile hydrogens, such as phenols [295–298], phenylene diamines [299], and acetylenes [300], can be oxidatively coupled in the presence of a metal complex to form polymeric compounds as represented by Eqs. (1–4):

are called oxidative coupling polymerizations. As an example, 2,6-disubstituted phenols with an amine–Cu complex catalyst produce poly(2,6-disubstituted phenylene oxide)s through C–O coupling [301–303]. Poly(2,6-dimethylphenylene ether) (PPE) (Scheme 4.43) and poly(2,6-diphenylphenylene ether) (PPPE) (Scheme 4.44) are commercially produced from 2,6-dimethylphenol (XOH) and 2,6-diphenylphenol, respectively, and have excellent performances as engineering plastics.

In the Cu-complex-catalyzed oxidative polymerization of phenols, the substrate (phenol) coordinates to the Cu(II) complex and is then activated. In the next stage the activated phenol then selectively couples. The Cu complex effectively acts as a catalyst at concentrations of $0.2-2$ mol% compared with phenols. The oxidation proceeds rapidly at room temperature under an air atmosphere to give PPE with a molecular weight of $2-10 \cdot 10^4$ Da in a quantitativ yield. Moreover, the polymerization follows Michaelis-Menten-type kinetics [304, 305]. This catalytic behavior is similar to that of the copper oxidases. Indeed, enzymatic oxidation of phenols is an important pathway in the biosynthesis of lignin in plants [306], which is catalyzed by a metalloenzyme.

The oxidative polymerization of phenols or dithiols has been proposed to proceed via a radical coupling, which might involve coupling of neutral radicals for a long time. The neutral radical is formed by a one-electron transfer after dissociation of a proton from the monomer (Scheme 4.45).

In the case of an asymmetric monomer such as a phenol, the coupling fashion needs to be controlled on order to obtain the polymer. The phenoxy anion is oxidized to the phenoxy radical through a one-electron transfer. The radical is localized on the oxygen atom and the carbon atoms in the *para* and *ortho* positions. The C–O coupling (binding of the oxygen atom with a carbon atom in the *para* position) needs to proceed preferentially for the formation of PPE (Scheme 4.46). In contrast, C–C coupling (binding of carbon atoms to each other) products yield nonpolymerizable 3,3′5,5′-tetramethylbiphenoquinone (Scheme 4.46) as a side product. Methyl substituents in the *ortho* position and orientation of the phenoxy anion in the coordination sphere of the metal-complex catalyst retard the side reaction (C–C coupling) [307, 308]. The Cu-complex catalyst not only enhances the rate of polymerization, but also has a significant effect on the coupling reaction.

Activation of the dimer during the oxidative polymerization of phenols is believed to occur after the consumption of most of the monomer by coupling, because the complex formation constant of the monomer with Cu ion is much larger than the dimer. The mechanism would indicate a stepwise reaction. Of

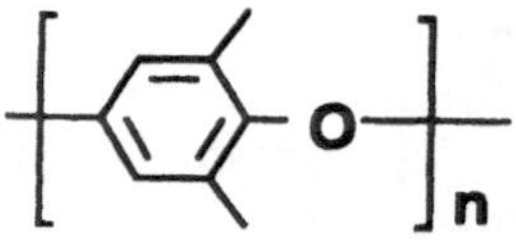

Scheme 4.43

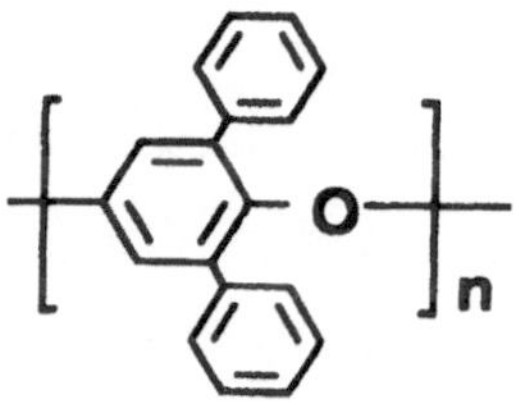

Scheme 4.44

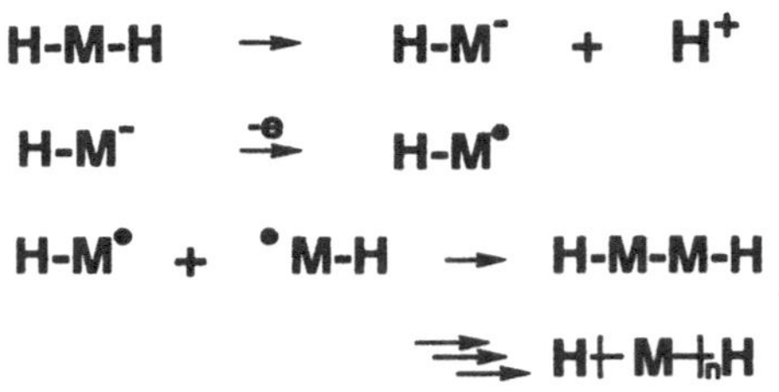

H-M-H   →   H-M⁻ + H⁺

H-M⁻ $\xrightarrow{-e}$ H-M•

H-M• + •M-H → H-M-M-H

          $\overset{}{\Longrightarrow}$ H+M+ₙH   **Scheme 4.45**

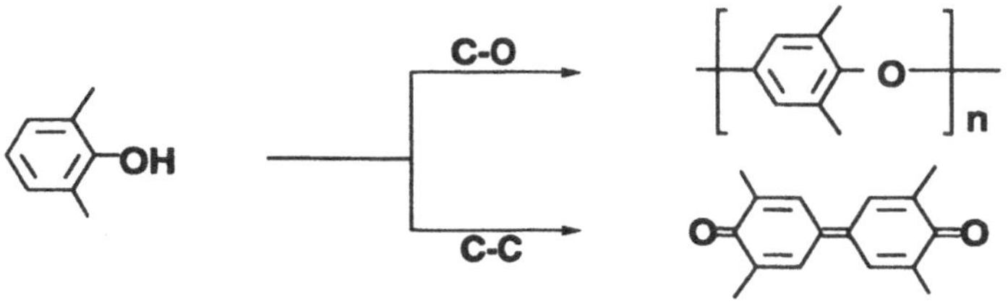

Scheme 4.46

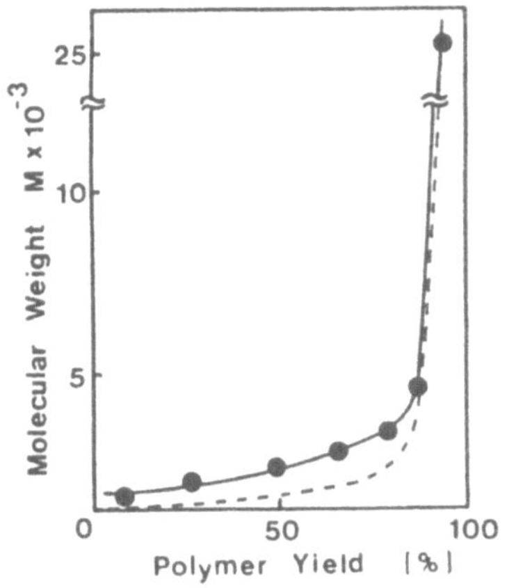

Fig. 4.36. Relationship between molecular weight and conversion in oxidative polymerization of phenols

course, the relationship between the conversion in the reaction and the molecular weight of the polymer supports the stepwise polycondensation (Fig. 4.36) [309, 310]. The oligomer, although it less readily binds to the catalyst, is oxidized more easily than the monomer because of its lower oxidation potential (Table 4.8). The formed oligomers grow to PPE through continuous repeating of the reoxidation and coupling reactions.

4.5.3.2 Catalytic Mechanism of Copper–Amine Complexes

The catalytic mechanism of the Cu-complex during the oxidative polymerization of phenol has been with the understanding that the polymerization proceeds via a radical coupling mechanism through the one-electron transfer.

Table 4.8 Oxidative polymerization of phenols

Monomer	PPE yield (wt. %)	Mw ($\cdot$ 10^{-3})	Oxidative Peak potential[a]	
			PhOH[b]	PhO^{-}[c]
⬡-OH	85	5.8	1.6	0.5
⬡-O-⬡-OH	88	6.1	1.2	0.2

PPE   poly (2,6-dimethylphenylene ether).
[a] V vs Ag/AgCl.
[b] In CH_2Cl_2.
[c] In MeOH-KOH.

In the first step the substrate coordinates to the Cu(II) complex (S1) and a one-electron transfer from the substrate to the Cu(II) ion (S4) occurs. The activated substrate then dissociates from the catalyst (S3), and the reduced catalyst is reoxidized to the origial Cu(II) complex by dioxygen (S4) (Scheme 4.47). The catalytic mechanism is based on the fact that the quantitative formation of the phenoxy radical was observed by ESR measurement during the electro- and chemical oxidation using nonpolymerizable 2,4,6-tri-tert-butylphenol [311]. Furthermore, the phenoxy anion is electropolymerized to yield PPE in methanol at a low applied potential of 0.5 V [312]. Although the one-electron transfer mechanism is reasonable, the formation of water through the four-electron transfer reduction of oxygen cannot be explained very well. It is difficult to study these elementary reactions in order to determine the complicated catalytic mechanism. Many investigations of the metal-complex-catalyzed reactions have been reported [313–315], however, the details of the catalysis are still not completely understood because of its complexity.

Recently, Chen et al. [316–319] and Challa et al. [320] reported that the oxidative polymerization of phenol proceeds through the one-step two-electron transfer. The catalysis obeys the mechanism shown in Scheme 4.48.

In the absence of phenol, the [Cu(L)$_4^{2+}$ complex (1 in Scheme 4.48) is predominantly formed. The phenol itself can hardly substitute OH$^-$ or Cl$^-$ in (bridged) copper complexes, but the phenoxy anion can, because shorter induction times and higher reaction rates are observed when adding the hydroxide. With a weak base, such as Et_3N, the rats are smaller. The phenoxy anion (PhO$^-$) is coordinated to Cu(II) to form a complex (2 in Scheme 4.48). The overall reaction rates obeyed Michaelis-Menten kinetics in phenol. The coordination of phenol yields complex (2 in Scheme 4.48) under all conditions after binding of the catalytic amounts of phenoxy anion.

At very low concentrations of Cu^{II}, the dimerization of the copper complex quantitatively takes place to form compound 3 in Scheme 4.48, because the overall reaction rate is proprotional to [Cu(II)$_o$. Nearly the same rates were observed even by using a nonbridging anion, such as Cu (Cl$_4$)$_2$, as a catalyst. This indicates that the phenoxy anion is the bridging ligand as proposed by

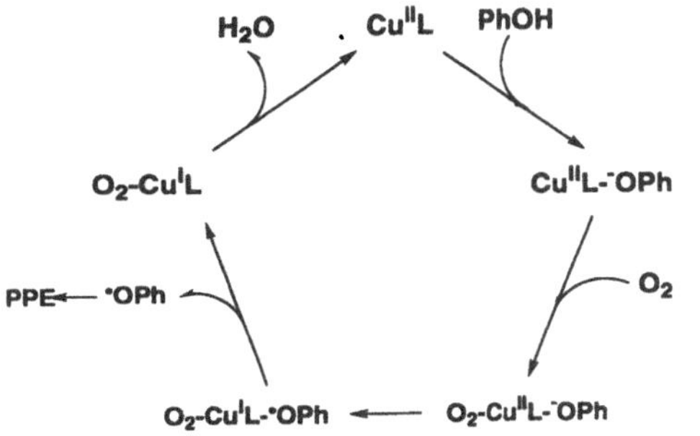
H₂O
Cu^II L
PhOH
O₂-Cu^I L
Cu^II L-⁻OPh
PPE ← •OPh
O₂
O₂-Cu^I L-•OPh
O₂-Cu^II L-⁻OPh

Scheme 4.47

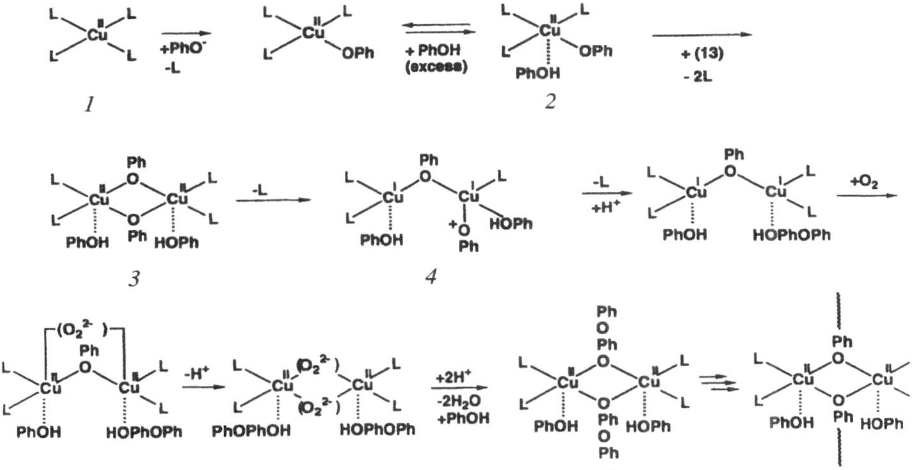
+PhO⁻
-L
+ PhOH
(excess)
+ (13)
- 2L
-L
+H⁺
+O₂
-H⁺
+2H⁺
-2H₂O
+PhOH
PhOH
HOPh
HOPhOPh
PhOPhOH

1 2 3 4 5 6 7

L = N-Substituted Imidazole
PhOH = 2,6-Dimethylphenol

Scheme 4.48

Karlin et al. [321] in the dinuclear complexes that act as the actual catalyst in the active state.

The electron transfer step is estimated to be rate-determining, because the complex conformation is significantly changed [Cu^{II}-Cu^{II}(square planar) = Cu^{I}-Cu^{I}(tetrahedral)]. A phenoxonium cation formed through a one-step two electron transfer from one phenoxy anion is preferred, because one phenoxy anion can then keep a bridge intact *4* (in Scheme 4.48) in between the two Cu(I) ions, which are known to prefer dinuclear structures.

Because electrophilic reactions of the cation species are well known to be affected by basic species, an intramolecular process should be taken into account; e. g., electrophilic substitution at the *para* position of the coordinated phenol by the neighboring "side-on" coordinated phenoxonium cation [322]. The specificity is determined by the way the phenoxonium cation attacks the benzene ring of the phenol. It is interesting that Reedijk and Challa claim that in the triplet state, the phenoxy anion follows a path yielding the dimer via C–O coupling, and in the single state, path finally producing the side product (Scheme 4.46) via C–C coupling [323].

The reoxidation step of the complex takes place after coordination of O_2 by dinuclear Cu complexes (5 in Scheme 4.48). The μ-dioxo complex is estimated to be formed on the basis of Karlin's model [321]. In the next step the phenoxy anion is oxidized by two electron transfers to O_2^{2-} via Cu(II) followed by coupling of the resulting phenoxonium cation with phenol according to Scheme 4.38. The resulting complex (6 in Scheme 4.48) is similarly transformed into compound 7 in Scheme 4.48. The catalytic center is regenerated through the four-electron transfer oxidation with oxygen.

In the following oxidation cycles, phenolate anions of the dimer, trimer, etc. are preferentially oxidized because of their lower oxidation potential compared with phenol. This is the current catalytic mechanism of the oxidative polymerization of phenol with copper–amine complexes. Here, it is important to form the polynuclear complexes and to undergo the one-step multielectron transfer. After all, the polyphenylene formation takes place through multielectron transfer.

4.5.4 Oxidative Polymerization of Disulfides

Aromatics for oxidative polymerization usually show high oxidation potentials. The direct oxidation of these monomers by O_2 does not proceed because of the large potential gap between the monomers and O_2 (diphenyl disulfide: ca. 1 V; phenol: ca. 0.5 V). The multielectron transfer shifts the reduction potential of O_2 to the positive side. The potential gap in the case of a multielectron transfer is thus smaller than that for the one-electron process.

Vanadium complexes are of considerable interest not only in complex chemistry, but also as redox catalysts and in molecular oxygen chemistry. Vanadium (III, IV, and V) complexes bridge a large potential gap between oxygen and aromatic compounds with a high oxidation potential such as diphenyl disulfide, pyrrole, thiophene, and benzene. The vanadium complexes are excellent electron mediators, because V(III) and V(V), whose redox potentials are located in the potential gap, can act as powerful reducing and oxidizing agents, respectively.

A novel catalysis of vanadyl complexes is applicable to the synthesis of oligo(p-phenylene sulfide) (OPS) containing an S–S bond by the oxygen oxidative polymerization of diphenyl disulfide in which the multielectron transfer process [324] plays an important role (Eq. (5)):

$$n/2 \; \langle \rangle\text{-S-S-}\langle \rangle + n/4 \, O_2 \longrightarrow \text{--}[\langle \rangle\text{-S}]_n\text{--} + n/2 \, H_2O \qquad (5)$$

This is the first example of a vanadium binuclear complex having a reversible two-electron transfer process that is applicable to the molecular conversion [325, 326].

4.5.4.1 Oxidative Polymerization

In the presence of VO (acac)$_2$, the polymerization is accompanied by a quantitative oxygen uptake. Oxygen is essential for the polymerization of diphenyl disulfides. The initial rate of oxygen uptake (V_O) was measued at various partial oxygen pressures during the polymerization of diphenyl disulfide (0.05 mol/L). The dependence of VO on the O$_2$ pressure over the range 0–0.5 atm is shown in Fig. 4.37. First-order disulfide substrate kinetics were observed at O$_2$ pressures above 0.6 atm. Zero order was found below 0.5 atm. VO became constant above 0.5 atm and was proportional to the oxygen partial pressure below 0.5 atm. These results support the belief that the rate-determining step in this catalytic cycle under atmospheric pressure is the oxidation process in which diphenyl disulfide is oxidized by the high-valence species of the vanadyl complex. The observed zero-order [O$_2$] dependence is consistent with this, because the activated vanadyl species is a much better oxidant than O$_2$. This is also consistent with the fact that a two-electron transfer occurs in the catalytic system as indicated by the electrochemical measurements.

One intriguing aspect of these VO-catalyzed oxidations was the observed effect of substituents in a series of methyl-substitued diphenyl disulfides on the observed rate of the reaction (Fig. 4.37). The VO-catalyzed oxidation of bis(3,5-dimethylphenyl) disulfide displays zero-order substrate kinetics. Not only a linear plot of oxygen partial pressures was obtained over the range of 0.33–1 atm, but also the initial rate agrees with that of diphenyl disulfide at

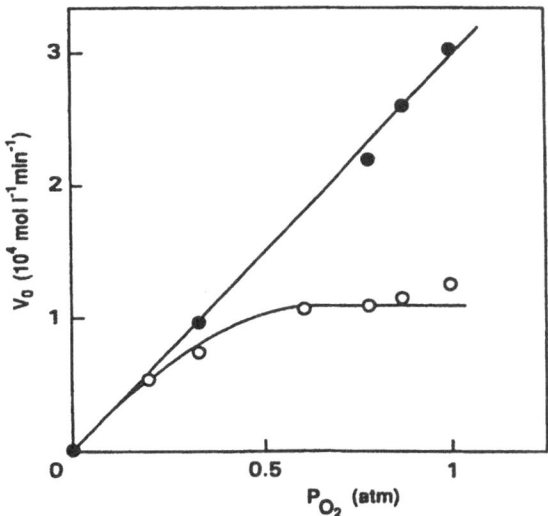

Fig. 4.37. Effect of oxygen partial pressure on oxygen uptake rate in VO-catalyzed oxidative polymerization of diphenyl disulfides

the low partial pressure (0.33 atm). These results do not exclude the possibility, that the rate-determining step is the reoxidation of the lower-valence VO species (three-valence) formed by the disproportionation with oxygen.

Bis(3,5-dimethylphenyl) disulfide is oxidized in this system three times more rapidly than the nonsubstituted diphenyl disulfide system. Although this may be due at least in part to the fact that bis(3,5-dimethylphenyl) disulfide is approximately 0.2 V easier to oxidize than the diphenyl disulfide, it is not clear at this point if other factors (e. g., steric) affecting the interaction with high-valence VO species may be operative.

A variety of other metal complexes were utilized under the same conditions. However, all of them are not effective as a catalyst for the oxygen oxidative polymerization of diphenyl disulfide (Table 4.9). Although an equimolar amount of 2,3-dichloro-5,6-dicyano-p-benzoquinone (DDQ) and lead tetraacetate can oxidize disulfide to yield OPS, due to the high oxidizing ability, the oxidants do not act as a catalyst for the efficient formation of OPS. The reduced species of the oxidant is not reoxidized with oxygen. The other complexes, such as $Cu(acac)_2$, $Fe(acac)_2$, and VO(salen), cannot oxidize disulfide, due to this lower redox potential.

In contrast, $VO(Bzac)_2$ with the same redox potential as $VO(acac)_2$ is effective as a catalyst during the polymerization. Even in the presence of the 0.2 % VO(TPP) complex as a catalyst the polymerization quantitatively proceeds for 120 h accompanied by oxygen uptake, although the polymerization rate is slow.

Equimolar vanadium(V) pentaoxide reacts with diphenyl disulfide to yield PPS (>90 %), even under an oxygen-free atmosphere. $VO(acac)_3$ (trivalent

Table 4.9. Oxygen polymerization of diphenyl disulfide with catalyst

Catalyst	[Cat] [Monomer]	E1/2 (V)[a]	Polymer yield (wt. %)
DDQ[b]	0.1	1.2[c] (Q/OH)	0
$Pb(CH_3COO)_4$	0.1	1.5[c](3+/4+)	0
$Cu(acac)_2$	0.1	−0.3(1+/2+)	0
$Fe(acac)_3$	0.1	−0.65(2+/3+)	0
VO(salen)	0.1	0.5(4+/5+)	0
$VO(Bzac)_2$	0.1	1.1(4+/5+)	95
$VO(acac)_2$	0.1	1.1(4+/5+)	92
$VO(acac)_2$	0.01	1.1(4+/5+)	68
VOTPP[d]	0.01	1.5(4+/5+)	95
$V(acac)_3$	0.1	0.8(3+/4+)	85
$V_2O_5^e$	2.0	–	97

[a] VS Ag/AgCl.
[b] 2,3-Dichloro-5.6-dicyano -p-benzoquinone.
[c] Oxicative peak potential.
[d] Reaction time 120 h.
[e] Five-valence under N_2.

species) is easily oxidized to VO(acac)$_2$ with oxygen. The redox potentials of VO(acac)$_2$ and VO(acac)$_3$ are 0.9 and 1.1 V (vs Ag/AgCl), respectively. The activated VO(acac)$_2$ in the presence of acid exhibits the properties of both vanadium(III) and vanadium(V) species in the catalytic system.

The OPS yield also depends on the acidity of the mixture. The oxidative polymerization does not proceed in the absence of acids. Strong acids, such as trifluoromethanesulfonic acid and trifluoroacetic acid, are effective in the VO-catalyzed polymerization. A 100-times smaller amount of trifluorome-thanesulfonic acid is sufficient to efficiently form PPSs in comparison with trifluoroacetic acid (TFA), because the acidity of trifluoromethanesulfonic acid is more than 100 times greater than that of trifluoroacetic acid. The oxidative polymerization of diphenyl disulfides is facilitated by the high oxidizing ability of the activate VO(acac)$_2$ produced by the acid. Diphenyl disulfides are not oxidized with only oxygen or only an equimolar amount of VO(acac)$_2$ in the absence of acid, due to the high oxidation peak potential (1.6 V vs Ag/AgCl) of diphenyl disulfide [327]. The VO catalyst is estimated to an excellent electron mediator through activation by acid to promote electron transfer between diphenyl disulfide and molecular oxygen.

The intermediate species formed from diphenyl disulfide during this type of oxidation have been studied using nonpolymerizable dimethyl disulfide as a model compound. Methylbis(methylthio)sulfonium cation salt has been quantitatively isolated by the oxidation of diphenyl disulfide. The detailed structure has been confirmed by ^{1}H-NMR spectra and elemental analysis [328–330]. It is believed that the phenylbis(phenylthio)sulfonium cation is an active species during this polymerization, and the polymerization mechanism (Scheme 4.49) has been previously discussed [331–333]. Diphenyl disulfide is oxidized to yield the phenylbis(phenylthio)sulfonium cation by the activated vanadium species through electron transfer. The electrophilic attack of the sulfonium cation proceeds on the benzene ring of the monomer or oligomer. The oxidation and the electrophilic reaction are continuously repeated to form OPS.

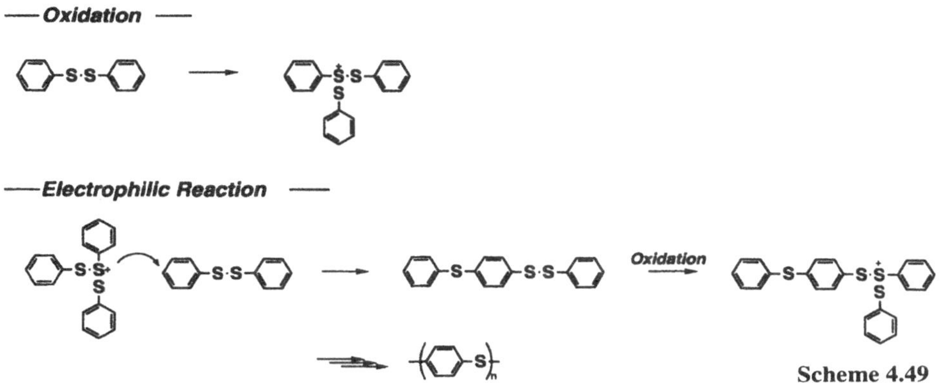

Scheme 4.49

4.5.4.2 Effect of Substituents on Polymerization Behaviors

The VO-catalyzed oxidative polymerization was applied to the synthesis of poly(arylene sulfide)s [334, 335]. The oxidative polymerization of alkyl-substituted diphenyl disulfides, such as bis(2-methylphenyl) disulfide, bis(3-methylphenyl) disulfide, bis(2,5-dimethylphenyl) disulfide, bis(2,6-dimethylphenyl) disulfide, and bis(3,5-dimethylphenyl) disulfide, were carried out under the same conditions (Table 4.10). The polymerization of the alkyl-substituted monomers was accompanied by the O_2 uptake, and the corresponding polymers were formed in high yield. V_{max} and Km (Michaelis-Menten constant) were calculted using the Lineweaver-Burk (L-B) plots based on the measurements of the initial rate of O_2-uptake rates during polymerization. The L-B plots of the alkyl-substituted monomers also showed linear relationships (Fig. 4.38). V_{max} and Km for the polymerization of bis(2,6-dimethylphenyl) disulfide were determined to be $1.3 \cdot 10^{-4}\,molL^{-1}min^{-1}$ and $7.4 \cdot 10^{-2}\,molL^{-1}$, respectively, from the L-B plots. Km for the polymerization of bis(2,6-dimethylphenyl) disulfide is larger than that for unsubstituted diphenyl disulfide (Table 4.10). It suggests that the steric hindrance in the case of the polymerization of bis(2,6-dimethylphenyl) disulfide may labilize the intermediate complex (Michaelis complex), which results in the larger Km.

Km values for the polymerization of bis(3-methylphenyl) disulfide and bis(2,5-dimethylphenyl) disulfide were not determined, because their oxovanadium-catalyzed oxidation showed zero-order substrate kinetics, i.e, the initial rate of O_2 uptake was independent of the concentration of the monomers. These results imply two possibilities: (a) that Km is close to zero, or (b) that the polymerization rate is independent of the monomer concentration. Bis(3-methylphenyl) disulfide, bis(2,5-dimethylphenyl) disulfide, and bis(3,5-dimethylphenyl) disulfide exhibit a lower oxidation potential compared with that of unsubstituted diphenyl disulfide or bis(2,6-dimethylphenyl) di-

Table 4.10. Kinetic parameters in oxidative polymerization of diphenyl disulfides[a]

Subsituents	Polymer yield (wt. %)	Mw ($\cdot 10^2$)	E^b (V)	Km $\cdot 10^{2c}$ (mol/dm³)	$V_{max} \cdot 10^{4d}$ (mol/dm³min)
Non	76	1.2	1.7	1.6	1.5
2-Methyl	85	2.6	1.6	–	–
3-Methyl	97	4.6	1.6	–	–
2,5-Dimethyl	100	2.0	1.5	–	–
3,5-Dimethyl	93	9.2	1.5	–	–
2,6-Dimethyl	73	1.6	1.7	7.4	1.3
Phenol[e]				73	2.5

[a] [Disulfide] = 0.1. [VO] = 0.5 x 10^{-3} (molL^{-1}).
[b] Oxidative potential vs Ag/AgCl.
[c] Michaelis constant.
[d] Initial rate of oxygen uptake.
[e] Oxidative polymerization of 2.6-dimethylphenol with Cu-pyridine.

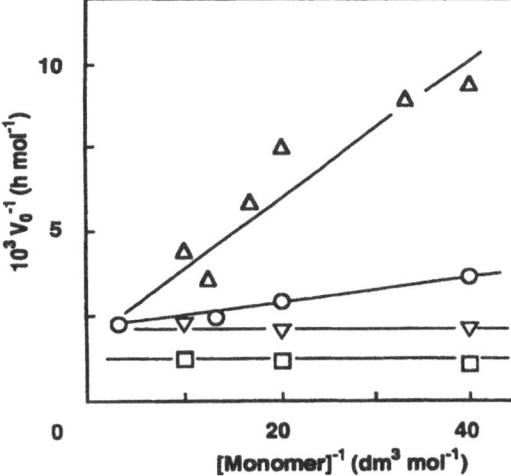

Fig. 4.38. Lineweaver-Burk plots for substituted diphenyl disulfides: diphenyl disulfide (*circles*), bis(2,6-dimethylphenyl) disulfide (*triangles*), bis(3-methylphenyl) disulfide (*triangles down*) bis(2,5-dimethylphenyl) disulfide (*squares*)

sulfide (Table 4.10), which results in an approximately 10 times larger rate of disulfide oxidation. The polymerization rate of bis(3,5-dimethylphenyl) disulfide increases in proportion to the partial pressue of O_2. Therefore, during the oxidative polymerization of monomers that exhibit lower oxidation potentials, the rate-determining step is estimated to be the reoxidation of the catalyst by O_2, as is the polymerization of unsubstituted diphenyl disulfide under low partial pressures of O_2.

4.5.4.3 One-Step, Two-Electron Transfer of Vanadium Complex

μ-Oxo-bis[(N,N'-ethylenebis(salicylideneaminato))vanadium(IV)] tetrafluoroborate, is synthesized from (N,N'-ethylenebis(salicylideneaminato))oxovanadium(IV) (VO(salen)) through an electrophilic reaction by triphenylmethyl tetrafluoroborate (Ph$_3$CBF$_4$) in dichloromethane under a dry argon atmosphere. The stoichiometry obeys Eq. (6). An electrophile, such as triphenylmethyl tetrafluoroborate (Ph$_3$CBF$_4$) or trifluoromethane sulfonic acid (CF$_3$SO$_3$H), is efficient for the deoxygenation of VO(salen).

$$2VO(salen) + 2Ph_3C^+ = [\{V(salen)\}_2(\mu\text{-}O)]^{2+} + Ph_3COCPh_3 \qquad (6)$$

[{V(salen)}$_2$(μ-O)](BF$_4$)$_2$ is isolated with > 95% yield as a black powder having the empirical formula of C$_{32}$H$_{24}$N$_4$V$_2$B$_2$F$_8$. The dimeric structure is confirmed by the fragments of 650, 333, and 315(+ 1) in FAB MAS spectroscopy, The electronic spectrum of [{V(salen)}$_2$(μ-O)](BF$_4$)$_2$ displays the absorption band at $\lambda_{max} = 579$ nm ($\varepsilon_{max} = 365$ mol dm^3 cm^{-1}) with 200 nm of half-band width. The typical V–O–V stretching band is observed at 655 cm^{-1}, which is assigned by using the V^{13}O(salen) complex. The spectroscopic data indicate the formation of a μ-oxo dimer, [{V(salen)}$_2$(μ-O)](BF$_4$)$_2$.

The cyclic voltammograms in Fig. 4.39 shows the redox of [{V(salen)}$_2$ (μ-O)](BF$_4$)$_2$ (0.5 mmolL^{-1}) in dichloromethane containing 0.5 molL^{-1} of tetrabutylammonium tetrafluoroborate using the platinum-disk electrode. A reversible redox wave is observed at 0.58 V (vs Ag/AgCl) [336] with a potential separtion (ΔEp) of 75 mV between the cathodic and anodic peaks, which is smaller than that for VO(salen) (85 mV) under the same conditions. The redox potential of the μ-oxo shifts to a more positive potential than the one-electron oxidation potential of VO(salen) (0.56 V vs Ag/AgCl) [337].

As an additional check on the system the rotating-disk voltammetry of VO(salen) and [{V(salen)}$_2$(μ-O)](BF$_4$)$_2$ were measured under the same conditions. The rotating-disk voltammetry (Fig. 4.40) shows symmetric waves at $E_{1/2} = 0.58$ V vs Ag/AgCl, i.e., the cathodic current is equal to the anodic current regardless of the rotation rate. The liming currents, i$_{la}$ and i$_{lc}$, are identified to be the oxidation and reduction of the V^{IV}–O–V^{IV} complex, respectively. A plot of the plateau current (i$_{la}$ + i$_{lc}$) vs the square root of the rotation rate (Levich plot) yields a straight line. The electrode potential at i = 0 was equal to the half-wave potential regardless of the rotation rate. The plateau current for [{V(salen)}$_2$(μ-O)]$^{2+}$ was approximately two times larger than the one-electron oxidation (V^{IV}/V^V) current of VO(salen), which indicates two-electron transfer of the dinuclear complex. The coulomb titration (prepartive electrolysis) also supported the fact that the complex exhibits a reversible two-electron transfer in a single voltammogram ($n = 1.98$).

These results support the idea that the following disproportionation takes place (Eq. (7), which is caused by the fact that $E_{1/2[III–IV/IV–IV]}$ and $E_{1/2[IV–IV/IV–V]}$ are located at close potentials:

$$2[V^{IV}–O–V^{IV}] = [V^{III}–O–V^{IV}] + [V^V–O–V^{IV}] \tag{7}$$

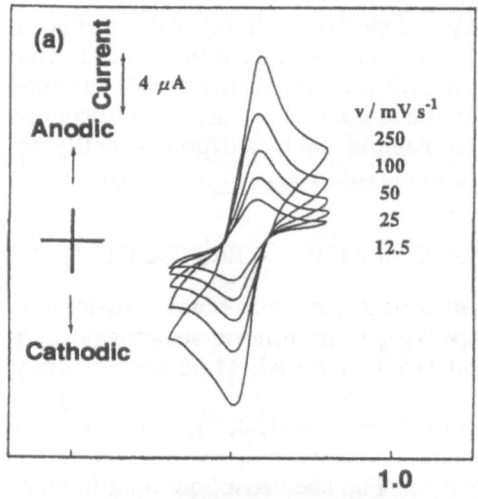

Fig. 4.39. Cyclic voltammograms of [{V(salen)}$_2$(μ-O)](BF$_4$)$_2$

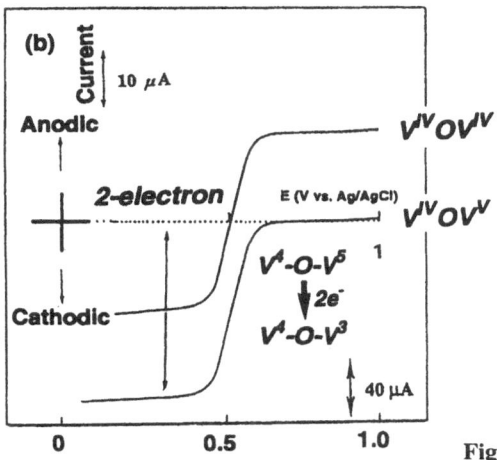

Fig. 4.40. Rotating-disk voltammograms of $[\{V(salen)\}_2(\mu\text{-}O)](BF_4)_2$ in dichloromethane

The disproportionation of the μ-oxo binuclear complex ($V^{IV}\text{–}O\text{–}V^{IV}$) results in the transfer of two electrons as the net redox reaction of the system.

4.5.5 Oxidative Polymerization of Benzene

Poly(p-phenylene) (PPP) is an insoluble, infusible material that is resistant to oxidation, radiation, and thermal degradation [338]. The polymer has attracted recent attention because of its high electrical conductivity when doped [339]. Poly(p-phenylene) is synthesized through the cationic oxidative polymerization of benzene by electrolysis (Eq. (8)):

$$n \; \langle \bigcirc \rangle \;+\; n/2 \; O_2 \longrightarrow \; {\Big[}{\langle\bigcirc\rangle}{\Big]}_n \;+\; n \; H_2O; \qquad (8)$$

or with aluminum-cupric chloride ($AlCuCl_4$) [340]. Research has also been directed toward elucidation of its mechanistic pathway.

This oxidative polymerization proceeds via a cationic mechanism. The oxidation that the deuterated polymer is of lower molecular weight has mechanistic consequences for the benzene–$AlCl_3$–$CuCl_2$ system. Benzene is oxidized to its cation radical with cuprous chloride. A recent report postulates a unique mechanism that involves the formation of $C_6H_6^+$ followed by coordination with the benzene nuclei in a propagative manner [341, 342]. When the chain build up reaches a certain stage, the radical cation character on the terminal members is too small to induce further propagation. Subsequent C–C bonds are generated. The intermediates then undergo competing proton loss and dealkylation.

The radical-containing end group in the hydropolymer may lose on electron in the presence of an oxidizing agent such as $CuCl_2$ (Scheme 4.50). The

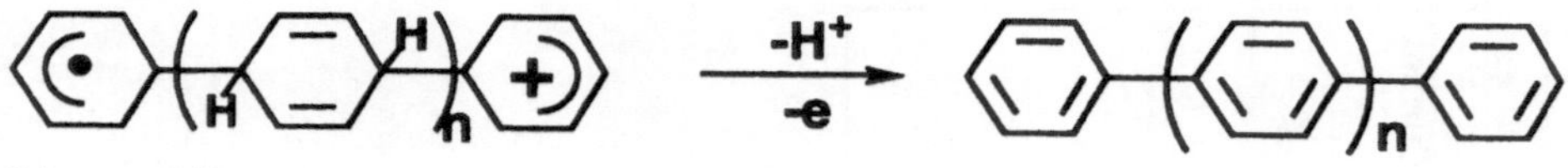

Scheme 4.50

rearomatization of the hydropolymer occurs through the oxidation by cupric chloride and leads to the formation of a poly(p-phenylene) chain.

The cationic oxidative polymerization is strongly influenced by the acidity of the reaction mixture. The formation of poly(p-phenylene) does not occur in a weaker acidic medium, i.e., CF_3COOH-$CuCl_2$ and $MgCl_2$. In contrast, in the presence of a strong acid, such as trifluoromethane sulfonic acid (CF_3SO_3H) or $AlCl_3$, the polymerization efficiently yields PPP with cupric chloride ($CuCl_2$) [343–345]. The effect of a strong acid, such as CF_3SO_3H and $AlCl_3$ is remarkable in this polymerization process in comparison with that of weak acids. This suggests a specific role for the strong acid during this polymerization process. Figure 4.41 shows cyclic voltammograms of benzene in nitromethane. The virgin oxidation of benzene is initiated at 1.7 V, and the oxidation peak potential (2.3 V) shifts toward a less anodic potential in the presence of CF_3SO_3H compared with voltammograms in the absence of acid (Epa = 2.8 V). Benzene was also oxidized above 1.7 V in the presence of $AlCl_3$. Upon the addition of benzene to strong acid solutions, the mixture turned from colorless to light brown ($\lambda = 380$ nm). Benzene is electropolymerized in the presence of the e acids at a potential lower than in the absence.

In the presence of a weak acid, such as trifluoroacetic acid, the oxidation potential does not shift cathodicaly. The voltammogram is the same as in the absence of a weak acid and the mixture remains colorless. The remarkable decrease in the oxidation potential of benzene in the presence of the strong acids can be explained as follows: Benzene forms a complex with a strong

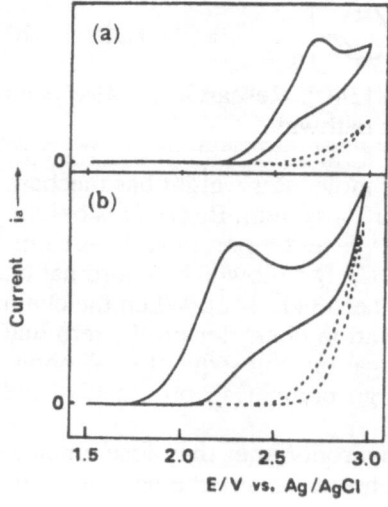

Fig. 4.41. Cyclic voltammograms of benzene in nitroemthane (**a**) and in the presence of CF_3SO_3H (**b**)

acid or Lewis acid, such as CF_3SO_3H and $AlCl_3$, and the resonance stabilization of the benzene ring is reduced, i.e., the oxidation potential shifts slightly toward less anodic potential in the presence of these strong acids, which corresponds to the current efficiency in the polymerization. The strong acid forms a cationic complex with benzene, which promotes the oxidation reaction and an electrophilic attack of the cation radical on benzene, and increases the polymerization efficiency.

Poly(phenylene) is prepared by the cationic oxidative polymerization under dioxygen by using a complex salt of cuprous chloride and aluminum chloride ($AlCuCl_4$) as a catalyst [346]. Because this complex is soluble by complex formation with benzene, the complex system provides the O_2-oxidative polymerization of benzene (Scheme 4.51). The polymerization reaction proceeds without any evolution of hydrogen chloride, although the conventional chemical method results in a huge amount of hydrogen chloride gas from slurry reaction mixtures.

The subsequent addition of excess aluminum chloride drastically improves the yields. The increase in the acidity results in a high polymerization efficiency. The O_2-oxidative polymerization of benzene is achieved by means of a decrease in the potential gap between oxygen and benzene, because the oxidation potential of benzene is reduced by the formation of a complex.

4.5.6 Conclusion

A molecular conversion system based on a two-electron transfer was accomplished during the O_2 oxidative polymerization of diphenyl disulfide using vanadyl complexes as the electron mediator. The geared cycle, which consists of V^{III}–V^{IV} and V^{IV}–V^V via disproportionation, enables the electron transfer from disulfide (oxidation potential: 1.6 V) to molecular oxygen. This is the first example of a one-step multielectron transfer electron mediator, which was applied to the efficient molecular conversion system. Interestingly, recent research reveals that the oxidative polymerization of phenol with oxygen also obeys the one-step multielectron transfer with the polynuclear copper complexes. Multielectron transfer is also applied to the epoxidation of olefines using ruthenium porphyrin [347–349]. The efficient redox cycle is formed

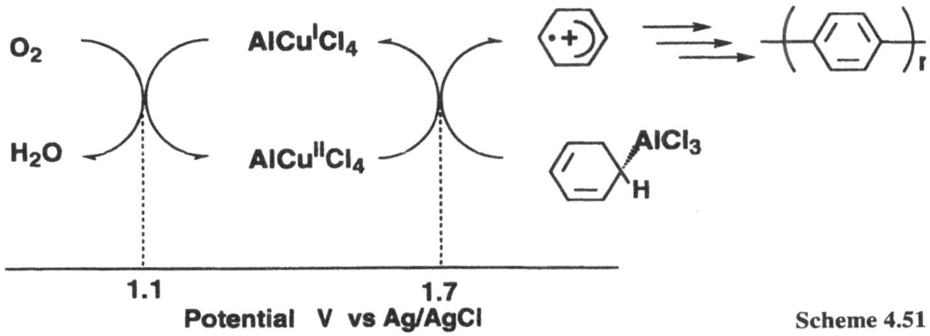

Scheme 4.51

in this system based on the disproportionation of ruthenium(IV) to (II) and (VI) (Fig. 4.42). The detailed structure of the active species has not been elucidated. The disproportioantion is estimated to proceed, due to the change in the coordination structure, which induces a shift in the redox potential.

The multielectron oxidation of substrates is also promising in terms of producing new types of active species, which could be applicable to the effective design of novel synthtic systems. Macromolecular metal complexes should offer effective systems for multielectron transfer where the entropy effect or the cooperative interaction is applicable as characteristic properties of macromolecules. Multielectron transfer is one of many efficient processes as a general method for molecular conversion that may allow an artificial system for the entire photosynthesis process, or catalytic systems applicable for a wide range of redox processes.

It is noteworthy that the one-step multielectron transfer is accomplished only by polynuclear complexes, some of which are profoundly related to biological reactions, such as photosynthesis or nitrogen fixation, which are difficult to achieve in artificial systems presently [350]. To elucidate the chemistry of a multielectron transfer is, therefore, excepted to realize new molecular conversions that provide efficient redox processes not only for the activation of small molecules, such as H_2O, CO_2, and N_2, but also for energy conversion.

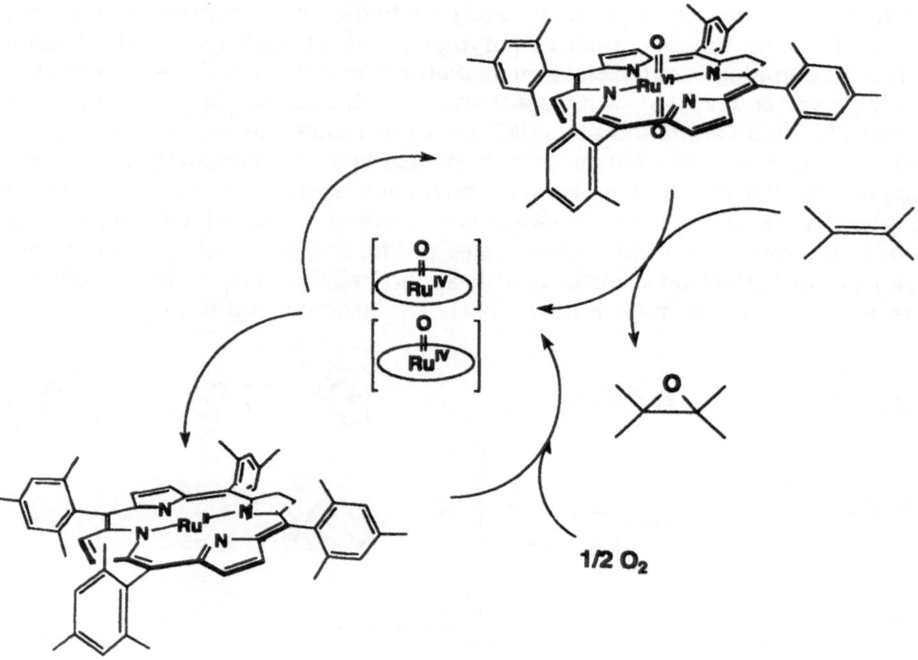

Fig. 4.42. Molecular conversion catalyzed by ruthenium porphyrin through disproportionation

Acknowledgements. This work was partially supported by a Grant-in-Aid for Scientific Research (nos. 06226279, 0455223, and 05650865) from the Ministry of Education, Science, and Culture, Japan.

References

1. Tubandt C (1921) Z Anorg Allgen Chem 115:105
2. Takahashi T, Yamamoto O, Yamada S, Hayashi S (1979) J Electrochem Soc 126:1954
3. Weber N, Kummer JT (1967) Proc Ann Pawer Sources Conf 21:37
4. Goodenough JB, Hong HY-P (1976) Mat Res Bull 11:203
5. Collongues R, Thery J, Boilot JP (1978) In Hagenmuller P, van Gool W (eds) Solid Electrolyte. Academic Press, New York
6. Wright PV (1975) Br Polym J 7:319
7. Cowie JMG, Sadaghianizadeh K (1990) Solid State Ionics 42:243
8. Wang YP, Feng HY, Fu ZS, Tsuchida E, Takeoka S, Ohta T (1991) Polym Adv Technol 2:295
9. Tsunemi K, Ohno H, Tsuchida E (1983) Electrochim Acta 28:833
10. Abraham KM, Alamgir M (1990) J Electrochem Soc 137:1657
11. Croce F, Brown SD, Greenbaum SG, Slane SC, Salomon M (1993) Chem Mater 5:1268
12. Bohnke O, Frand G, Rezrazi M, Rousselot C, Truche C (1993) Solid State Ionics 66:97
13. Tsuchida E, Ohno H, Tsunemi K, Kobayashi N (1983) Solid State Ionics 11:227
14. Takeoka S, Horiuchi K, Yamagata S, Tsuchida E (1991) Macromolecules 24:2003
15. Papke BL, Ratner MA, Shriver DF (1981) J Phys Chem Solid 42:493
16. Chatani Y, Okamura N (1987) Polymer 28:1815
17. Armand MB, Chabagno JM, Duclot MJ (1979) In: Vashishta P et al. (eds) Fast ion transport in solids, p. 131
18. Robitaille CD, Marques S, Boils D, Prud'homme J (1987) Macromolecules 20:3203
19. Nicholas CV, Wilson DJ, Booth C, Giles JRM (1988) Br Polym J 20:289
20. Nagaoka K, Naruse H, Shinohara I, Watanabe M (1984) J Poly Sci Polym Lett Ed 22:659
21. Kobayashi N, Uchiyama M, Shigehara K, Tsuchida E (1985) J Phys Chem 89:987
22. Cowie JMG, Martin ACS (1987) Polymer 28:267
23. Yamada A, Shigehara K, Kurata Y (1985) Polym Prepr 34:903
24. Hall PG, Davis GR, McIntyre JE, Ward IM, Bannister DJ, Le Brocq KMF (1986) Polym Commun 27:98
25. Fish D, Khan IM, Smid J (1986) Macromol Chem Rapid Comm 7:115
26. Fish D, Khan IM, Smid J (1988) Br Polym J 20:281
27. Blonsky PM, Shriver DF, Austin P, Allcock HR (1984) J Am Chem Soc 106:6854
28. Killis A, Le Nest JF, Cheradame H (1980) Macromol Chem Rapid Comm 1:595
29. West K, Christiansen BZ, Jacobsen T (1988) Br Polym J 20:243
30. Watanabe M, Nagano S, Sanui K, Ogata N (1986) Solid State Ionics 18, 19:338
31. Bannister DJ, Davies GR, Ward IM (1984) Polymer 25:1291
32. Hardy LC, Shriver DF (1985) J Am Chem Soc 107:3823
33. Kobayashi N, Uchiyama M, Tsuchida E (1985) Solid State Ionics 17:307
34. Tsuchida E, Kobayashi N, Ohno H (1988) Macromolecules 21:96
35. Ganapathiappan S, Chen K, Shriver DF (1988) Macromolecules 21:2299
36. Zhou G, Khan IM, Smid J (1989) Polym Commun 30:52
37. Lc Ncst J-F, Gandini A, Cheradamc H, Cohen-Added J-P (1987) Polym Commun 28:302
38. Watanabe M, Nango S, Sanui K, Ogata N (1988) Solid State Ionics 28, 30:911
39. Tsuchida E, Ohno H, Kobayashi N, Ishizaka H (1989) Macromolecules 22:1771
40. Takeoka S, Maeda Y, Ohno H, Tsuchida E (1990) Polym Adv Technol 1:201
41. Papke BL, Ratner MA, Shriver DF (1982) J Electrochem Soc 129:1694
42. Saito S, (1986) Koubunshi 17:672
43. Barker RE Jr (1976) Pure Appl Chem 46:157
44. Porter CH, Boyd RH (1971) Macromolecules 4:589

45. Albinsson I, Mellander B-E, Stevens JR (1993) Solid State Ionics 60:63
46. Torell LM, Schantz S (1989) In: MacCallum JR, Vincent CA (eds) Polymer electrolyte reviews 2, Elsevier Applied Science, London, p. 1
47. Kakihana M, Schantz S, Torell LM, Stevens JR (1990) Solid State Ionics 40, 41:641
48. Kakihana M, Schantz S, Torell LM (1990) J Chem Phys 92:6271
49. Petersen G, Brodin A, Torell LM, Smith M (1994) Solid State Ionics 72:165
50. Armand M, Gorecki W, Andre'ani R (1990) In: Serosati B (ed) Second international symposium on polymer electrolytes. Elsevier Applied Science, London, p. 91
51. Brown SD, Greenbaum SG, McLin MG, Wintersgill MC, Fontanella JJ (1994) Solid State Ionics 67:257
52. Benrabah D, Baril D, Sanchez JY, Armand M, Gard GG (1993) J Chem Soc Faraday Trans 89:355
53. Webber A (1991) J Electrochem Soc 138:2586
54. Nagasubramanian G, Distefano S (1990) J Electrochem Soc 137:3830
55. Yeh TF, Okamoto Y, Skotheim TA (1990) Mol Cryst Liq Cryst 190:205
56. Watanabe M, Yamada S, Sanui K, Ogata N (1993) J Chem Soc Chem Commun 929
57. Angell CA, Liu C, Sanchez E (1993) Nature 362:137
58. Angell CA, Fan J, Liu C, Lu Q, Sanchez E, Xu K (1994) Solid State Ionics 69:343
59. Miyamoto T, Shibayama K (1973) J Appl Phys 44:5372
60. Killis A, Le Nest JF, Cheradame H, Gandini A (1982) Makromol Chem 183:2835
61. Williams ML, Landel RF, Ferry JD (1955) J Am Chem Soc 77:3701
62. Watanabe M, Itoh M, Sanui K, Ogata N (1987) Macromolecules 20:569
63. Kobayashi N, Ohno H, Tsuchida E (1986) Nippon Kagaku Kaishi 1986:441
64. Robitaille C, Marques S, Borls D, Prud'homme J (1987) Macromolecules 20:3023
65. Souquet JL, Levy M, Duclot M (1994) Solid State Ionics 70, 71:337
66. Ohno H, Sasayama H (1991) Polym Prep Jpn 40:596
67. Takeoka S, Sakai H, Tsuchida E (1990) Chem Lett 1990:1539
68. Ohno H, Kobayashi N, Takeoka S, Ishizaka H, Tsuchida E (1990) Solid State Ionics 40, 41:655
69. Frensdorff HK (1971) J Am Chem Soc 93:600
70. Takeoka S, Maeda Y, Tsuchida E, Ohno H (1990) Polym Adv Technol 1:201
71. Ohta T, Takeoka S, Tsuchida E, Feng HY, Fu ZS, Wang YP (1992) Polym Adv Technol 3:433
72. Hopper A (1988) Chem Indust 17:198
73. Tofield BC, Dell RM, Jensen J (1984) AERE Harwell Report, 11261
74. Scrosati B (1992) J Electrochem Soc 139:2776
75. Gauthier M, Fauteux D, Vassort G, Belanger A, Duval M, Ricoux P, Chabagno J-M, Muller D, Rigaud P, Armand MB, Deroo D (1985) J Electrochem Soc 132:1333
76. Hooper A, North JM (1983) Solid State Ionics 9, 10:1161
77. Hammou A, Hammouche A (1988) Electrochim Acta 33:1719
78. Capuano F, Croce F, Scrosati B (1992) J Power Sources 37:369
79. Abraham KM, Alamgir M, Perrotti SJ (1988) J Electrochem Soc 135:535
80. Liu M, Visco SJ, De Jonghe LC (1991) J Electrochem Soc 138:1891
81. Vassort G, Gauthier M, Harrey PE, Brochu F, Armand MB (1988) In: Gabano JP, Takehara Z, Bro P (eds) Proceedings symposium on lithium batteries, vol 88–86. Electrochemical Society, p. 780
82. Abraham KM, Alamgir M (1991) Chem Mater 3:399
83. Bonino F, Croce F, Panero S (1994) Solid State Ionics 70, 71:654
84. Croce F, Passerini S, Scrosati B (1994) J Electrochem Soc 141:1405
85. Chiang CK (1981) Polymer 22:1454
86. Flouletier M, Degott P, Armand MB (1983) Solid State Ionics 8:165
87. Nagatomo T, Kakehata H, Ichikawa C, Omoto O (1985) Jpn J Appl Phys 24:L397
88. Nova'k P, Inganas O (1988) J Electrochem Soc 135:2485
89. Deb SK (1968) Proc Roy Soc A304:211
90. Gottesfield S, McIntyre JE, Beni G, Shay JL (1978) Appl Phys Lett 33:208
91. Itaya K, Shibayama K, Akahoshi H, Toshima S (1982) J Appl Phys 53:804

92. Akahoshi H, Toshima S, Itaya K (1981) J Phys Chem 85:818
93. Ohsaka T, Kunimura S, Oyama N (1988) Electrochim Acta 33:639
94. Gusfafsson JC, Liedberg B, Inganas O (1994) Solid State Ionics 69:145
95. Nishikawa M, Ohno H, Kobayashi N, Tsuchida E, Hirohashi R (1988) Nippon Shashin Gakkai-shi 51:184
96. Oyama N, Ohsaka T, Menda M, Ohno H (1989) Denki Kagaku 57:1172
97. Watanabe M, Wooster TT, Murray RW (1991) J Phys Chem 95:4573
98. Watanabe M, Longmire ML, Murray RW (1990) J Phys Chem 94:2614
99. Shi G, Ohno H (1991) J Electroanal Chem 314:59
100. Oliver BN, Egekeze JO, Murray RW (1988) J Am Chem Soc 110:2321
101. Ohno H, Yamaguchi N (992) Kagaku 47:360
102. Reed RA, Geng L, Murray RW (1986) J Electroanal Chem 208:185
103. Geng L, Reed RA, Kim M-H, Wooster TT, Oliver BN, Egekeze J, Kennedy RT, Jorgenson JW, Murray RW (1989) J Am Chem Soc 111:1619
104. Jernigan JC, Murray RW (1987) J Am Chem Soc 109:1738
105. Ohno H, Yamaguchi N, Watanabe M (1993) Polym Adv Technol 4:133
106. Ohno H (1992) Electrochim Acta 37:1649
107. Ohno H, Tsukuda T (1992) J Electroanal Chem 341:137
108. Arbizzani C, Mastragostino M, Meneghello L, Hamaide T, Guyot A (1992) Electrochim Acta 37:1631
109. Sirkar KK, Wo WS (ed) (1992) Membrane handbook. Van Nostrand Publishers, New York
110. Comyn J (ed) (1985) Polymer Permeability. Elsevier Applied Science Publishers, London
111. Tsuchida E, Nishide H (1986) Topics Curr Chem 132:64
112. Tsuchida E, Nishide H (1977) Adv Polymer Sci 24:1
113. Nishide H, Tsuchida E (1991) Multiple interaction in molecule transport through macromolecular complexes. In: Tsuchida E (ed) Macromolecular complexes: dynamic interactions and electronic processes, chap 6. VCH Publishers, New York
114. Way JD, Noble RD (1992) Facilitated transport. In: Kirkar KK, Wo WS (eds) Membrane handboock, chap 44, Van Nostrand Publishers, New York
115. Johnson BM, Baker RW, Matson SL, Smith KL, Roteman IC, Tuttle ME, Lonsdale L (1987) J Mebr Sci 31:31
116. Wöhrle D, Bohlen H, Blum IK (1986) Makromol Chem 187:2081
117. Nishide H, Tsuchida E (1992) Polymer complex membranes for gas separation. In: Toshima N (ed) Polymers for gas separation, chap 6, VCH Publishers, New York
118. Nishide H, Ohyanagi M, Okada O, Tsuchida E (1986) Macromolecules 19:494
119. Ohyanagi M, Nishide H, Suenaga K, Tsuchida E (1988) Macromolecules 21:1590
120. Nishide H, Ohyanagi M, Okada O, Tsuchida E (1988) Macromolecules 21:2910
121. Nishide H, Kawakami H, Suzuki T, Azechi T, Tsuchida E (1990) Macromolecules 23:3714
122. Tsuchida E, Nishide H, Ohyanagi M, Kawakami H (1987) Macromolecules 20:1970
123. Nishide H, Kawakami H, Kurimura Y, Tsuchida E (1989) J Am Chem Soc 111:7175
124. Nishide H, Kawakami H, Tsuchida E, Kurimura Y (198) J Macromol Sci Chem A25:1339
125. Nishide H, Kawakami H, Toda S, Kamiya Y, Tsuchida E (1990) Macromolecules 23:4325
126. Tsuchida E, Nishide H, Ohyanagi M (1988) J Phys Chem 92:641
127. Nishide H, Ohyanagi M, Tsuchida E (1987) Polymer J 19:839
128. Nishide H, Suzuki T, Soejima Y, Tsuchida E (1994) Makromol Chem Symp 80:145
129. Nishide H, Kawakami H, Suzuki T, Azechi Y, Soejima Y, Tsuchida E (1991) Macromolecules 24:6306
130. Kurimura Y, Ohta F, Gohta J, Nishide H, Tsuchida E (1982) Macromol Chem 183:2889
131. Nishide H, Ohyanagi M, Okada O, Tsuchida E (1987) Macromolecules 20:417
132. Barrer RM (1984) J Membr Sci 18:2
133. Nishide H, Suzuki T, Kawakami H, Tsuchida E (1994) J Phys Chem 98:5084
134. Krieger IM, Mulholland GW, Dickey CS (1967) J Phys Chem 71:1123

135. Suzuki T, Soejima Y, Nishide H, Tsuchida E (1995) Bull Chem Soc Jpn 68:333
136. Kapoor A, Yong RT (1991) Chem Eng Sci 45:3261
137. Nishide H, Suzuki T, Nakagawa R, Tsuchida E (1994) J Am Chem Soc 116:4503
138. Franklin B, Bernard GF (1987) In: Dyson J (ed) Hemoglobin: molecular, genetic and clinical aspects. Saunders, Philadelphia, p. 37
139. Imai K, Yonetani T (1975) J Biol Chem 250:7093
140. Mathis P, Rutherford AW (1987) In: Amesz J (ed) Photosynthesis. Elsevier, New York, p. 63
141. Deisenhofer J, Epp O, Miki K, Huber R, Michel H (1985) Nature 318:618
142. Koening DF (1965) Acta Cryst 18:663
143. Roberet W, Lee YJ (1987) Struct Bond 64:1
144. Caughey WS, Ibers JA (1977) J Am Chem Soc 99:6639
145. Byrn MP, Curtis CJ, Khan SI, Sawin PA, Tsurumi R, Strouse CE (1991) J Am Chem Soc 113:2501; Byrn MP, Burtis CJ, Goldberg I, Hsiou Y, Khan SI, Sawin PA, Tendick SK, Strouse CE (1991) J Am Chem Soc 113:6549
146. Guilard R, Kadish KM (1988) Chem Rev 88:1121
147. O'Keeffe DH, Barlow CH, Smythe GA, Fuchsman WH, Moss TH, Lilienthal HR, Caughey WS (1975) Bioinorg Chem 5:125
148. Hoffman AB, Collins DM, Day VW, Fleischer EB, Srivastava TS, Hoard JL (1972) J Am Chem Soc 94:3620
149. LaMar GN, Eaton GR, Holm RH, Walker FA (1973) J Am Chem Soc 95:63
150. Schultz H, Lehmann H, Rein M, Hanack M (1991) Struct Bond 74:41
151. Hanack M (1984) Mol Cryst Liq Cryst 105: 133; Hanack M, Degar S, lange A, Ziplies T (1986) Synth Metals 15:207; Hanack M, Keppeler V, Lange A, Hirsch A, Dieing R (1993) In: Phthalocyanines, properties and applications, vol 2. VCH Publishers, Weinheim, p. 43
152. Collman JP, McDevitt JT, Yee GT, Leidner CR, McCullough LG, Little WA, Torrance JB (1986) Proc Natl Acad Sci USA 83:4581; Collman JP, McDevitt JT, Leidner CR, Yee GT, Torrance JB, Little WA (1987) J Am Chem Soc 109:4606
153. Gunther MJ, McLaughlin GM, Berry KJ, Murray KS (1984) Inorg Chem 23:283
154. Fleischer EB, Schacher AM (1991) Inorg Chem 30:3763
155. Anderson HL (1994) Inorg Chem 33:972
156. White WI (1978) In: Dolphin D (ed) The porphyrins 5:303
157. Alexander AE (1937) J Chem Soc 1813
158. Gallagher WA, Elliot WB (1973) Ann NY Acad Sci USA 206:463
159. Caughey WS, Eberspaeche H, Fuchsman WH, McCoy S, Alben JO (1969) Ann NY Acad Sci USA 153:722
160. Inamura I, Uchida K (1991) Bull Chem Soc Jpn 64:2005
161. Fuhrhop J-H, Demoulin C, Boettcher C, Koening J, Siggel U (1992) J Am Chem Soc 114:4159
162. Fisher JRE, Rosenbach-Belkin V, Scherz A (1990) Biophys J 58:461
163. Guilard R, Senglet N, Liu YH, Sazou D, Findsen E, Faure D, Courieres TD, Kadish KM (1991) Inrg Chem 30:1898; Guilard R, Senglet N, Liu YH, Sazou D, Findsen E, Faure D, Courieres TD, Kadish KM (1991) Inorg Chem 30:1898
164. Fuhrhop J-H, Koening J (1994) In: Membranes and molecular assemblies: the synkinetic approach. Royal Chemistry Society, Cambridge, p. 98
165. Fuhrhop J-H, Bindig U, Siggel U (1994) J Chem Soc Chem Commun 1583
166. Fuhrhop J-H, Bindig U, Siggel U (1993) J Am Chem Soc 115:11036
167. Tsuchida E, Komatsu T, Kumamoto S, Toyano N, Nishide H (1993) J Chem Soc Chem Commun 1731
168. Komatsu T, Arai K, Nishide H, Tsuchida E (1993) Chem Lett 1949
169. Chang CK, Traylor TG (1973) Proc Natl Acad Sci USA 70:2647; Traylor TG (1981) Acc Chem Res 14:102; Traylor RG, Tsuchiya S, Campbell D, Mitchel M, Stynes D, Koga N (1985) J Am Chem Soc 107:604
170. Collman JP, Gagne RR, Halbert TR, Marchon J-C, Reed CA (1973) J Am Chem Soc 95:7868; Collman JP, Gagne RR, Reed CA, Halbert TR, Lang G, Robinson WT (1975)

J Am Chem Soc 97:1427; Collman JP, Brauman, Iverson BL, Sesler JL, Morris RM, Gibson QH (1983) J Am Chem Soc 105:3052; Collman JP, Herrmann PC, Boitrel B, Zhang X, Eberspacher TA, Fu L (1994) J Am Chem Soc 116:9783

171. Almog J, Baldwin JE, Dyer RL, Peters M (1975) J Am Chem Soc 97:226; Budge JR, Ellis PE Jr, Jones RD, Linard JE, Basolo F, Baldwin JE (1979) J Am Chem Soc 101:4760

172. Momenteau M (1986) Pure Appl Chem 58:1493; Momenteau M, Mispelter J, Loock B, Bisagni E. J Chem Soc Perkin Trans 1:189

173. Chang CK, Ward B, Young R, Kondylis MP (1988) J Macromol Sci Chem 25:1307

174. Battersby AR, Hamilton AD (1980) J Chem Soc Chem Commun 117; Battersby AR, Bartholomew SA, Nitta T (1983) J Chem Soc Chem Commun 1291

175. Suslick KS, Fox MM (983) J Am Chem Soc 105:3507; Suslick KS, Fox MM, Reinert TJ (1984) J Am Chem Soc 106:4522

176. Baldwin JE, Cameron JH, Crossley MJ, Dagley IJ, Hall SR, Klose T (1984) J Chem Soc Dalton Trans 1739

177. Komatsu T, Hasegawa E, Nishide H, Tsuchida E (1990) J Chem Soc Chem Commun 66; Komatsu T, Hasegawa E, Jumamoto S, Nishide H, Tsuchida E (1992) J Chem Soc Dalton Trans 3281; Tsuchida E, Komatsu T, Nakata T, Hasegawa E, Nishide H (1991) J Chem Soc Dalton Trans 3285; Komatsu T, Kumamoto S, Nishide H, Tsuchida E (1993) Bull Chem Soc Jpn 66:1640; Tsuchida E, Komatsu T, Arai K, Nishide H (1993) J Chem Soc Dalton Trans 2465; Komatsu T, Kumamoto S, Ando K, Nishide H, Tsuchida E (1995) J Chem Soc Perkin Trans 2 (in press)

178. Komatsu T, Nakao K, Nishide H, Tsuchida E (1993) J Chem Soc Chem Commun 728

179. Komatsu T, Tsuchida E, Sieggel U, Fuhrhop J-H (unpublished results)

180. Chang CK, Traylor TG (1975) Proc Natl Acad Sci USA 72:1166; Geibel J, Cannon J, Campbell D, Traylor TG (1978) J Am Chem Soc 100:3575; Traylor TG, Chang CK, Geibel J, Berzinis GA, Mincey T, Cannon J (1979) J Am Chem Soc 101:6716

181. Esch J, Roks MFM, Nolte RJM (1986) J Am Chem Soc 108:6093

182. Groves JT, Neumann R (1987) J Am Chem Soc 109:5045; Groves JT, Neumann R (1989) J Am Chem Soc 111:2900; Groves JT, Fate GD, Lahiri J (1994) J Chem Soc 116:5477

183. Nango M, Mizusawa T, Miyake T, Yoshinaga J (1990) J Am Chem Soc 112:1640

184. Tsuchida E, Nishide H, Yuasa M, Sekine M (1983) Chem Lett 473; Tsuchida E, Nishide H (1986) Top Curr Chem 132:64; Tsuchida E, Nishide H, Yuasa M (1985) J Chem Soc Dalton Trans 275; Tsuchida E, Nishide H, Yuasa M, Hasegawa E, Eshima K, Matsushita Y (1989) Macromolecules 22:2103

185. Tsuchida E, Komatsu T, Arai K, Nishide H (1993) J Chem Soc Chem Commun 730; Tsuchida E, Komatsu T, Arai K, Yamada K, Nishide H, Fuhrhop JH (1994) Langumuir (submitted)

186. Schick GA, Schreiman IC, Wagner RW, Lindsey JS, Bocian DF (1989) J Am Chem Soc 111:1344

187. Esch JH, Peters AMP, Nolte RJM (1990) J Chem Soc Chem Commun 638

188. Barber DC, Freitag-Beestron RA, Whitten DG (1991) J Phys Chem 95:4074

189. Nagata T, Osuka A, Maruyama K (1990) J Am Chem Soc 112:3054

190. Kasha M (1963) Rad Res 20:55; Kasha M, Rawls HR, El-Bayoumi MA (1965) Pure Appl Chem 11:371

191. Emerson ES, Conlin MA, Rosenoff AE, Norland K, Rodriguez H, Chin D, Bird GR (1967) J Phys Chem 71:2396

192. Sharma VS, Schmidt MR, Ranney HM (1976) J Biol Chem 251:4267

193. Steinmier RC, Parkhuvst LJ (1975) Biochemistry 14:1564

194. Brunori M, Schster TM (1969) J Biol Chem 244:4046; Antonini F, Brunori M (1970) In: Hemoglobin and myoglobin and their reactions with ligands. North Holland, Amsterdam, p. 220

195. Schuberth O, Wretlind A (1961) Acta Chir Scand (Suppl) 278:1

196. Hallberg D (1965) Acta Physiol Scand 64:306

197. Untracht SH (1982) Biochim Biophys Acta 711:176

198. Redgrave TG, Maranhao RC (1985) Biochim Biophys Acta 835:104
199. Davis SS, Washington C, West P, Illum L, Liversidge G, Sternson L, Kirsh R (1987) Ann NY Acad Sci 507:75
200. Riess JG, Dalfors JL (1992) Biomat Art Calls Immob Biotech 20:839
201. Komatsu T, Matsubuchi E, Nishide H, Tsuchida E (1992) Chem Lett 1325; Tsuchida E, Komatsu T, Kawai N, Nishide H, Kakizaki T, Kobayashi K (1993) Artif Organs Today 3: 137; Komatsu T, Muramutsu Y, Nishide H, Tsuchida E, Kakizaki T, Kobayashi K (1994) Artif Organs Today (in press)
202. Pittman CU Jr (1982) In: Wilkinson G, Stone FGA, Abel G (eds) Comprehensive organo-metallic chemistry, vol 8. Oxford, p. 533
203. Ciardelli F, Braca G, Carlini C, Sbrana G, Valentini G (1982) J Mol Catal 14:1
204. Tsuchida E (1982) Macromol Rev 16:397
205. Pracejus H East German Patent 92:031
206. Kinting A, Dobler C, Kreuzfeld H-J, Krause H (1985) Wiss Z Karl-Marx-Univ Leizpig, Math-Naturwiss R, 34, 169 and references therein
207. Bradley JS, Hill E, Lenowicz ME, Witzke H (1987) J Mol Catal 41:59
208. Ciardelli F, Carlini C, Pertici P, Valentini G (1989) J Macromol Sci Chem A26:327
209. Ozin GA (1985) Chemtech:488
210. Ciardelli F, Altomare A, Conti G, Arribas G, Mendez B, Ismayel A (1994) Macromol Symp 80:29
211. Ozin GA, Steele MR (1994) Macromol Symp 80:45
212. Pomogailo AD (1992) Russian Chem Rev 61:133
213. Carlini C, Braca G, Ciardelli F, Sbrana G (1972) J Mol Catal 2:379
214. Lawrence SA, Sermon PA, Feinstein-Jaffe I (1989) J Mol Catal 51:117
215. Gokak DT, Ram RN (1989) J Mol Catal 49:285
216. Diversi P, ingrosso G, Lucherini A, Minutillo A (1987) J Mol Catal 40:359
217. Terreros P, Pastor E, Fierro JLG (1989) J Mol Catal 53:359
218. Innorta G, Modelli A, Scagnolari F, Foffani A (1980) J Organometal Chem 185:403
219. Valentini G, Sbrana G, Braca G (1981) J Mol Catal 11:383
220. Cao SK, Huang MY, Jiang YY (1989) J Macromol Sci Chem A26:381
221. Uematzu T, Nakazawa Y, Akutsu F, Shimazu S, Miura M (1987) Makromol Chem 188:1085
222. Lito J, Prochazka M, Arnold DB, Gates BC (1985) J Mol Catal 31:89
223. Rybak WK, Ziolkowski JJ (1981) J Mol Catal 11:365
224. Dobson A, Robinson SD (1977) Inorg Chem 16:1321
225. Valentini G, Cecchi A, Di Bugno C, Braca G, Sbrana G (1986) In: Yermakov Y, Lik-holobov V (eds) Homogeneous and heterogeneous catalysis. VNU Science Press, Utrecht, The Netherlands, p. 765
226. Shah HN, Gokak DT, Ram RN (1990) J Mol Catal 60:141
227. John J, Ram RN (1994) Polymer Int 34:369
228. Peuckert M, Keim W (1984) J Mol Catal 22:289
229. Braca G, Di Girolamo M, Raspolli-Galletti AM, Sbrana G (1992) J Mol Catal 74:421
230. Cermák J, Soukupová L, Chvalovsky V (1993) J Mol Catal 80:181
231. Sherrington DC, Tang H (1994) Macromol Symp 80:193
232. Tang H, Sherrington DC (1994) J Mol Catal 94:7
233. van Leeuwen PWNH, Jongsma T, Challa G (1994) Macromol Symp 80:241
234. Pan C, Zong H (1994) Macromol Symp 80:265
235. Tsuchida E, Nishide H, Ohyanagi M, Kawakami H (1987) Macromolecules 20:1907
236. Nishide H, Yuasa M, Hasegawa E, Tsuchida E (1987) Macromolecules 20:1913
237. Bedioui F, Moisy F, Devynck J, Salmon L, Bied-Charreton C (1989) paper presented at VISHHC, Pisa, 24–29 September
238. Ganeshpure PA, Satish S, Sivaram S (1989) J Mol Catal 50:L1
239. Challa G (1981) Makromol Chem (Suppl) 5:70
240. Yamashita K, Okada L, Suzuki Y, Tsuda K (1988) Makromol Chem Rapid Commun 9:705

241. Miyazawa T, Endo T (1988) J Mol Catal 49:L31
242. Muralidharan S, Freiser H (1989) J Mol Catal 50:181
243. Tsuchida E (1994) Macromol Symp 80:17
244. Bhatia RK, Rao GN (1994) J Mol Catal 93:29
245. Collman JP, Hegedus LS, Cooke MP, Norton JR, Dolcetti G, Marquardt DN (1972) J Am Chem Soc 94:1789
246. Efendiev AA (1994) Macromol Symp 80:289
247. Andrews MP, Ozin GA (1989) Chem Mater 1:174
248. Pertici P, Vituli G, Carlini C, Ciardelli F (1981) J Mol Catal 11:353
249. Ciardelli F, Pertici P (1985) Z Naturforsch 40b:133
250. Wilke G (1963) Angew Chem Int Ed Engl 2:195
251. Collman JP Kosydar KM, Bressan M, Lamanna W, Garret T (1984) J Am Chem Soc 106:2569
252. Toshima N, Taranishi T, Saito Y (1991) Macromol Symp 59:327
253. Monterra C, Zecchina A, Costa G (1989) Structure and reactivity of surfaces, Elsevier, Amsterdam
254. Yermakov YI, Likholobov V (1987) Homogeneous and heterogeneous catalysis, VNU Science press, Utrecht, The Netherlands
255. Iwazawa Y, Gates BC (1989) Chemtech:173
256. Aramendia MA, Bran V, Jimenez C, Marinas JM, Rodero F (1987) Gazz Chim Ital 117:39
257. Deganello G, Duca D, Liotta LF, Martorana A, Venezia AM (1994) Gazz Chim Ital 124:229
258. Yermakov YuI (1986) Catal Rev 13:77
259. Marks T (1992) Acc Chem Res 25:57
260. Schock LE, Marks T (1988) J Am Chem Soc 110:7701
261. Collins S, Kelly WM, Holden DA (1992) Macromolecules 25:1780
262. Soga K, Kaminaka M (1993) Makromol Chem 194:1745
263. Chien JCW, He D (1991) J Polym Sci (Part A) Polym Chem 29:1603
264. Ciardelli F, Altomare A, Arribas G, Conti G, Masi F, Menconi F (1994) In: Soga K, Terano M (eds) Catalyst design for tailor-made Polyolefins. Kodansha-Elsevier, Tokyo, p. 257
265. Conti G, Arribas G, Altomare A, Ciardelli F (1994) J Mol Catal 89:41
266. Engelhardt G, Michel D (1987) High resolution solid-state NMR of silicates and zeolites. Wiley, New York
267. Klinoswki J (1991) Chem Rev 91:1459
268. Sishta C, Hatorn RM, Marks TJ (1992) J Am Chem Soc 114:1112
269. Nesterov GA, Fink G, Zakarov VA (1989) Makromol Chem Rapid Common 10:669
270. Nesterov GA, Zakarov VA, Fink G, Fenzi W (1991) J Mol Catal 69:129
271. Nesterov GA, Zakarov VA, Fink G, Fenzi W (1991) J Mol Catal 66:367
272. Soga K, Koide R, Uozumi T (1993) Makromol Rapid Commun 14:511
273. Buck T, Wöhrle D, Schulz-Ehloff G, Anoheev A (1991) J Mol Catal 70:259; ibidem (1993) J Mol Catal 80:253
274. Kim DH, Britt RD, Klein MP, Sauger K (1990) J Am Chem Soc 112:9389
275. Kirk ML, Chan MK, Armstrong WH, Solomon EI (1992) J Am Chem Soc 114:10432
276. Collman JP, Rothrock RK, Finke RG, Moore EJ, Munch FR (1982) Inorg Chem 21:161
277. Lintvedt RL, Lynch WE, Zehetmair JK (1990) Inorg Chem 29:3009
278. Ghosh MC, Gelerinter E, Gould ES (1991) Inorg Chem 31:702
279. Astruc D, Lacoste M, Toupet L (1991) J Chem Soc, Chem Commun:558
280. Lintvedt RL, Schoenfelner BA, Rupp KA (1986) Inorg Chem 25:2704
281. Sudha C, Mandal SK, Chakravarty AR (1993) Inorg Chem 32:3801
282. Sasaki Y, Suzuki M, Nagasawa A, Tokiwa A, Ebihara M, Yamaguchi T, Kabuto C, Ochi T, Ito T (1991) Inorg Chem 30:4903
283. Collman JP, Denisevich P, Konai Y, Marrocco M, Koval C, Anson FC (1980) J Am Chem Soc 102:6027
284. Durand RR Jr, Bencosme CS, Colknan JP, Anson FC (1983) J Am Chem Soc 105:2710

285. Liu H-Y, Weaver MJ, Wang CB, Chang CK (1983) J Electroanal Chem Interfacial Electrochem 145:439
286. Ni C-L, Abdalmuhdi L, Chang CK, Anson FC (1987) J Phys Chem 91:1158
287. Collman JP, Kim K (1986) J Am Chem Soc 108:7847
288. Shi C, Anson FC (1991) J Am Chem Soc 113:9564
289. Zhang J, Anson FC (1992) J Electroanal Chem 341:323
290. Shi C, Anson FC (1990) Inorg Chem 29:4298
291. Shi C, Anson FC (1992) Inorg Chem 31:5078
292. Steiger B, Shi C, Anson FC (1993) Inorg Chem 32:2107
293. Matsumoto K, Sakai K, Nishino K, Tokisue Y, Ito R, Nishide T, Shichi Y (1992) J Am Chem Soc 114:8110
294. Matsumoto K, Matiba N (1988) Inorg Chim Acta 142:59
295. Hay AS, Blanchard HS, Endres GF, Eustance JW (1959) J Am Chem Soc 81:6335
296. Hay AS (1959) Adv Polymer Sci 4:496
297. Hay AS, Dana DE (1989) J Polymer Sci, Polymer Chem 27:873
298. Tsuchida E, Kaneko M, Nishide H (1972) Makromol Chem 151:221
299. Tsuchida E, Kaneko M, Kurimura Y (1970) Makromol Chem 132:209, 215
300. Hay AS (1960)(1962) J Org Chem 25: 1275; 27:3320
301. Hay AS (1962) Encycl Polymer Sci Technol 10:92
302. Nishide H, Minakata T, Tsuchida E, Yamada S (1982) Makromol Chem 183:1989
303. Hyun SH, Nishide H, Tsuchida E, Yamada S (1988) Bull Chem Soc Jpn 61X:1319
304. Tsuchida E, Kaneko H, Nishide H (1972) Makromol Chem 151X:235
305. Tsuchida E, Kaneko M, Nishide H (1973) Makromol Chem 164:203
306. Taylor WL, Battersby AR (1968) Oxidative coupling of phenols. Marcel Dekker, New York
307. Challa G (1981) Macromol Chem (Suppl) 5:70
308. Schouten AJ, Noordegraaf D, Jekel AP, Challa G (1979) J Mol Catal 5:5331
309. Viersen FJ, Challa G, Reedijk J (1990) Polymer 31:1368
310. Tsuchida E, Yamamoto K (1993) Bioinorganic catalysis In: Reedijk J (ed) Marcel Dekker, New York, p. 29
311. Tsuchida E, Nishide H, Maekawa T (1985) Reactive oligomer. InL Harms FW, Spinelli HJ (eds) ACS Symp. Series, Washington, DC, p. 175
312. Yamamoto K, Nishide H, Tsuchida E (1987) Macromol Chem Rapid Commun 8:11
313. Verlaan JPL, Bootsma JPC, Koning CE, Challa G (1983) J Mol Catal 18:159
314. Meinders HC, Challa G (1980) J Mol Catal 7:321
315. Koning CE, Brinkhuis R, Wevers R, Challa G (1987) Polymer 28:2310
316. Chen W, Challa G (1990) Eur Polym J 26:1211
317. Chen W, Challa G (1990) Polymer 31:2171
318. Chen W, Challa G (1991) React Polym 14:63
319. Chen W, Boven G, Challa G (1991) Macromolecules 24:3982
320. Challa G, Chen W, Reedijk J (1992) Makromol Chem Macromol Symp 59:59
321. Karlin KD (1984) J Am Chem Soc 106:2121
322. Uechi T, Yamaguchi H, Ueda I, Yasukouchi K (1980) Bull Chem Soc Jpn 53:3483
323. Li Y, Abramovitch RA, Houk KN (198) J Org Chem 54:2911
324. Tsuchida E, Yamamoto K, Oyaizu K, Iwasaki N, Anson FC (1994) Inorg Chem 33:1056
325. Yamamoto K, Tsuchida E, Nishide H, Jikei M, Oyaizu K (1993) Macromolecules 26:3432
326. Tsuchida E, Yamamoto K, Jikei M, Nishide H (1989) Macromolecules 22:4138
327. Yamamoto K, Yoshida S, Nishide H, Tsuchida E (1992) J Electrochem Soc 369:2401
328. Tsuchida E, Yamamoto K, Nishide H, Yoshda S, Jikei M (1990) Macromolecules 23:2101
329. Capozzi G, Lucchini V, Modena G, Rivietti F (1974) J Chem Soc Perkin Trans 2:900
330. Gibin AS, Smitl WA, Bogdanov VS (1980) Tetrahedron 21:383
331. Tsuchida E, Yamamoto K, Nishide H, Jikei M (1990) Macromolecules 23:930
332. Yamamoto K, Jikei M, Tsuchida E (1994) Macromolecules 27:4312
333. Tsuchida E, Yamamoto K, Shouji E (1994) J Makromol Sci Chem A34:1579

334. Yamamoto K, Jikei M, Oyaizu K, Suzuki F, Nishide H, Tsuchida E (1994) Bull Chem Soc Jpn 67:251
335. Yamamoto K, Jikei M, Katoh J, Nishide H, Tsuchida E (1992) Macromolecules 25:2698
336. Yamamoto K, Oyaizu K, Iwasaki N, Tsuchida E (1993) Chem Lett 1993:1223
337. Bnadies JA, Carrano CJ (1986) J Am Chem Soc 108:4088
338. Viersen FJ, Challa G, Reedijk J (1990) Polymer 31:1368
339. Kovacic P, Koch FW (1971) J Macromol Sci Rev Macromol Chem C5:259
340. Speigh JG, Ivory DH, Millar GG,Sowa JM, Shackellete LW, Chace RR, Baughman RH (1979) J Chem Phys 71:1501
341. Kovacic P, Kyriakis A (1963) J Am Chem Soc 55:454
342. Brown CE, Jones MB, Kovacic P (1980) J Polymer Sci, Polym Lett 18:653
343. Yamamoto K, Asada T, Nishide H, Tsuchida E (1988) Polymer Bull 19:533
344. Tsuchida E, Yamamoto K, Asada T, Nishide H (1987) Chem Lett 1987:1541
345. Yamamoto K, Asada T, Nishide H, Tsuchida E (1988) Bull Chem Soc Jpn 61:1731
346. Toshima N, Komaki K, Koshirai A, Hirai H (1988) Bull Chem Soc Jpn 61:2551
347. Groves JT, Quinn R (1985) J Am Chem Soc 107:5790
348. Che C, Lau K, Lau T, Poon C (1990) J Am Chem Soc 112:5176

5 Photoinduced Electron Transport of Macromolecular Metal Complexes

M. Kaneko and D. Wöhrle

Electron transport as an important process for macromolecule–metal complexes (MMC) to function as active centers in catalysis is discussed in Chap. 4.4 and 4.5. This behavior is directly related to energy levels of metals, metal ions, or metal clusters in the special environemnt of a ligand and/or high molecular compound to lavor thermodynamically the electron interaction and to allow quick and selective electron transport. Electrical conduction of solid polymer metal complexes as discussed in Chap. 5.3.1 is related to the charge conduction in the polymer bulk either by transport in valence and conduction band levels or by charge hopping between metal complex centers often enhanced by doping with small molecules or ions. Interaction of small molecules with metal ions in the core of the complexes also as a result of transfer of electron density is part of Chaps. 4.2 and 4.3.

Irradiation of macromolecule–metal complexes, for example in the visible region of light if colored MMC are employed, produce the excited states that can be used for electron transport in photochemical reactions, photoelectrochemical devices, photovoltaic cells, sensors, etc. The most important photoinduced electron transport exists in photosynthesis of green plants providing almost all the energy sources for biological activities [1, 2]. In recent years electron transport in biological systems has attracted a great deal of attention [3–8] especially for metalloenzymes (see also Chap. 3). In these studies photoexcited-state electron transfer reactions of metal complexes are often utilized in order to analyze the rapid ns-to-ps order reaction rate by laser flash photolysis techniques.

The aim of this chapter is not to review the extended area of research on biological and its model systems, but to focus on our approach to photoinduced electron transport of macromolecule–metal complexes in order to elucidate especially the microenvironmetal effect of polymer solutions, polymer bulk, or molecular aggregates in the photoprocesses. This view is important not only for fundamental sciene, but also for the future development of devices based on photoinduced electron transport. On one side conversion of solar energy (solar photochemistry, photoelectrochemistry, photovaltaic), and on the other side use of artificial light (photoconduction, optoelectronic, information storage), must be considered. Most of these subjects are in the stage of fundamental research, and few of them have found the way for broad application as has photoconduction.

The photoinduced electron transport of macromolecule–metal complexes in solution is described in Chap. 5.1, and at the solid/liquid interface, in Chap. 5.2. In many cases the metal complex itself is the photoexcitation center that is coupled with another electron donor or acceptor, but in some cases the metal complex works as an electron acceptor of donor, or as a catalyst with another photoexcitation center such as a semiconductor. The situation in solid macromolecular–metal complexes without and under irradiation is reviewed briefly in Chap. 5.3. Besides fundamentals in conduction, processes under irradiation, such as photoconduction, nonlinear optics, and hole burning as information storage, are considered.

5.1 Photoinduced Electron Transport in Solution

5.1.1 Determination of Rate and Mechanism

In order to study excited-state electron transfer reactions, laser flash photolysis is usually used by which formation and/or decay rate of the reaction intermediate or product can be obtained from transient absorption measurement. When the photoexcited state of the excitation center emits light (fluorescence or phosphorescence), the electron transfer causes quenching of the photoluminescence: therefore, the rate can be obtained by studying the quenching rate, and in this case the electron acceptor or donor is called a quencher.

When the electron transfer takes place by diffusion and collision of the excited-state molecule and a quencher (electron acceptor or donor) in a solution, the mechanism is called dynamic. In this case the second-order rate constant $(k_{q,2})$ can be determined by the dependence of the relative emission intensity (I_0/I) in the absence (I_0) and presence (I) of the quencher (Q) upon the quencher concentration [Q] according to Eqs. (1) and (2):

$$I_0/I = 1 + k_{sv}[Q] \tag{1}$$

$$k_{q,2} = k_{sv}/\tau_0, \tag{2}$$

where Eq. (1) is called the Stern-Volmer equation [8], k_{sv} Stern-Volmer constant, and τ_0 emission lifetime of the photoexcited state in the absence of the quencher. In this dynamic quenching the relative emission lifetime (τ_0/τ) is equal to I_0/I, namely the Stern-Volmer plots based on I_0/I and τ_0/τ fall on the same line. In the dynamic mechanism the rate is diffusion-limited so that the maximal second-order constant is below the order of approximately 10^{10} dm^3 mol^{-1} s^{-1}.

When there is an interaction between the excited center and the quencher (acceptor or donor) before excitation, or when the quencher exists in a quenching sphere around the excitation center molecule where quenching takes place without diffusion, photoinduced electron transfer can take place immediately after excitation without diffusion of the molecule. This mechanism is called "static," and the Stern-Volmer plots of I_0/I sometimes show a curve. In a complete static reaction case the slope of τ_0/τ plots is always zero, i.e., the emission lifetime does not change by the presence of the quencher. In a system of complete static reaction, the I_0/I vs [Q] plots may be upward-deviating when [Q] is high enough, following the so-called Perrin plots [9] expressed by Eq. (3), where V is the quenching sphere volume and N_A is Avogadro's number:

$$\ln(I_0/I) = VN_A[Q]. \tag{3}$$

From this relation the quenching sphere volume is calculated in an entire static quenching reaction.

In the static reaction the rate can exceed diffusion limit, and therefore can be very rapid. In biological systems electron transport often takes place without diffusion of the molecules, which allows a very rapid process exceeding diffusion limit.

Also, in polymer systems there exists often interaction between excitation center and quencher because of polymer environment, which often leads to anomalously enhanced electron transport. However, the mechanism in polymer systems is often not simple and involves both dynamic and static processes. In addition, the interaction of the excitation center and the quencher may not be simple, eventually showing multistep equilibria of the interaction. These problems are described in the following sections.

5.1.2 Photoinduced Electron Transport of Polymer-Pendant Ru(bpy)$_3^{2+}$

Tris (2,2′-bipyridine) ruthenium(II) complex Ru(bpy)$_3^{2+}$ (Scheme 5.1) is attracting a great deal of attention as a photoexcitation center for photochemical energy conversion, because this complex is capable theoretically of photolyzing water [1, 2, 10, 11].

Unfortunately, there still exists difficulty in using this sensitizer as a real and reproducible photoexcitation center for water cleavage, although there have been some reports to claim water photolysis with this complex. Despite these circumstances, a lot of papers have appeared about fundamental excited state reactions of this sensitizer, because electron transfer reactivity of its excited state (charge transfer from metal to ligand) is very high. Note now that this complex emits phosphorescence at approximately 600 nm excited by 450 nm light, and the emission is quenched by electron transfer, so that the photoinduced electron transfer can be studied by its phosphorescence. It was of interest to study how the microenvironment of this complex affects the photoinduced electron transport relevant to biological systems in which the protein part plays an important role for the event. However, very few works have been done in this regard.

The present authors have prepared various copolymer-pendant Ru(bpy)$_3^{2+}$ complexes (Scheme 5.2) for the first time [12, 13], and studied the effect of the polymer in the photochemical electron transfer reactions (see also Chaps. 2.2.3 and 2.2.6).

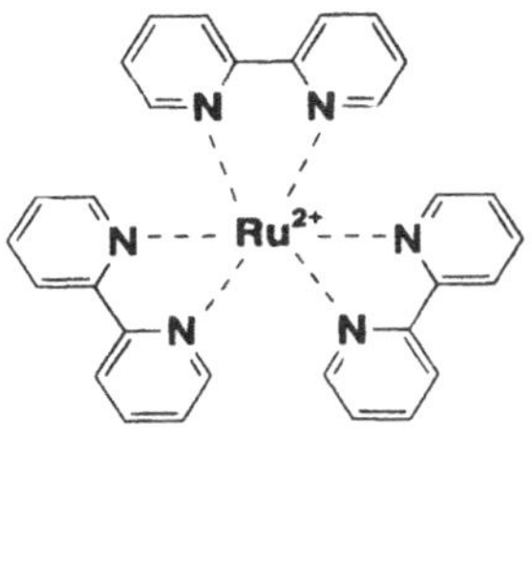

Scheme 5.1

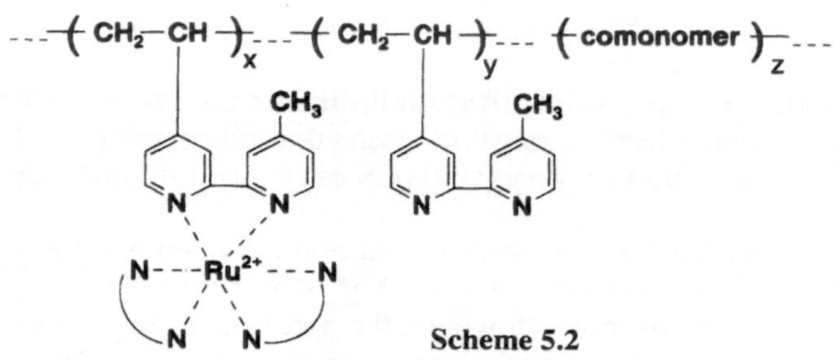

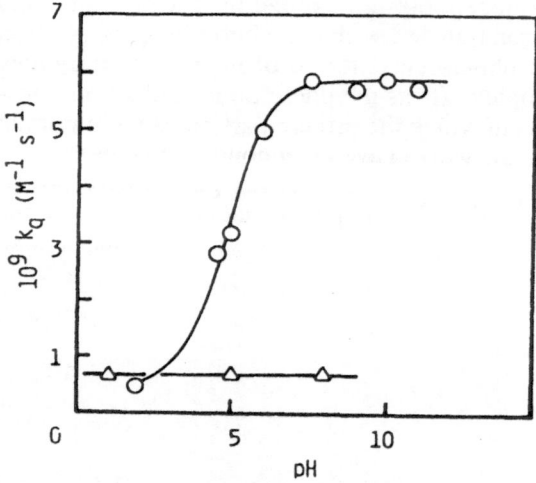

Scheme 5.2

1,1′-dimethyl-4,4′-bipyridinium dichloride (methylviologen, MV^{2+}) was usually used as an electron acceptor for the photoexcited $Ru(bpy)_3^{2+}$. This acceptor is often used because its reduced form is capable of reducing protons to hydrogen H_2 in the presence of some catalyst. The electron transfer is shown by Eq. (4):

$$Ru(bpy)_3^{2+*} + MV^{2+} \rightarrow Ru(bpy)_3^{3+} + MV^{+}. \tag{4}$$

Because back electron transfer is rapid, this reaction is studied only by laser flash photolysis or by emission from the excited Ru complex. When the comonomer is acrylic acid in the copolymer complex (Scheme 5.2), the apparent $k_{q,2}$ obtained by Eqs. (1) and (2) in an aqueous solution for the reaction of Eq. (4) was remarkably higher than the monomeric complex $Ru(bpy)_3^{2+}$ under neutral and alkaline conditions as shown in Fig. 5.1 [14].

The pH of approximately 5 where the curve for the polymer complex increases steeply indicates that dissociation of the acrylic acid ($pK_a = 5.08$ for poly(acrylic acid)) is responsible for the effect. The addition of NaCl to the

Fig. 5.1. pH dependence of the electron transfer rate constants from the photoexcited polymer complex (Scheme 5.2, comonomer = acrylic acid (*circles*)) and monomeric (Scheme 5.1 (*triangles*)) to MV^{2+} in water at 30 °C. Ru unit 10 μm

alkaline solution decreased the rate of the same value as the monomeric complex showing that the effect is electrostatic. It was concluded that the anionic domain formed by the dissociated carboxylate incorporates the cationic acceptor (MV^{2+}) to enhance the electron transfer.

It is of interest that the emission quenching of the $Ru(bpy)_3^{2+*}$ induced by the electron transfer was strongly dependent the molecular weight of the polymercomplex. The Stern-Volmer plots are shown in Fig. 5.2 [15]. When the molecular weight is low (2100), the plots show a line indicating that the process can be understood as a dynamic one. However, the higher molecular weight complexes (4400 and 13300), show higher quenching rate, i.e., higher electron transfer rate, and the curved plots suggest involvement of static quenching.

In methanol solution the slope and the profile of the Stern-Volmer plots are strongly dependent on the additives such as neutral salt, acid, and base as shown in Fig. 5.3 [16] indicating a different mechanism dependent on the conditions.

These results could not be understood by a simple interaction mechanism of the substrate and the polymer domain: therefore models have been proposed based on multistep equilibria of the interaction between the substrate and the polymer domain that surround the Ru complex. The most fitted interaction mo-

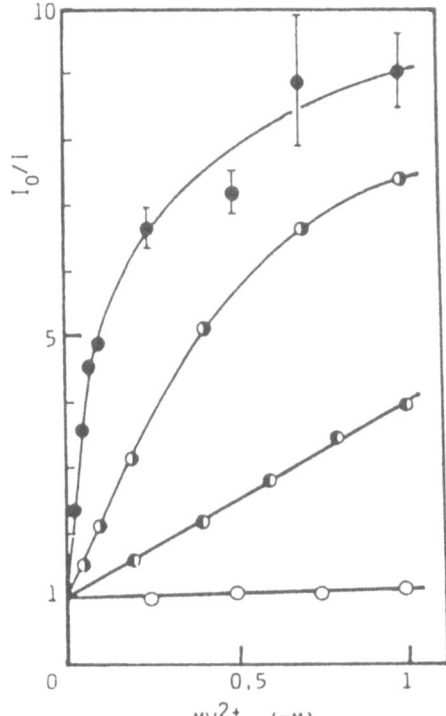

Fig. 5.2. Stern-Volmer plots for the electron transfer quenching of the photoexcited polymercomplex (Scheme 5.2) (mol. wt.: *left-shaded circles* 2100, *right-shaded circles* 4400; *filled circles* 13300) and of the monomeric complex (Scheme 5.1) (*open circles*) in 10 mM NaOH aqueous solution

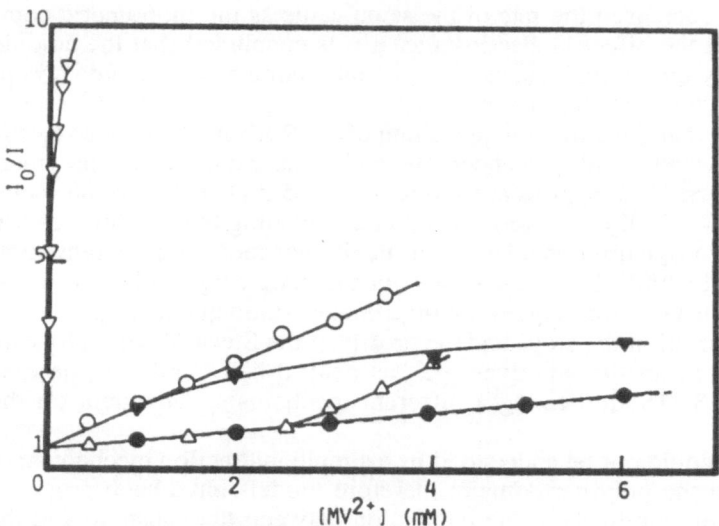

Fig. 5.3. Stern-Volmer plots for the electron transfer quenching of the photoexcited poly-mercomplex (Scheme 5.2) to MV^{2+} in methyl alcohol. Ru complex unit 10 μm. Additives: *filled triangles* none; *filled circles* 0.5 mM HCl; *open circles* 0.5 mM HCl and 0.1 M Et_4NBr; *open triangles down* 1 mM NaOH; *open triangles up* 0.5 mM HCl and 0.1 M H_2NCSNH_2; *dotted line* Ru $(bpy)_3^{2+}$ (Scheme 5.1)

del is the one that considers the stepwise incorporation of quencher Q (MV^{2+}) into the domain the Ru complex (denoted as [Ru]) as follows [15, 16], where K_i is each equilibrium constant:

$$[Ru] \qquad + Q \underset{k_{-a}}{\overset{k_a}{\rightleftharpoons}} [Ru\text{-}Q_1] \quad K_1 = k_a/k_{-a} \tag{5}$$

$$[Ru\text{-}Q_1] \quad + Q \underset{2k_{-a}}{\overset{k_a}{\rightleftharpoons}} [Ru\text{-}Q_2] \quad K_2 = K_1/2 \tag{6}$$

$$[Ru\text{-}Q_{i-1}]] + Q \underset{ik_{-a}}{\overset{k_a}{\rightleftharpoons}} [Ru\text{-}Q_1] \quad K_i = K_1/i \tag{7}$$

Under these equilibria, reaction mechanism models from 1 to 4 have been pro-posed, where k_d is rate constant for radiationless transition from the excited state to ground one, k_f rate constant for the emission, and k_q, 1 the first-order static quenching rate constant. When the static quenching rate is proportional to the number of Q incorporated into the domain, the rate constant should be $ik_{q,1}$.

Model 1

$$[Ru] \overset{h\nu}{\rightarrow} [Ru^*] \quad \left|\begin{array}{l} \xrightarrow{k_{q,2}[Q]} [Ru] \\ \xrightarrow{k_d} [Ru] \\ \xrightarrow{k_f} [Ru] \end{array}\right. \qquad \begin{array}{l}(8)\\(9)\\(10)\end{array}$$

$$[\text{Ru-Q}_i] \xrightarrow{h\nu} [\text{Ru}^*\text{-Q}_i] \begin{cases} \xrightarrow{k_{q,1}} [\text{Ru-Q}_i] & (11) \\ \xrightarrow{k_d} [\text{Ru-Q}_i] & (12) \\ \xrightarrow{k_f} [\text{Ru-Q}_i] & (13) \end{cases}$$

Model 2

$$[\text{Ru}] \xrightarrow{h\nu} [\text{Ru}^*] \begin{cases} \xrightarrow{k_d} [\text{Ru}] & (14) \\ \xrightarrow{k_f} [\text{Ru}] & (15) \end{cases}$$

$$[\text{Ru-Q}_i] \xrightarrow{h\nu} [\text{Ru}^*\text{-Q}_i] \begin{cases} \xrightarrow{k_{q,1}} [\text{Ru-Q}_i] & (16) \\ \xrightarrow{k_d} [\text{Ru-Q}_i] & (17) \\ \xrightarrow{k_f} [\text{Ru-Q}_i] & (18) \end{cases}$$

Model 3

$$[\text{Ru}] \xrightarrow{h\nu} [\text{Ru}^*] \begin{cases} \xrightarrow{k_{q,2}[Q]} [\text{Ru}] & (19) \\ \xrightarrow{k_d} [\text{Ru}] & (20) \\ \xrightarrow{k_f} [\text{Ru}] & (21) \end{cases}$$

$$[\text{Ru-Q}_i] \xrightarrow{h\nu} [\text{Ru}^*\text{-Q}_i] \begin{cases} \xrightarrow{ik_{q,1}} [\text{Ru-Q}_i] & (22) \\ \xrightarrow{k_d} [\text{Ru-Q}_i] & (23) \\ \xrightarrow{k_f} [\text{Ru-Q}_i] & (24) \end{cases}$$

Model 4

$$[\text{Ru}] \xrightarrow{h\nu} [\text{Ru}^*] \begin{cases} \xrightarrow{k_d} [\text{Ru}] & (25) \\ \xrightarrow{k_f} [\text{Ru}] & (26) \end{cases}$$

$$[\text{Ru-Q}_i] \xrightarrow{h\nu} [\text{Ru}^*\text{-Q}_i] \begin{cases} \xrightarrow{ik_{q,1}} [\text{Ru-Q}_i] & (27) \\ \xrightarrow{k_d} [\text{Ru-Q}_i] & (28) \\ \xrightarrow{k_f} [\text{Ru-Q}_i] & (29) \end{cases}$$

In model 1 both dynamic and static quenchings are considered, the former being induced by the bulk Q and the latter by the incorporated Q. In model 2 only static quenching is considered. Model 3 considers both dynamic and static quenchings, but the static quenching rate here is proportional to the number of the incorporated quencher i. In model 4 only the static quenching takes place whose rate is proportioanl to i.

According to the equilibria shown in Eqs. (5)–(7), the number of Q molecules bound to the domain obeys the Poisson distribution. The total concentrations of the domain $[\text{Ru}]_t$, the concentration of Q incorporated into the domain $[\text{Q}]_b$, and the free Q are represented by Eqs. (30)–(32), respectively.

$$[Ru]_t = \sum_{i=0}^{\infty} [Ru\text{-}Q_i] = [Ru] \sum_{i=0}^{\infty} \frac{(K_1[Q])^i}{i!}$$
$$= [Ru] \exp(K_1[Q]) \tag{30}$$

$$[Q]_b = \sum_{i=0}^{\infty} i[Ru\text{-}Q_i] = [Ru]K_1[Q] \exp(K_1[Q]$$
$$= [Ru]_t K_1[Q] \tag{31}$$

$$[Q] = [Q]_t/(K_1[Ru]_i + 1) \tag{32}$$

Calculation of the I_0/I values for models 1–4 in a steady state showed that the profiles of the I_0/I vs $[Q]_t$ plots are as in Fig. 5.4a (for details see [15, 16].

The emission decay profile for each model is shown in Fig. 5.4b.

According to these profiles of Fig. 5.4, it can be decided which model is applicable to the reaction within the Q concentration studied.

From the I_0/I vs $[Q]_t$ plots and the emission decay profiles, it was revealed that the results of Figs. 5.1 and 5.2 studied in an aqueous solution are explained either by a conventional dynamic model (for mol. wt. 2100) or by model 1 (for mol. wt. 4400 and 13 300), which involves both the dynamic and static quenchings [15]. It means that when the molecular weight is lower, the accelerating effect of the polymer is explained only by a concentration effect of the acceptor into the polymer domain around the Ru complex. However, when the molecular weight becomes higher, with parallel to the dynamic quenching, specific static quenching takes place in the domain based on interactions between the excited Ru complex and the acceptor, which causes a much higher electron transfer rate of the system.

In methanol solution, because electrostatic interaction is more enhanced than in water, reaction behavior is entirely different from that in water. The results of the analysis are summarized in Table 5.1 [16] (see also Fig. 5.3).

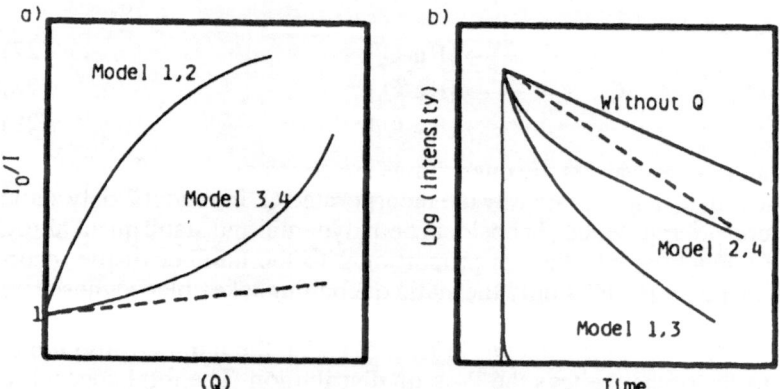

Fig. 5.4. Profiles of the **a** Stern-Volmer plots and **b** emission decays for the models 1–4 and for the conventional dynamic model (*dotted line*)

Table 5.1. Quenching mechanism of the excited copolymer-pendant $[Ru(bpy)_3]^{2+}$ (Scheme 5.2) by MV^{2+} in methyl alcohol

Model	Rates of quenching		Stern-Volmer plot based on emission intensity	Additive
	Bulk Q	Bound Q		
1	$k_{q,2}[Q]$	$k_{q,1}$	Downward curve	None
2	–	$k_{q,1}$	Downward curve	NaOH
3	$k_{q,2}[Q]$	$ik_{q,1}$	Upward curve	$HCl + H_2NCSNH_2$
4	–	$ik_{q,1}$	Upward curve	
Conventional dynamic model	$k_{q,2}[Q]$	–	Linear relationship	HCl or $HCl + Et_4NBr$

Without any additive, static quenching occurs to increase the rate with conventional dynamic quenching. The addition of NaOH enhances the rate anamalously by more than two orders or magnitude, the mechanism being only static.

The effect of the polymer domain in enhanced reactions of polymer–metal complexes have usually been explained by a conventional concentration effect [17]. However, the previously described analyses show that specific interactions between the reactions compounds arise depending on the conditions, which brings about a rapid reaction path for the process.

5.1.3 Photoinduced Electron Transport and Energy Transfer in Polymer-Pendant Porphyrin Complexes

Uncharged, negatively charged, and positively charged polymers containing moities of covalently bound porphyrins, phthalocyanines, and naphthalocyanines were prepared (Schemes 5.3–5.7) [18–20]. (For a general description of polymer binding of metal chelates see Chap. 2.2.3.) With these polymer–metal complexes the electron [18] and photoinduced electron transport [20] were investigated in solution in order to study the influence of the charge of the polymer on these processes. The porphyrins are intensively absorbing ($\varepsilon > 10^5$ l/mol cm) in the visible region of light. Examples of prepared polymers with polymer-bound zinc(II)-5,10,15,20-tetrakis(4-amino)phenylporphyrin are shown in polymers of Schemes 5.3–5.5. Polymers in Schemes 5.6 and 5.7 contain viologen or combined porphyrin and viologen units. The reason that the metal-containing porphyrins were employed instead of metal-free ones is explained by better photophysical properties of the Zn(II) chelates compared with metal-free porphyrins [21].

The excited triplet lifetimes were measured in under inert gas water at 300 K. For all polymers in Schemes 5.3–5.5 triplet lifetimes between 1.0 and 1.5 ms were found [20]. Analogous negatively and positively charged low molecular weight Zn(II) porphyrins exhibit triplet lifetimes of 1.5 ms. This important finding means that the excited state is not affected by different polymer

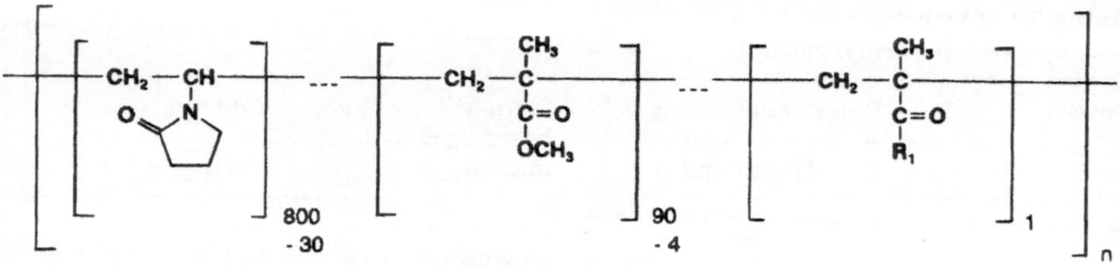

Scheme 5.3

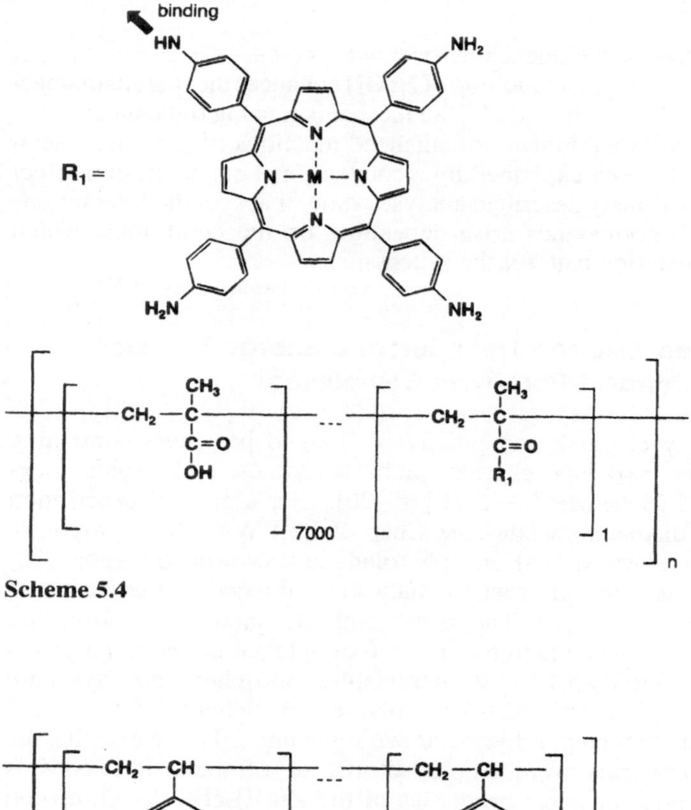

$R_1 =$

Scheme 5.4

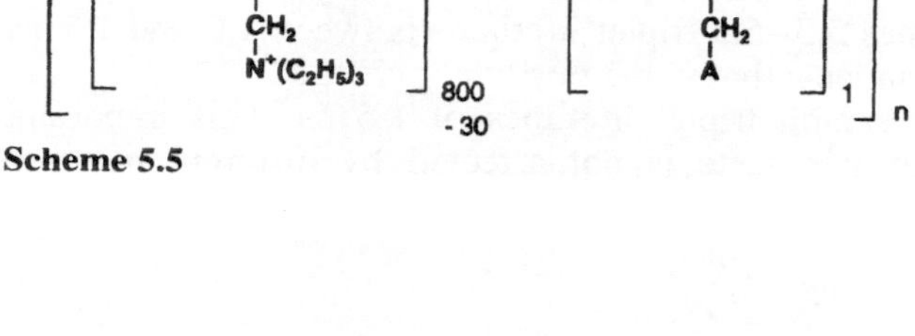

Scheme 5.5

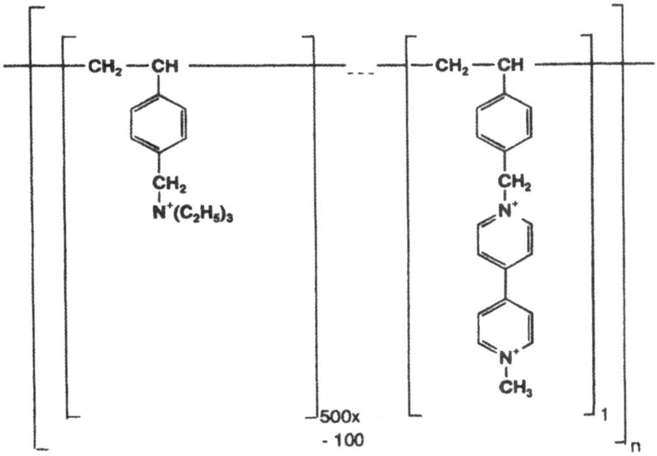

Scheme 5.6

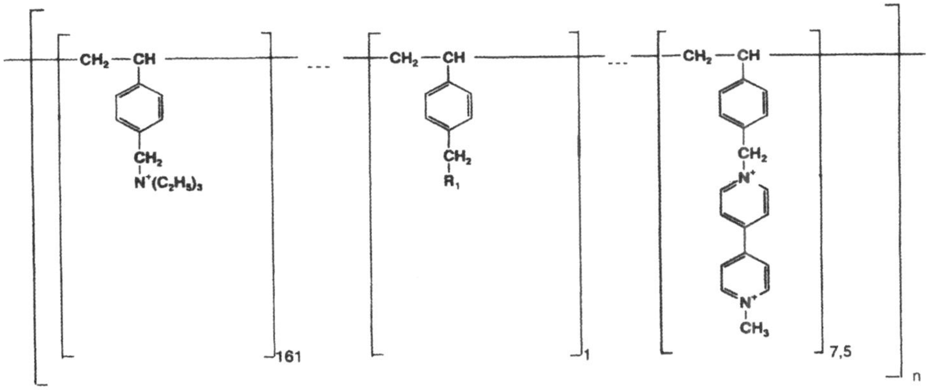

Scheme 5.7

environments. In dimethylformamide (DMF) water mixture the triplet lifetimes increase from 0.08 ms to 1.9 ms from low molecular weight porphyrins to polymer in Scheme 5.5. A specific polymer may shield the Zn(II) in the porphyrin core against coordination of the polar DMF and/or against bimolecular processes of triplet–triplet annihilation. For polymer in Scheme 5.7 with porphyrin and viologen (acceptor) units, no lifetime higher than 10^{-8}s could be measured, because of efficient quenching (see Chap. 5.1.2). Similar to the previously mentioned Ru complexes, also the porphyrin complexes are efficient photoredox catalysts [21]. This was investigated in solution in the presence of an excess of low molecular weight or polymer-bound viologens as acceptor, and 2-mercaptoethanol as donor [20], by irradiation with visible light. In water the photoredox activities using a low molecular weight porphyrin as standard (100%) are as follows: negatively charged polymer (Scheme 5.4) 81%, uncharged polymer

(Scheme 5.3) 110% and positively charged polymer (Scheme 5.5) 172%. As shown in Chap. 5.2.1 the charge of the polymer has a dominating influence on acceleration or retardation of electron transfer in comparison with low molecular wight electrolytes. Details for the polymer-pendant porphyrins in this direction are discussed in [20]. Investigating these photon-driven processes with polymer-bound viologen, the following was found using polymer in Scheme 5.5 in the presence of low molecular weight MV^{2+} (and thiol as donar) as standard (100%): polymer porphyrin in Scheme 5.5 and polymer-bound viologen in Scheme 5.6 165%, polymer in Scheme 5.7 with bound porphyrin and viologen 260%. The advantage of having acceptor and sensitizer moieties in a fixed neighbourhood seems of general importance – as shown in photosynthesis and direct viologen-linked porphyrins ([22] and references cited therein) – to advance photoelectron transfer reactions. Also, hydrogen evolution in the described system using hydrogenase as a catalyst is running efficiently with the polymer-pendant porphyrins [20]. After the results the mentioned reactions run according to Eqs. (33)–(35):

$$\text{Sens} \xrightarrow{h\nu} \text{Sens}^* \xrightarrow{MV^{2+}} [^T\text{Sens}^* \cdots MV^{2+}] \rightarrow \text{Sens}^{+\cdot} + MV^{+\cdot} \tag{33}$$

$$\text{Sens}^{+\cdot} + \text{RSH} \rightarrow \text{Sens} + (\text{RSH})_{ox} \tag{34}$$

$$MV^{+\cdot} + H_2O \xrightarrow{\text{catalyst}} MV^{2+} + (1/2)H_2 + OH^- \tag{35}$$

Another photoinduced process that must be considered is the energy transfer. Under irradiation in the presence of the sensitizers porphyrin and the acceptor, oxygen energy transfer from the excited triplet state of the sensitizer to triplet oxygen under formation of singlet oxygen is thermodynamically possible. The quantum yields of such a process is 20–40%, and therefore very efficient [21, 23–25].

It is important to consider that singlet oxygen can oxidize various substrates: model reactions for solar photochemistry [24] and as photodrugs in the photodynamic therapy of cancer [25, 26]. As a photochemical reaction the photooxidation of thiols was investigated. Whereas in the dark metal phthalocyanines can oxidize – in the presence of O_2 – thiols catalytically only to the corresponding disulfides (MEROX process in the petroleum industry), irradiation photooxidation gives sulfonic acids and sulfate [24]. Zn(II) phthalocyanines covalently bound at silica gel of low surface area exhibit comparable activities to low molecular weight dissolved phthalocyanines. Organic polymers as carriers for sensitizers are photooxidized over time by singlet oxygen. Silica gel is one example of a stable high molecular weight support for metal complexes to use as a heterogeneous sensitizer. The reactions for the photooxidation of thiols are summarized in Eqs. (36) and (37) (details are described in [24]):

$$\text{Sens} \xrightarrow{h\nu} {}^T\text{Sens}^* \xrightarrow{{}^3O_2} \text{Sens} + {}^1O_2 \tag{36}$$

$$2RS^{-} + 3\,{}^1O_2 \rightarrow 2\,RSO_3^{-} \text{ (also } SO_4^{2-}) . \tag{37}$$

Another application of porphyrin derivatives is for the photodynamic therapy of cancer. Porphyrins and phthalocyanines were covalently bound at methoxy-poly(ethylene glycol) over carboxylic acid or phenolic groups to get, for example, MP (substituent-$(OCH_2CH_2)_nOCH_3)_{2-4}$ [27]. According to in vivo experiments the tumor accumulation after intravenous application of polymer-bound sensitizers is up to 30 % higher in comparison with commericial hematoporphyrin derivatives. Another direction uses successful incorporation of long wavelength absorbing sensitizers, such as naphthalocyanines ($\lambda \approx 770$ nm), incorporated into liposomes, in the photodynamic therapy of cancer [25].

5.1.4 Photoinduced Electron Transport Between Ru(bpy)$_3^{2+}$ and Colloidal Particles

Visible light cleavage of water has been achieved with Ru(bpy)$_3^{2+}$ and Prussian Blue (PB) colloids in an aqueus solution [28, 29]. Prussian Blue (Chap. 2.3.2.1) is a mixed valent iron cyanide complex whose repeating unit is represented by the following and forms a colloidal aqueous solution:

$$Fe_4^{3+}[Fe^{II}(CN)_6]_3^{4-} \ .$$

It is a high molecular weight polymeric complex with $\overline{M}$ over 5 million as studied by electron microscopy. In this water cleavage system it has been found that the emission from the excited Ru complex is quenched by PB by a static mechanism following typical Perrin type [29, 30]. This emission quenching takes place also by electron transfer from the excite Ru complex to PB as is clear from the redox reaction products. The Stern-Volmer plots shown in Fig. 5.5 with an upward-deviating curve indicate clearly the involvement of a static quenching mechanism. The Stern-Volmer plots based on the relative emission lifetime (τ_0/τ) showed no

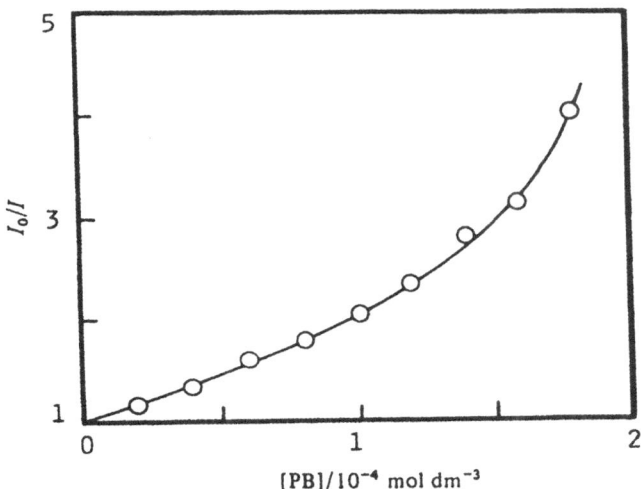

Fig. 5.5. Stern-Volmer plots for the electron transfer quenching of the photoexcited Ru (bpy)$_3^{2+}$ (Scheme 5.1) by Prussian Blue (PB) colloids. Ru complex 20 μM, KCl 0.5 M, pH 2

slope, meaning that the quenching follows an entirely static mechanism. It is reasonable to think about electrostatic interaction, because the PB colloids are anionially charged.

The most fit model for the quenching was stepwise interaction of the Ru complex and the repeating unit of PB as shown in Eqs. (38)–(40):

$$Ru + PB \underset{k_{-a}}{\overset{k_a}{\rightleftharpoons}} RuPB_1, \qquad K_1 = k_a/k_{-a} \tag{38}$$

$$RuPB_1 + PB \underset{2k_{-na}}{\overset{k_a}{\rightleftharpoons}} RuPB_2, \qquad K_2 = K_1/2 \tag{39}$$

$$RuPB_{i-1} + PB \underset{ik_{-na}}{\overset{k_{na}}{\rightleftharpoons}} RuPB_i, \qquad K_i = K_1/i. \tag{40}$$

This is again a Poisson type interaction model: the probability distribution P_i is expressed by Eqs. (41) and (42) for which m is the average number of bound PB units $[PB]_b$ per total number of the Ru complex $[Ru]_t$. The calculation has led to Eq. (43) [30], where K_1 is the equilibrium constant for the first step of the interaciton (Eq. (38)).

$$P_i = (m^i/i!) \exp(-m) \tag{41}$$

$$m = [PB]_b/[Ru]_t \tag{42}$$

$$\ln(I_0/I) = K_1[PB] = \frac{K_1}{1 + K_1[Ru]_t} [PB]_t. \tag{43}$$

When the results of Fig. 5.5 were plotted according to Eq. (43), it gave a linear relationship, which led to a K_1 value of $8.57 \cdot 10^3$ dm^3 mol^{-1}.

The distribution curve of the number of the adsorbed Ru complex on an average size (23 nm diameter) colloidal PB particle was depicted by calculating the results using this K_1 value. In a most typical concentration conditions of $1 \cdot 10^{-5}$ M Ru(bpy)$_3^{2+}$ and 1 mM PB repeating unit, it was calculated that the colloidal PB surface was covered by a monolayer Ru complex.

In this system, the interaction is observed only by a small shift in the nuclear magnetic resonance (NMR) spectrum of the Ru complex, and not observed by ultraviolet (UV) visible absorption or emission spectrum. The present analysis by emission quenching provides useful methodology to obtain information about such interaction.

5.1.5 Photoinduced Electron Transport in Molecular Assemblies

Molecular assemblies, such as micelles liposomes, and LB films, provide useful models for studying biological electron transfer (Chap. 4.2). Metal porphyrin is an excellent photoexcitation center with high-excited-state reactivity, and has been intensively studied as a model for the photosynthetic excitation center molecule. Amphiphilic zinc porphyrin (ZnP) containing four amphiphilic chains (Fig. 5.6) was synthesized and incorporated into liposomes made of dipalmitoyl-phosphatidylcholine (DPPC). It has been shown that the porphyrin ring plane is oriented parallel to the wall surface [31] as shown in Fig. 5.7.

The distance from the ZnP plane to the outer surface of the liposome was changed by changing the alkyl chain length, and photoinduced electron transfer from the excited ZnP embedded into the liposome to MV^{2+} present at the outer aqueous phase was studied in an aqueous solution as a function of the distance [32]. The cationic viologen exists at the anionic surface of the liposomes. The singlet excited state of ZnP was not effective for the electron transfer, and only the triplet state transferred electrons to MV^{2+}. The formation of $MV^{+\cdot}$ (603 nm) and the decay of T–T absorption (470 nm) was studied by laser flash photolysis. It was interesting to find that the photoinduced electron transfer takes place only within 1.2 nm, and not beyond 1.7 nm. It should be noted that electron hopping distance in the biological system is often reported to be approximately 1.3 nm, which is close to the value obtained in the present work. The electron transfer followed second-order reaction, and the rate constant was a 10^9 ($dm^3 mol^{-1} s^{-1}$) order of magnitude.

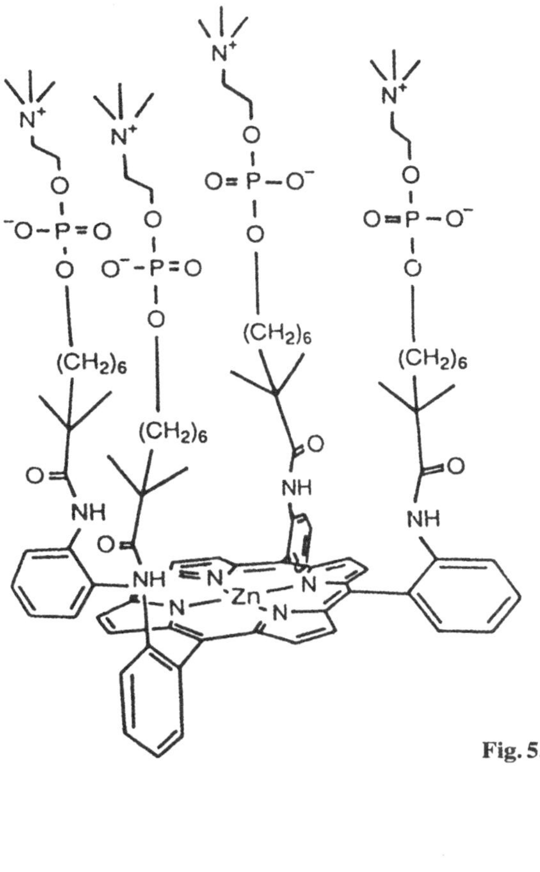

Fig. 5.6. Amphiphilic zinc porphyrin (ZnP)

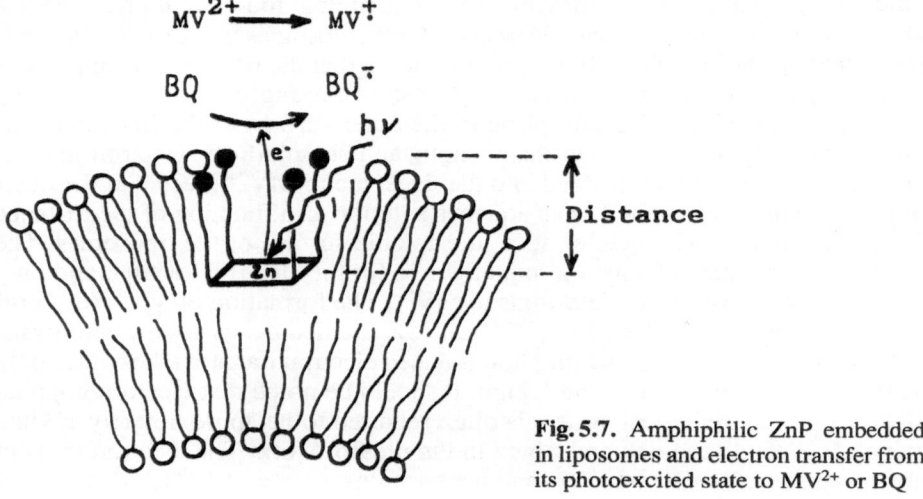

Fig. 5.7. Amphiphilic ZnP embedded in liposomes and electron transfer from its photoexcited state to MV^{2+} or BQ

When benzoquinone (BQ) was put in the outer aqueous phase and used as an acceptor in the above liposome embedded ZnP system, a very different type of electron transfer took place [33]. The single-excited ZnP also transferred electrons to BQ in parallel with the triplet excited state. The second order rate constant obtained from the ZnP^+ formation ($2.5 \cdot 10^9$ mol^{-1} dm^3 s^{-1}) was four times higher than the value obtained from the triplet decay ($6 \cdot 10^8$ M^{-1} s^{-1}). The quenching of the singlet ZnP by electron transfer to BQ was almost static showing that the acceptor is incorporated inside the liposome near ZnP and reacts there. Model analysis showed that a similar mechanism as the previously described $Ru(bpy)_3^{2+}$/PB system is applicable; BQ is incorporated into the quenching sphere around ZnP stepwise (Eq. (44)) as in Eqs. (5)–(7), and quenches singlet ZnP by a static mechanism:

$$[ZnP\text{-}BQ_{i-1}] + BQ \underset{k_{-a}}{\overset{k_a}{\rightleftharpoons}} [ZnP\text{-}BQ_i] \tag{44}$$

The quenching of the fluorescence by BQ was studied; the $\ln(I_0/I)$ vs $[BQ]_t$ plots according to Eq. (43) gave a linear relationship, and from the slope K_I was calculated as 52 dm^{3}mol^{-1}. From the K_I the distribution of BQ incorporated into the quenching sphere around ZnP was calculated and shown in Fig. 5.8 for various total BQ concentrations.

The UV/visible absorption and emission spectra did not give any information about incorporation of BQ into liposomes. Only through such analysis as the above could evidence be given for the incorporation of BQ into liposomes. One of the interesting features of this system is the reaction of the singlet excited state of ZnP. In a solution only triplet is capable of reacting, because singlet state is too short-lived (approximately 3 ns in the present case). In the photosynthetic photoinduced charge separation, singlet-state chlorophyll participates

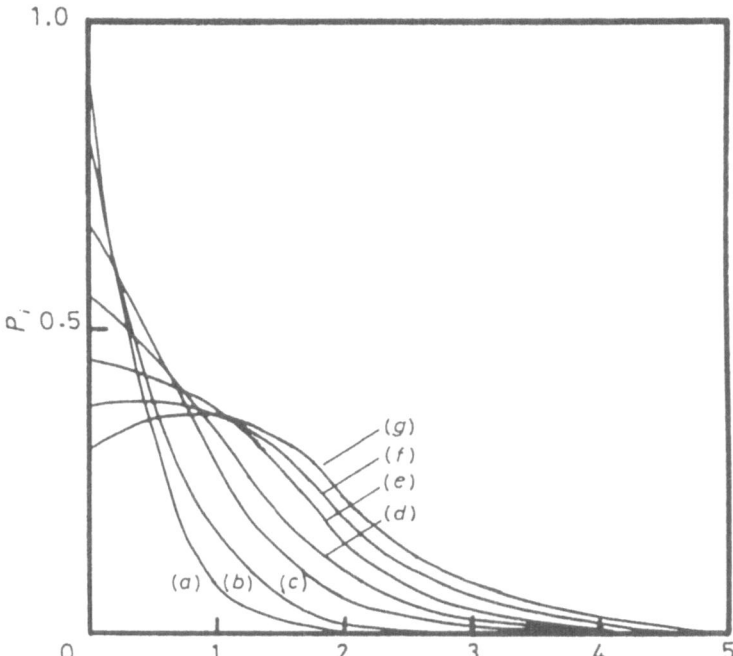

Fig. 5.8. Probability distribution of BQ incorporated into the quenching sphere of ZnP present in liposomes. ZnP 50 μM; BQ (mM) (**a**) 2, (**b**) 5, (**c**) 10, (**d**) 15, (**e**) 20, (**f**) 25, and (**g**) 30

in the electron transfer to make the reaction very rapid of the ps order, which would be possible only for molecularly arranged systems. In the present system also, by incorporation of the acceptor into the liposome, probably due to the hydrophobic nature of BQ, rapid static reaction is made possible in which singlet state is responsible.

5.2 Photoinduced Electron Transport at Solid/Liquid Interface

Photoinduced electron transport can take place also at the solid/liquid interface. In many cases the metal complex works as a photoexcitation center in the presence of another acceptor and/or donor molecule, but in some cases the metal complex works as an electron acceptor or donor coupled with another photoexcitation center such as organic sensitizer or semiconductor. This process is important especially for constructing photoenergy conversion systems and photosensing devices based on photoelectrochemistry.

As a photoexcitation center-metal complexes, such as $Ru(bpy)_3^{2+}$ and macrocyclic ligand-metal, was well as p-/n-type semiconductors, are used.

5.2.1 Photoelectrochemial Electron Transport with $Ru(bpy)_3^{2+}$ Excitation Center

Photoresponsive devices, such as solar cell and photosensor, are usually composed of semiconductors with pn or Schottky junction. Photochemical reactions can also be utilized for photoelectrochemical devices if at least one of the photochemical reaction products has a change to exchange electron(s) with an electrode. When the photochemical reaction product in a solution is long-lived enough, it can exchange electrons with an electrode dipped into the solution to give photocurrent. On the contrary, when the photochemical product is too short-lived, it has no chance to react to the electrode resulting in no photoresponse to the electrode. However, when the product is attached to the electrode surface, there is a chance to exchange electron with the electrode to give photocurrent [1].

Photocurrent has been successfully obtained with a $Ru(bpy)_3^{2+}/MV^{2+}$ photochemical reaction couple by coating polymer-pendant $Ru(bpy)_3^{2+}$ film on an electrode [33–36], for which photochemical reaction products of monomeric $Ru(bpy)_3^{2+}$ and $MV^{+\cdot}$ are too short-lived to induce photocurrent from their solution photochemistry. The acceptor, MV^{2+}, can either be present in solution in contact with the Ru-complex-coated electrode or as a second polymer layer on top of the polymer-Ru-complex film. The photoinduced electron transfer is shown in the following scheme, where BPG is a basal plane pyrolytic graphite.

Electrode
(BPG)
$\xrightarrow{e^-}$ $Ru(bpy)_3^{2+*}$
Polymer
layer
$\xrightarrow{e^-}$ MV^{2+}
In solution or
second polymer
layer
Electrolytic solution

Prussian Blue film coated on the top of the polymer-pendant $Ru(bpy)_3^{2+}$ layer worked as electron acceptor or donor, but because PB is also excited by visible light (λ_{max} ca. 700 nm) and induces photocurrent by itself [37], the photocurrent direction and major mechanism depend on the applied potential at the electrode as shown in the following schemes:

(Major process under cathodic bias)

Electrode
(BPG)
$\xrightarrow{e^-}$ $Ru(bpy)_3^{2+*}$
First
polymer
layer
$\xrightarrow{e^-}$ PB
Second
polymer
layer
Aqueous electrolytic solution

(Major process under anodic bias)

$$
\text{Electrode (BPG)} \quad \overset{|}{\underset{|}{\overset{e^-}{\longleftarrow}}} \quad
\begin{array}{c} Ru(bpy)_3^{2+*} \\ \text{First} \\ \text{polymer} \\ \text{layer} \end{array}
\quad \overset{e^-}{\longleftarrow} \quad
\begin{array}{c} PB^* \\ \text{Second} \\ \text{polymer} \\ \text{layer} \end{array}
\quad \Big| \quad \text{Aqueous electrolytic solution}
$$

In such photoelectrochemical systems O_2 works as a direct electron acceptor from the excited Ru complex to induce cathodic photocurrent [30].

$$
\text{Electrode (ITO)} \quad \overset{e^-}{\longrightarrow} \quad
\begin{array}{c} Ru(bpy)_3^{2+*} \\ \text{Polymer} \\ \text{layer} \end{array}
\quad \overset{e^-}{\longrightarrow} \quad O_2 , \quad \text{Aqueous electrolytic solution}
$$

This photoelectrochemical system shows that O_2 can work as a direct electron acceptor from the excited $Ru(bpy)_3^{2+}$, although the extent of the electron transfer in comparison with a parallel occurring energy transfer is not clear [38, 39].

The presence of the MV^{2+} second polymer layer in the $Ru(bpy)_3^{2+}/O_2$ system enhanced much the photocurrent for which MV^{2+} works as an efficient electron transport mediator from $Ru(bpy)_3^{2+*}$ to O_2 [40].

$$
\text{Electrode (ITO)} \quad \overset{e^-}{\longrightarrow} \quad
\begin{array}{c} Ru(bpy)_3^{2+*} \\ \text{First} \\ \text{polymer} \\ \text{layer} \end{array}
\quad \overset{e^-}{\longrightarrow} \quad
\begin{array}{c} MV^{2+} \\ \text{Second} \\ \text{polymer} \\ \text{layer} \end{array}
\quad \overset{e^-}{\longrightarrow} \quad O_2 , \text{Aqueous electrolytic solution}
$$

Great progress has been made in the conversion of solar light into electrical energy by a photoelectrochemical cell consisting of working electrode conducting glass covered with TiO_2/an adsorbed Ru (II) complex/iodide–triodide redox electrolyte/ conducting glass counter electrode [41–43]. The Ruthenium complex is a trinuclear one of the following composition: $[Ru(bpy)_2(CN)_2]_2Ru(bpy)(Coo)_2)_2^{2-}$. The TiO_2 film (10 µm thickness; average particle size 15 nm) is deposited from colloidal solutions. A monolayer of the Ru complex (λ_{max} in solution 530 nm) is deposited via electrostatic interactions of the TiO_2 surface with the carboxylic acid groups of the complex from solution. The layer shows an absorption onset shift of TiO_2 to 750 nm (spectral overlap between solar emission leads to 46% absorption of photons, for general description of metal chelates on the surface of oxides, see Chap. 2.2.3). Scheme 5.8 explains energetically the function of such a cell. TiO_2 is a semiconductor working as a ceramic membrane. It excepts electrons from the excited state of the bound sensitizer in a quantum yield of approximately 80%. The electrons cross a 5-µm-thick membrane within 2 µs (charge transport via a 5-nm photosynthetic membrane within 100 µs is longer). The electrons flow to the counter electrode and can reduce iodine that is diffusing to the working electrode to reduce the oxidized sensitizer. The overall light-to-

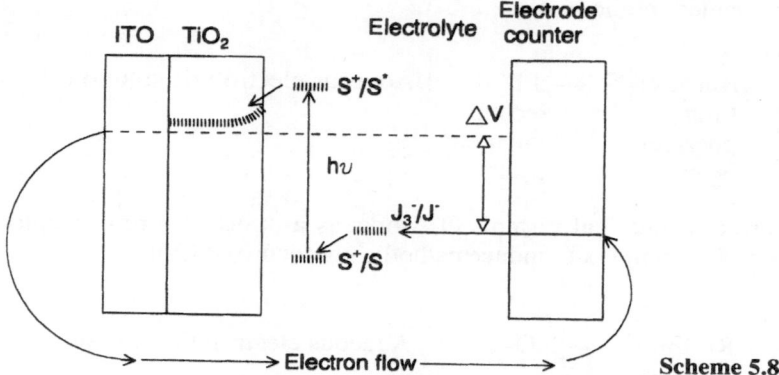

Scheme 5.8

electrical energy conversion yield with short circuit currents of $12\,mA/cm^2$ and open circuit voltages of $679\,mV$ is approximately 7.5% in simulated solar light.

5.2.2 Photoinduced Electron Transport with Macrocyclic Metal Complexes

Metal complexes with macrocyclic ligands such as metal–phthalocyanine (MPc; Scheme 5.9) and metal–porphyrin, e.g., metal–tetraphynylporphyrin (MTPP; Scheme 5.10), are also excellent photoexcitation centers for photoelectrochemical electron transfer.

Their films are easily coated on an electrode by vapor deposition or by casting their solution. Acceptors such as MV^{2+} and O_2, can also be used for the excited state of their films in contact with an electrolytic solution. Molecular oxygen works also as a primary acceptor for the photoexcited MPc [44].

Vapor deposited pure ZnPc film did not give stable photoelectrochemical characteristics, but when the complex was dispersed in a polymer film such as

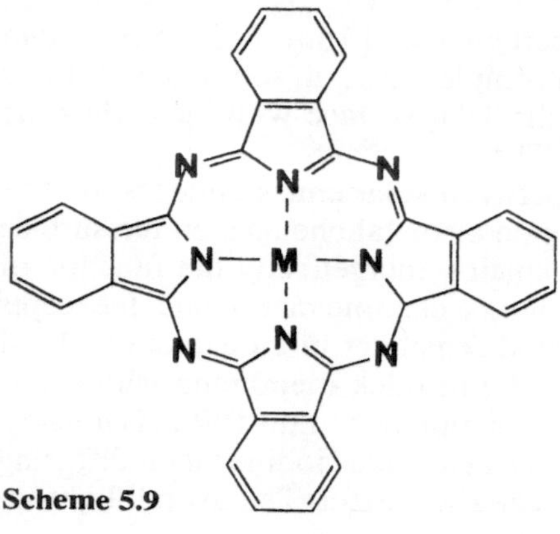

Scheme 5.9

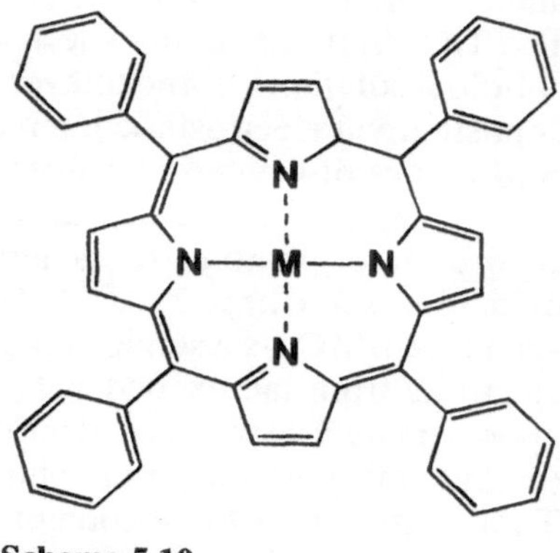

Scheme 5.10

poly(*N*-vinylcarbazole) (PVCz) or poly(vinylidene fluoride) (PVDF), stable and higher photocurrents were obtained. The current change induced by switching on and off the irradiation of ZnPc dispersed in a PVDF film (coated on an ITO electrode by mixture casting) in the presence of O_2 is shown in Fig. 5.9 [45, 46].

The quick response of the current change may suggest a mechanism based on a *p*-semiconductor property of the complex layer in comparison with a fairly slow change of the photoresponse based on photoresponse based on photochemical reactions of a photosensitizer [1].

The effect of the polymers in the following order might suggest that higher dielectric constant of the polymer favors charge separation resulting in the increase of the photocurrent:

PVDF > PVCz > Poly(vinyl chloride) >
Poly(acrylonitrile) > Poly(styrene)

For electrodes of ZnPc in poly(vinylidene fluoride) it was found that the rate-limiting step in the photoelectrochemical reduction of O_2 is the charge transfer at the electrode surface, and not the diffusion of O_2 [47]. Prior to the charge transfer the adsorption of oxygen at the electrode surface is explained by the Langmuir's adsorption isotherm. The photocurrent density induced by ITO/PVCz[ZnPc] increased with O_2 concentration in the aqueous solution, but approached saturation after approximately 0.5 mM [48]. This behavior was analyzed by interaction of O_2 with ZnPc dispersed in the PVCz film as shown in Eq. (45):

$$\text{PVCz[ZnPc]} + O_2 \underset{k_2}{\overset{k_1}{\rightleftharpoons}} \text{PVCz[ZnPc]} - O_2 \overset{k_3}{\underset{h\upsilon}{\rightarrow}} \text{PVCz[ZnPc]}^{+\cdot} + O_2^{-\cdot} \quad (45)$$

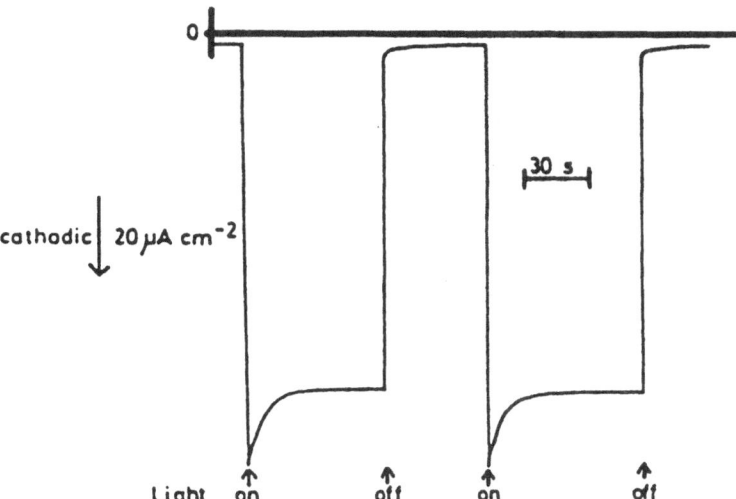

Fig. 5.9. Current changes induced by switching on and off of the irradiation (400 mW cm^{-2}) on a ZnPc/poly(vinylidene fluoride) film (weight ratio 1:1, thickness 250 nm) at −60 mV vs NHE

The affinity parameter K_I (= $k_1/k_2 + k_3$)) of O_2 and ZnPc was obtained as $6.06 \cdot 10^3 \, mol^{-1} \, dm^3$ from the $[O_2]$ dependene of the photocurrent [48].

The presence of methylbenzylviologen groups anchored on poly(methyl-β-phenethylsiloxane) (PPhS) enhanced much the photoresponse characteristics (V_{oc} 0.58 V; J_{sc} 5.5 μA cm^{-2}) as well as the Fill Factor (FF = 0.40). This is explained by an electron-mediating effect of the viologen groups as well as by an excellent diffusion property of O_2 into the polysiloxane matrix. There are still arguments whether such macrocyclic metal complexes work as a molecular sensitizer or as a semiconductor, which should be solved in the future [51].

Thin films of polymeric copper(II) phthalocyanines (see Chap 2.3.1.2) were prepared by in situ synthesis [49] from different oxy- and phenoxy-bridged diphthalonitriles with copper layers deposited on Ti plates and KCl [50]. Depending on the tetracarbonitrile the films exhibit different morphologies. The photoelectrochemical properties were investigated with the $Fe(CN)_6^{3-/4-}$ redox couple. Better structural uniformity of the polymer metal complexes of the films led to higher cathodic photocurrents of the materials.

5.2.3 Photoelectrochemical Electron Transport with Semiconductor Excitation Center Coupled with Metal Complexes Working as Catalysts or Charge Mediators

Utilization of semicondutor junctions, e. g., p/n or Schottky junction, as a photoexcitation and charge separation center is a promising approach to construct photoenergy conversion systems and photodevices [1, 2, 10, 11]. Metal complexes are important compounds as charge mediators or catalysts used in conjunction with such semiconductors. For this purpose metal complex is often used by being incorporated into a membrane coated on the semiconductor. The catalysis and charge transport characteristics of the metal complex in the membrane are especially important and are therefore described in this section with some applications.

5.2.3.1 Charge Transport in a Polymer Membrane Containing Metal Complex

Charge transport in a polymer membrane containing redox centers often takes place by charge hopping between the redox compounds, rather than by charge conduction, which is a usual process for conductive materials. Such charge transport by hopping has been studied in observing electrochemical charge propagation in a polymer membrane containing metal complex redox centers by using an in situ spectrocyclic voltammetry (SCV), which couples cyclic voltammetry with rapid-scan in situ UV/visible absorption spectroscopy [52].

As an example, charge propagation among the dispersed Ru(bpy)$_3^{2+}$ redox enters in a Nafion membrane coated on an ITO electrode has been studied by SCV (Fig. 5.10) [52]. The visible absorption spectral changes measured in situ under potential scanning from 0 to 1.3 V vs SCE (Fig. 5.10a) showed the oxidation of the 2^+ complex to 3^+ state by charge hopping. Almost 80% of the

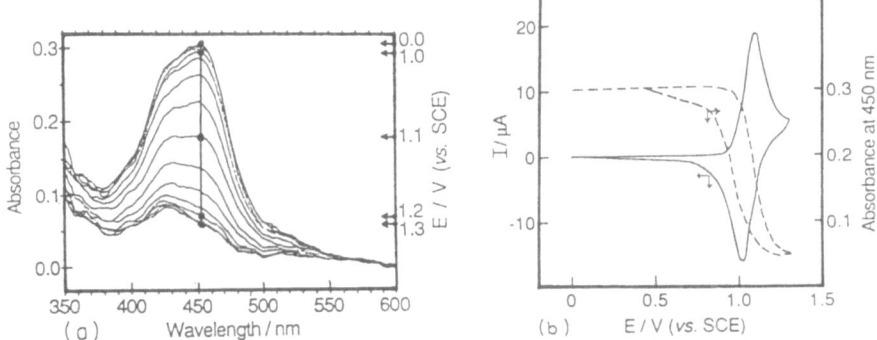

Fig. 5.10. *In situ* spectrocyclic voltammogram of Ru(bpy)$_3^{2+}$ incorporated into a Nafion membrane coated on ITO electrode dipped into 0.1 M KNO$_3$ aqueous solution measured at the scan rate of 2 mV s^{-1}. **a** Spectral change under 0–1.3 V anodic scanning. **b** Cyclic voltammogram and absorbance change at 450 nm

dispersed complex is electroactive, and on a reverse scan to 0 V, the spectrum recovered its original 2$^+$ complex state.

For the charge hopping between the redox centers, the distance between the molecules is important. The dispersion is usually random, so that the probability density P(r) of the distance between the nearest molecules (r) has been calculated by a modified Poisson distribution considering the excluded volume of the Ru complex as shown in Eq. (46), where N$_A$ is Avogadro's constant, c molar concentration of the complex in the membrane bulk, and s contact distance of the complex (0.82 nm).

$$P(r) = 4\pi r^2 N_A \cdot 10^{-24}c \exp\left[\frac{-4\pi (r^3 - s^3)N_A \cdot 10^{-24}c}{3}\right]. \tag{46}$$

The distance distribution of the nearest neighboring Ru(bpy)$_3^{2+}$ molecules dispersed in a polymer membrane is calculated and shown in Fig. 5.11 [1b].

From this distribution and the electrochemically active ratio of the complex in the membrane, the charge hopping distance was 0.96 nm when the scan rate was high. When the scan rate was low, the charge hopping distance was 1.5 nm, which involved bounded motion of the incorporated complex of approximately 0.6 nm[1c].

5.2.3.2 Catalytic Activities of Metal Complexes Incorporated into a Polymer Membrane to be used in Conjunction with Photoexcitation Center

In order to construct photochemical energy conversion systems it is important to couple a photoexcitation center with catalysts for designed reactions. In an artificial photosynthetic model, which is important for future new energy resources [1, 2, 17], water oxidation and CO$_2$ reduction catalysts are essential.

Electrocatalytic water oxidation to evolve O$_2$ (2H$_2$O $\rightarrow$ O$_2$ + 4e$^-$ + 4H$^+$) has been studied with metal complex catalysts incorporated into a polymer mem-

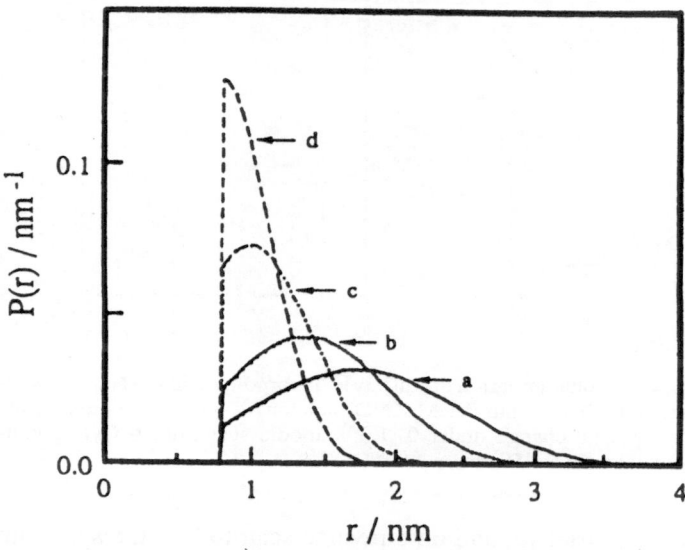

Fig. 5.11. Probability distribution of the intermolecular distance (r) (center to center) between the nearest neighboring Ru(bpy)$_3^{2+}$ dispersed in a polymer membrane. Concentration in M: (**a**) 0.05, (**b**) 0.1, (**c**) 0.261, and (**d**) 0.5

brane coated on an electrode as a site to be used for photoenergy conversion systems (see [1, 2, 17] for details). In the studies of their redox chemistry by SCV interesting phenomena have been observed. It has been found that catalytically inactive (for water oxidation) monomeric Ru-bpy complex, cis-[Ru(bpy)$_2$Cl$_2$], can be oxidatively dimerized to catalytically active dimeric complex as shown in Eq. (47) under anodic scanning of the electrode potential [53].

$$2 \; cis\text{-}[Ru(bpy)_2Cl_2] \xrightarrow{\text{oxidation potential}} [(bpy)_2Ru\text{-}O\text{-}Ru(bpy)_2]^{4+}. \tag{47}$$

The change of the monomeric complex (λ_{max} 480 nm) to the dimeric one (λ_{max} 645 nm) is clearly observed in the SCV spectral changes (Fig. 5.12) measured under cyclic voltammetric scanning from 0.4 to 1.4 V and back to 0.4 V vs SCE. This finding might be important for studying the photosynthetic water oxidation center in that an active catalyst structure is formed upon oxidation of an inactive structure.

It has also been found that an active trinuclear Ru complex (Ru-red) is stabilized when incorporated into a Nafion membrane against electrochemical oxidation [54]:

(Ru-red: [(NH$_3$)$_5$Ru-O-Ru(NH$_3$)$_4$-O-Ru(NH$_3$)$_5$]$^{6+}$).

In a catalytic water oxidation using organic ligand–metal complex, undesired oxidation of a neighboring complex ligand, due to the high oxidation potential of the active complex to result in its decomposition, is a problem to overcome. In an aqueus solution where two catalytic molecules diffuse and react with each

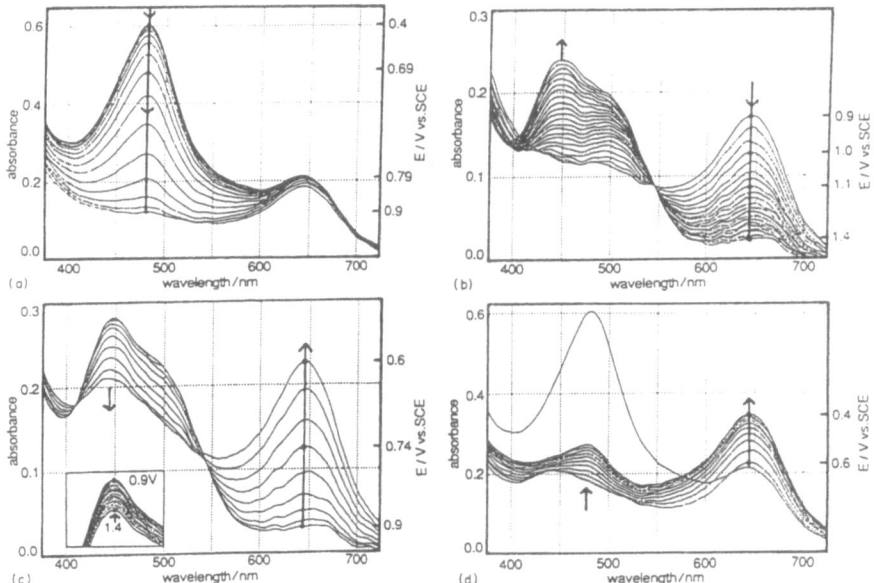

Fig. 5.12. *In situ* spectral changes under cyclic potential scanning between 0.4 and 1.4 V vs SCE at *cis*-[Ru(bpy)$_2$(H$_2$O)$_2$]$^{2+}$ incorporated into a Nafion membrane coated on ITO with scan rate of 0.5 mV s^{-1}. **a** scan from 0.4 to 0.9 V; **b** scan from 0.9 to 1.4 V; **c** scan from 0.9 to 0.6 V (inset shows the changes under 1.4 – 0.9 V scanning), and **d** scan from 0.6 to 0.4 V

other, such a side reaction often cannot be avoided. In a membrane or in a heterogeneous matrix where the complex is fixed, such a side reaction would be prohibited because of the immobilization of the catalyst.

Application of a polymer membrane to incorporate a metal complex catalyst is thus important not only for designing systems or devices, but also for realizing active and stabilized structure.

In order to couple a catalytic CO$_2$ reduction site with water oxidation in an artificial photosynthesis, it is essential to carry out the CO$_2$ reduction also in the water phase. Some metal complexes have been reported as CO$_2$ reduction catalysts, but they have been active only in an aprotic solvent, because proton reduction is usually much easier in water than CO$_2$ reduction. It has been discovered that selective electrocatalytic CO$_2$ reduction can be achieved in water by suppressing proton reduction when the complex catalyst (Re(bpy)(CO)$_3$Br [55] [Co(terpy)$_2$]$^{2+}$ [56]) is incorporated into a coated Nafion membrane. The hydrophobic region of the Nafion would prohibit proton reduction to make the CO$_2$ reduction selective.

Another important aspect of the CO$_2$ catalysis by the complex incorporated into a polymer membrane is that a one-electron reduced complex ([Co(terpy)$_2$]$^+$) is active [56], whereas in an aprotic solution two-electron reduced species [[Co(terpy)]^0) is required for the CO$_2$ reduction (Eq. (48)).

$$CO_2 + 2e^- + 2H^+ \rightarrow HCOOH \qquad (48)$$

In a polymer membrane there would exist catalytic sites where two molecules of the complex work together for one molecule of CO_2 to achieve the two-electron reduction. The activity of COCID-phthalocyanine in a polymer for CO_2 reduction was mentioned in Chap. 2.4.1.1 page 104.

The previously described water oxidation catalysts have been utilized for visible-light water cleavage [57] by combining them with a narrow bandgap n-semiconductor photoanode, which otherwise is decomposed under visible-light irradiation [1]. The degradative oxidation of a CdS photoanode by the holes formed in the depletion layer under irradiation has been successfully prohibited by coating with a Nafion membrane incorporating dispersed Ru-red water oxidation catalyst, and oxygen was produced in the membrane resulting in total water cleavage for H_2 and O_2 formation [57]:

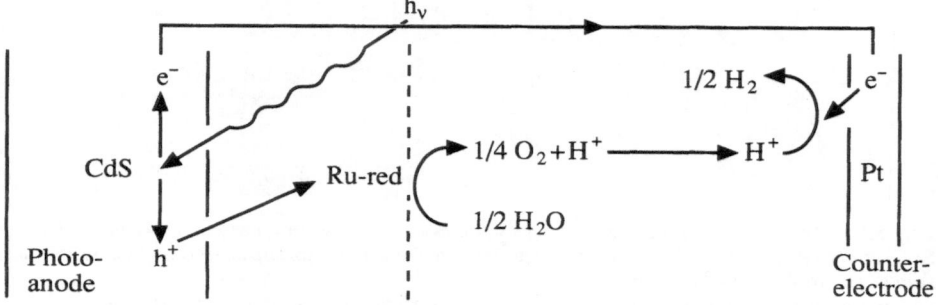

The combination of CO_2 reduction catalyst with a narrow bandgap p-semiconductor for achieving photochemical reduction of CO_2 is now under way.

5.3 Conduction and Photoinduced Behavior in Solid Macromolecular Metal Complexes

Solid macromolecular metal complexes, including metal complexes or metal clusters in a high molecular environment, exhibit the advantage of good stability and easy use. Processing, for example of thin film devices, is possible by different methods such coating from solution or dispersion, evaporation techniques, plasma polymerization and electrochemical polymerization of low molecular complexes (an overview on polymer coated electrodes is given in [1]).

Several fundamental physical solid state properties have been investigated. In the following results on electrical conductivity, photoconductivity, photoinduced charge separation, and hole burning of macromolecule metal complexes are given. Some advices for potential applications are included. An overview on applications of macromolecule complexes for advanced optoelectronic devices is described in [58]. The properties of metal clusters are not discussed here. Chapters 2.4.1.2 and 2.4.2.2 include some references. Recent reviews summarize electronic properties of metal clusters [59], also encapsulated in molecular sieves [60]. Some additional references consider electrical conductivity [61], quantum size effect [62], giant dipole resonance, or plasmon absorption [63].

5.3.1 Electrical Conductivity

Much progress has been achieved in the 1980s considering the preparation of low molecular weight and polymeric organic conducting materials. Examples are "molecular metals", such as donor–acceptor complexes, e. g., tetrathiofulvalene/tetracyanoquinodimethane, or polymers such as polyacetylene and polypyrrole. For "molecular metals" it is known that flat molecules with a conjugated π-electron system must crystallize in separate donor–acceptor staples with near distances in each staple including an incomplete population of the highest energy band (partial oxidation or reduction). Partial oxidation or reduction is also fundamental to high-conducting organic polymers. Besides, few examples of metalic-like conductivity, most conducting low molecular weight and polymeric organic materials exhibit characteristic semiconducting behavior of $\sigma \sim 10^{-1} - 10\,S \cdot cm^{-1}$.

Transport of electrons in these materials is of importance for shielding [64], sensors [66], energy storage [65], energy conversion (photovoltaic cells) [67], switching devices [68, 69], electrochromic devices [1, 70, 71] electroluminescence devices [58], nonlinear optical materials [72] and ferromagnetic materials [73]. Semiconducting behavior is also necessary for processes at electrodes in contact with an electrolyte either in the dark or under irradiation (Chaps. 4.1, 4.5 and 5.2). Therefore, it is necessary to discuss in this chapter the electrical conduction of macromolecular metal complexes.

Various metal complexes consisting of a transition metal ion and a ligand with donor atom and a π-electron system exhibit as low molecular weight or polymeric materials semiconducting behavior. In most cases the conductivity was determined at compressed powders either in the pure state or embedded in a polymer binder. Only in a few cases was anisotropic conduction in oriented compounds determined. Especially in polymeric metal complexes that are generally insoluble and not vaporizable is an oriented growth of a thin film difficult to achieve. The conductivity is sensitive against impurities that can act as traps or dopants. In most cases no care was taken regarding the purity of the materials and on the surrounding atmosphere (O_2 as an acceptor!).

Most intensively the conductivity of phthalocyanines either as part of a polymer network or in the cofacial stacked state (Chaps. 2.3.1 and 2.3.5) were investigated. Polymeric copper phthalocyanines ($[CuPc]_n$, Scheme 2.50 in Chap. 2.3.1.2) as compressed powders exhibit conductivities in the order of $10^{-3}-10^{-7}\,S \cdot cm^{-1}$ (Table 5.2), which are ten orders of magnitude higher in comparison with the low molecular copper phthalocyanine (but on the same order as copper 2,3,9,10,16,17,23,24-octacyanophthalocyanine) [74−76]. When the polymers are pure, meaning that uniform end groups and oxygen is excluded, these conductivities are "intrinsic" and stable over time. Real "intrinsic" conduction of these polymeric phthalocyanines grown as thin films were realized as follows [77]: Sputtered films of copper (thickness 1.5−30 nm) were reacted with 1,2,4,5-tetracyanobenzene from the gas phase at 400°C to obtain thin films of the polymer (thickness 40−1700 nm). Molecular images exhibit patterns with spacings of 1.34 nm in two directions that show periodicity in the plane. For the films σ of $10^{-2}-10^{-3}\,S \cdot cm^{-1}$ were determined. Electrochemical charge/-discharge properties of these films were investigated [77].

Table 5.2. Room temperature powder conductivities σ and thermal activation energies ΔE of phthalocyanines (undoped: "intrinsic", doped: "extrinsic")

Sample	σ (S cm^{-1})	ΔE (eV)	References
CuPc	10^{-11}–10^{-15}	1.4–1.8	68
[CuPc]$_n$	10^{-3}–10^{-7}	0.3–0.7	74–76
[CuPc]$_n$[a]	10^{-2}–10^{-3}	0.18	77
[Si(O)Pc]$_n$	$5 \cdot 10^{-6}$	0.29	78, 88, 89
[Si(O)Pc(J$_3$)$_{0.35}$]$_n$	$6 \cdot 10^{-1}$	0.0089	78, 88, 89
[Ge(O)Pc]$_n$	$2 \cdot 10^{-10}$		78, 88, 89
[Ge(O)Pc(J$_3$)$_{0.35}$]$_n$	$1 \cdot 10^{-1}$	0.034	78, 88, 89
[Al(F)Pc]$_n$	$< 10^{-7}$		75
[Al(F)Pc(J$_3$)$_{1.1}$]$_n$	5	0.017	79, 90
[Fe(pyz)$_2$Pc]	$3 \cdot 10^{-12}$		79, 80
[Fe(pyz)Pc]$_n$	$1 \cdot 10^{-6}$		79, 80
[Fe(pyz)Pc(J$_3$)$_{0.83}$]$_n$	$2 \cdot 10^{-1}$	0.045	79, 80, 91
[Ru(dib)Pc]$_n$	$2 \cdot 10^{-6}$		79, 92
[Ru(dib)Pc(J$_3$)$_{0.5}$]$_n$	$4 \cdot 10^{-3}$	0.1	79, 93
[Ru(tz)$_2$Pc]	$< 10^{-11}$		94
[Ru(tz)Pc]$_n$	$1 \cdot 10^{-2}$		94
[FeNc(tz)]$_n$	$3 \cdot 10^{-1}$		79
[Co(CN)Pc]$_n$	$2 \cdot 10^{-2}$		79

pyz pyrazine; dib p-diisocyanobenzene; tz tetrazine; CN cyanide; pc phthalocyanine; ne naphthalocyanine.
[a] Measured at oriented thin films between 77 and 93 K.

Cofacial stacked polymeric macrocycles (Chap. 2.3.5) were intensively investigated for their electrical conductivity. It is seen that some of the polymers exhibit high conductivity in the undoped "intrinsic" case, whereas for other polymers the conductivity increases by doping to the "extrinsic" state. The characteristics of the doped polymers had to be compared with doped low molecular weight macrocycles such as H$_2$Pc(J$_3$)$_{0.33}$ or NiPc(J$_3$)$_{0.33}$ [78]. Only very few examples of the investigated macrocycles reviewed in [75, 78–80] are mentioned here in (Table 5.2). Examples of polymers with covalent/covalent bonds between the central metal ions of the macrocycles (Scheme 2.68, Chap. 2.3.5.1) are [Si(O)Pc]$_n$ and [Ge(O)Pc]$_n$. The conductivities are enhanced by doping with J$_3^-$, BF$_4^-$ ClO$_4^-$, 2,3-dichloro-5,6-dicyano-p-benzoquinone and others. The doping with halogens as shown in (49) decreases the interplanar distances of the macrocycles: [Si(O)Pc]$_n$ 0.353 nm, [Si(O)Pc(J$_3$)$_{0.35}$]$_n$ 0.33 nm, [Ge(O)Pc]$_n$ 0.382 nm and [Ge(O)Pc(J$_3$)$_{0.35}$]$_n$ 0.348 nm (for comparison: NiPc 0.332 nm, NiPc(J$_3$)$_{0.35}$ = 0.34 nm). The electrical conductivity increases with decreasing distance for the macrocycles in the stacks: doped [Sn(O)Pc]$_n$<[Ge(O)Pc]$_n$<[Si(O)Pc]$_n$<NiPc.

$$[M(O)Pc]_n + 0.53 \, nX_2 \rightarrow [M(O)Pc(X_3)_y]_n \tag{49}$$
$$M = Si, Ge, Sn; \ X = J, Br; \ y \sim 0.35$$

Gradual addition of the oxidizing agents (acceptor) produces increasing stoichiometric oxidation with change of the crystal structure from o-rhombic to tetragonal [78]. A maximum degree of oxidation with different oxidation agents is achieved with a stoichiometry of 0.20–0.35 (or a band population of 0.20–0.35). The undoped and doped polymers are p-type conductors. Figures 5.13 and 5.14 show the dependence of the electrical conductivity and the thermoelectrical power as a function of the degree of doping. The conduction at $y > 0.20$ is explained by fluctuation-induced tunneling (switch on of high conducting state at $y = 0.2$). Localized charge carriers at $y < 0.2$ are delocalized at $y > 0.2$. The doping (from the gas phase from solution or electrochemically) is reversible. Electrochemically, also reductive doping is observed [78]. Electrochemical doping is interesting for chemiresistors [69]. Doping with organic acceptors of low vapor pressure, such as 2,3-dichloro-5,6-dicyano-p-benzoquinone or 2,3,4,6-tetrafluorotetracyanoquinodimethane, lead to more stable devices [77, 81, 82].

Polymeric metal complexes with covalent/coordinative bonds between a central metal ion, such as $[Al(F)Pc]_n$ (Chap. 2.3.5.1, Scheme 2.69), exhibit the advantage that films can be prepared by vapor deposition of the Al(F)Pc (e.g., epitaxial growth). Doping with various acceptors results in excellent conducting materials (Table 5.2) [77, 79, 80, 83, 84]. $[Ga(F)Pc]_n$ crystallize in stacks of nearly eclipsed macrocycles connected by linear Ga-F-Ga-F bridges (Ga-F-Ga distances 0.387 nm) [83]. The conductivity of this doped material is lower compared with doped $[Al(F)PcJ_y]_n$, due to greater interplanar distances of the Ga compounds and less π-orbital overlap. The conductivity of $[Al(F)Pc]_n$ doped with AsF_5 is lower (10^{-5} S $\cdot$ cm^{-1}) compared with the J_2 doped Al compounds (5 S $\cdot$ cm^{-1}). The AsF_5 rotates the eclipsed AlPc rings to a staggered conformation [85].

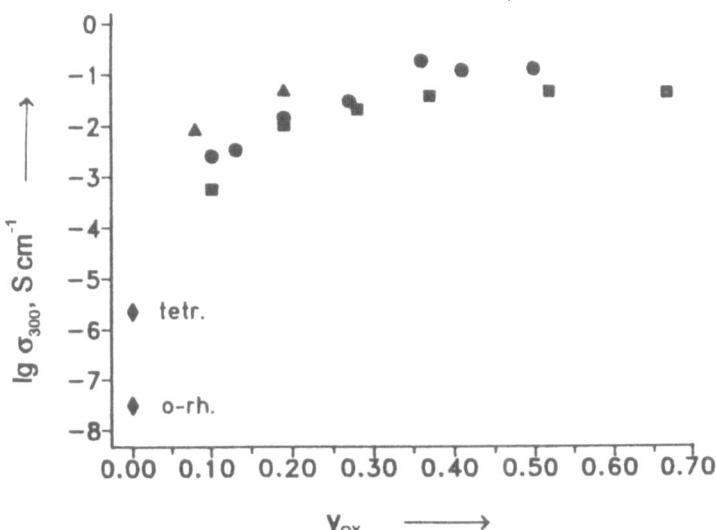

Fig. 5.13. Electrical conductivity log σ at RT of $[Si(O)PcX_y]_n$ from the degree of oxidation y_{ox}. filled circles = BF_4^-; filled squares = ToS$^-$; filled triangles = SO_4^{2-}

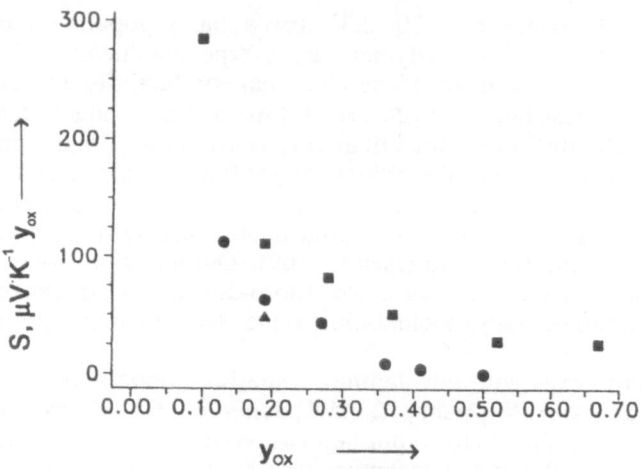

Fig. 5.14. Thermoelectrical power S at RT of $[Si(O)PcX_y]_n$ from the degree of oxidation y_{ox}. filled circles = BF_4^-; filled squares = ToS^-; filled triangles = SO_4^{2-}

Various other cofacial stacked macrocycles connected via covalent/coordinative bonds (Chap 2.3.5.2, Scheme 2.69) or coordinative/coordinative bonds (Chap. 2.3.5.3, Scheme 2.70) were investigated (Table 5.2) ([79, 80] and references cited therein). Some polymers, such as $[Co(CN)Pc]_n$ or $[Ru(tz)Pc]_n$, exhibit high "intrinsic" conductivities. For others the conductivity increases by doping. The conductivity increases also when going from a low molecular weight complex, such as $[Fe(pyz)_2Pc]$, to the corresponding polymer $[Fe(pyz)Pc]_n$ (Table 5.2). The conductivity is influenced by the kind of macrocycles. As confirmed by theoretical calculations [86] and redox potentials [87] destabilization of the HOMO level and narrowing of the HOMO–LUMO energy gap results in better σ. The high conductivity of tetrazine and other bridged polymers without doping is assigned by charge transfer process from the metallomacrocycle to the π^*-orbital of the bridging ligand (quasi-one-dimensional chain structure [80]).

A higher structural order in cofacial stacked macrocycles is obtained by preparing from suitable starting materials Langmuir-Blodgett films (LB), or by self-organization as discotic crystalline liquids. Tetra(methoxy)tetra(octyloxy)phthalocyaninatopolysiloxane forms stable monolayers on water that were transferred by LB with the polymers aligned parallel to the dipping direction to carriers such as ITO [95]. Conductivity of 10^{-7} S $\cdot$ cm^{-1} was measured of undoped films of 20 monolayers.

Stacks of metallomacrocycles were realized in liquid crystalline discotic phases (Chap. 2.3.5.4, Scheme 2.71). Electrical condutivities of such samples were determined by impedance spectroscopy [96, 98]. High "intrinsic" conduction (σ at 180 °C of $3 \cdot 10^{-2}$–$5 \cdot 10^{-5}$S $\cdot$ cm^{-1} with ΔE of approximately 0.5 eV) was observed in dependence on the alkali metal ion in bearing four hexaoxahexamethylene crown ether rings at the benzene annelands [96–99]. Doping with iodine increases the conductivities (σ at RT 0.7–0.02 S $\cdot$ cm^{-1} with ΔE approximately

0.07 eV). The high conductivities are explained by parallel packing of the Pcs at a distance of 0.34 nm, and caused by the substituents a higher flexibility allowing the formation of more tightly bound polaron states is given [98]. The authors tried to measure the conductivity of individual columns in a special set up above the phase transition temperature [98]. Metallic conductivity is expected. Tetra(crown ether)phthalocyaninatopolysiloxanes were synthesized and investigated [98, 100]. Conductivities of $10^{-4}-10^{-5}$ S $\cdot$ cm^{-1} with $\Delta E = 0.1-0.2$ eV were observed. The change of the conductivity of Cu-phthalocyanines bearing different crown ether substituents against ppm concentrations of NO_2 and NH_3 was determined [101]. The sensor characteristics decrease with increasing size of the crown ether rings. The gas-sensing properties enhance by treatment of the films with KCl.

Coordination polymers containing thiolate groups (Chap. 2.3.2.2) can exhibit high electrical conductivities at room temperature: poly(metal tetrathiooxalate). $\sigma < 20$ S $\cdot$ cm^{-1} [102]; poly(metal tetrathiofulvalene tetrathiolate or thiosquarate): $\sigma < 30$ S $\cdot$ cm^{-1} [103–105]; poly(metal benzene-1,2,4,5-tetrathiolate $\sigma < 0.2$ S $\cdot$ cm^{-1} [106]; poly(metal naphthalene tetrathiolate): $\sigma < 10^{-6}$ S $\cdot$ cm^{-1} [107]. The electronid structures of these polymers were calculated [108]. Dark-colored poly(arylenecobaltacylopentadienylene) (Chap. 2.3.6) are π-conjugated polymers with σ of 10^{-9} S $\cdot$ cm^{-1} [109]. Disadvantageous is that nearly all of these polymers are insoluble and can be used only as compressed tablets.

A hybrid electrode film consisting of metal phthalocyanines (as catalyst), polyvinylchloride (as binder) and polypyrrole (as conducting material) was obtained by electrochemical polymerization of pyrrole in DMF in the presence of suspended MPcs and PVC [110a]. Electrodes with 50 wt. % MPcs/PVC (1:1) and 50 wt. % polypyrrole show highest conductivities and catalytic activties for O_2 reduction. 50 wt. % octacyanophthalocyanine in polymer binders is necessary to achieve good particle contact for electron transport [110b]. The electrical properties of poly(methylmethacrylate) and polystyrene doped with metal salts ($FeCl_3$, CuCl, $AlCl_3$), and poly(vinyl chloride) doped with dimethylglyoxime metal complexes were investigated [111]. The conductivity increases several orders of magnitude by few percentage of doping. The a–c conductivities of alkali metal salt-polyethylene glycol complexes described by several authors were summarized recently (Chap. 4.1) [112]. The high conductivities up to 10^{-3} S $\cdot$ cm^{-1} are important of electrolytes in solid state batteries.

5.3.2 Photoconductivity

As with low molecular weight metal complexes, macromolecular metal complexes (polymeric metal complexes or low molecular ones dispersed in a polymer) are also intensively – in the case of polymeric π-conjugated system or of a good intermolecular electrical contact – up to black colored. Therefore, in an applied electrical field, irradiation in the absorption region of the material should, in addition to the conductivity, lead to photoconductivity. The transport of these additional formed charge carriers can be described by a band model and/or a hopping mechanism controlled by bandwidths, applied energy, and spatial distribution of defects and traps. It is found for molecular organic semiconductors that lower conducting materials show a higher additional contribution under irradiation than higher conducting materials.

Low molecular weight phthalocyanines were most intensively investigated [68, 79, 113, 114] and find broad technical application as xerographic photo-receptors of bilayer devices [115]. Relatively few contributions are available on the photoconduction of macromolecular metal complexes. The photocon-ductivities of cofacial stacked polymeric metal complexes were investigated [79, 116]. For polycrystalline samples of different linked stacked polymers (Chap. 2.3.5), such as $[Ge(O)Pc]_n$, $[Fe(pyz)Pc]_n$, $[Ru(tz)Pc]_n$, $[Co(CN)Pc]_n$ and others, the quantum yield at λ_{max} in an applied electrical field of 2000–2500 V $\cdot$ cm^{-1} is low (10^{-2}–10^{-7}). Also, the additional contribution of the photocurrent to the dark current seems to be only one or two orders of magnitude. No general rule of the photocurrent on the kind of stacking (covalent, coordinative) can be given. The photocurrent I_{ph} increase with light intensity φ according to $I_{ph} \sim \varphi^a$. For the intensity parameter a value between 0.6 and 1.0 was found. Values lower than 1.0 can be understood if the photocurrent is limitd by space-charge effects at higher light intensities. It is interesting to note that stacked polymeric macro-cycles exhibit after photocurrent action spectra peaks in the near-infrared region at approximately 1100–1200 nm (Fig. 5.15), which in comparison to the visible region are more intensive than those of low molecular weight phthalocyanines [79, 116, 117]. The photoconduction mechanism of stacked polymeric macro-cycles is discussed in [116]. Other macromolecular metal complexes were also investigated. Complexes of quinizarin and Al in poly(N-vinylcarbazol) exhibit photoconducting properties [118].

A complex system consisting of an ion-exchanged polymer (blend of poly(vinylpyridine-co-styrene) and a random ternary copolymer with cationic

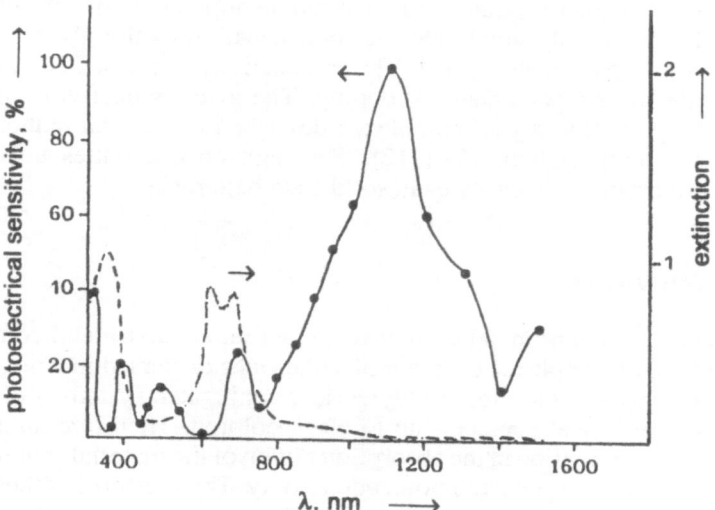

Fig. 5.15. Photoaction spectrum of polycrystalline $[Ge(O)Pc]_n$ (*solid line*) and the VIS absorption spectrum (*dotted line*)

sites) containing approximately 20 wt. % of Zn(II)-5,10,15,10-tetraphenylpor-phyrin was studied in a cell configuration SnO_2/dye, polymer (5 µm)/Au [119]. Whereas the dark current is not influenced by the incorporated porphyrin, (determined by the polymer), the photocurrent initated by the light absorption through the metal complex is dictated by the presence of a narrow distribution of traps that are believed to be the cationic sites of the polymer. The results shown very slow photocurrent-time response and also a low photoconductive gain.

In xerographic photoreceptor devices a high external electrical field is applied, which reduces the recombination of photoinduced charge carriers by their rapid separation [115]. For commercial devices the metal phthalocyanines are applied in approximately 50 wt. % in a polymer binder such as poly(vinylbutyral) or a polyester in the charge generation layer (CGL; Fig. 5.16). The polymers seem to be important for a good adhesion in the device. By 50 wt. % of Pcs a conducting pathway in the polymer is realized. Devices are prepared by solution casting with a thickness of the CGL between 0.2 and 2 µm. The photoreceptor in the CGL must have a low dark conductivity and become then photoconductive under irradiation. The highest quantum yield of photogeneration with 90 % as determined by xerographic photodischarge is measured at a field strength of 30 V/µm with 50 wt. % Y-TiOPc in poly(vinylbutyral) [115, 120]. Details on photogeneration, separation and effect of molecular architecture are discussed in [115]. This includes also the importance of precise stacking arrangement on the photogeneration efficiency, e. g., different phases of a substituted VOPc in polystyrene by solvent vapor treatment. Memory storage in an organopolysilane-based photoreceptor (Chap. 2.3.3) are described in [121]. The photoconducting behavior of the conjugated ethynyl–metal compounds $(R-C{\equiv}C-M)_n$ or $(C{\equiv}C-R-C{\equiv}C-M)_n$ described in Chap. 2.3.4 have been reviewed recently and are not discussed here [122].

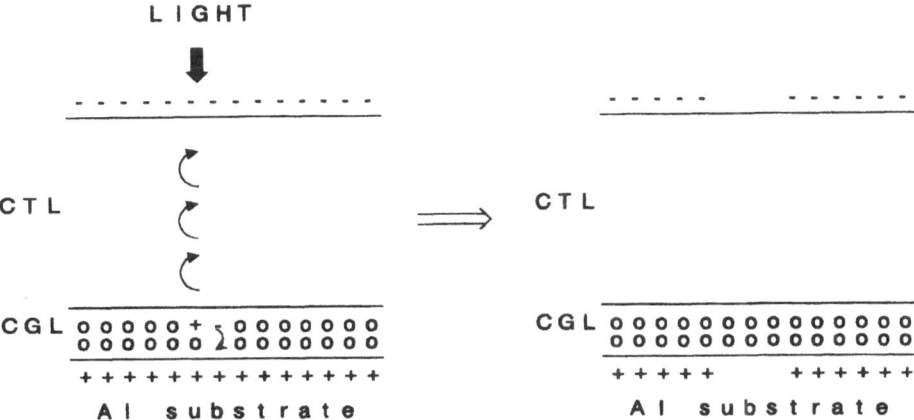

Fig. 5.16. The configuration and photodischarge process of a bilayer photoreceptor (CGL charge generation layer; CTL charge transporting layer)

5.3.3 Photovoltaic Devices

Low molecular weight metal complexes and also colored organic compounds exhibit in the solid state characteristics as molecular organic semiconductors [123]. They can be classified according to measurements of the thermoelectrical power, electrochemical/photoelectrochemical behavior [123] and junction characteristics [123–125] as p- or n-conductors. Typical examples of p-conductors are phthalocyanines 9, 5,10,15,20-tetraphenylporphyrins and for n-conductors, 5,10,15,20-tetrapyridylporphyrins, tetrapyridinotetraazaporphyrins and perylenetetracarboxylic acid diimide derivatives.

Contact of an organic p-conducting material to a metal of low work function, such as Al, In, or an organic n-conducting material, to a metal of high work function, such as Au or the organic p/n contact, results as usual in electrical rectification in the voltage/current (V/I) characteristics. Under irradiation with light in the absorption region of the colored materials typical photovoltaic characteristics are observed [67, 124–126]. Overall efficiencies between 0.5 and 1.5% in cells of ITO/n-conductor/p-conductor/Au or Ag have been achieved (Table 5.3), whereas the values in Schottky devices are lower [67]. With decreasing layer thickness the quantum yield of the photocurrent increased (but absorption of light decreases), and in 10-mm-thick devices this value can each 80% [124]. Epitaxial growth of the molecular organic semiconductors increases the photocurrent as shown in electrochemical devices [127]. The reason for both facts is explained by reduced recombinations of excited states/charge carriers. The question is whether polymer–metal complexes can lead to progress that means a combination of efficient charge separation (reduced recombination) and high absorption in thicker films with the result of a good efficiency. Two examples of cells consisting of n/p contacts of low molecular organic semiconductors are shown in Table 5.3 (nos 1 and 2). To demonstrate the function of an organic photovoltaic cell, Fig. 5.17 contains energy levels of used materials.

Only few reports on photovolatic devices with high molecular metal complex semiconductors are known. Polymeric phthalocyanines (Chap. 2.3.1.2) ob-

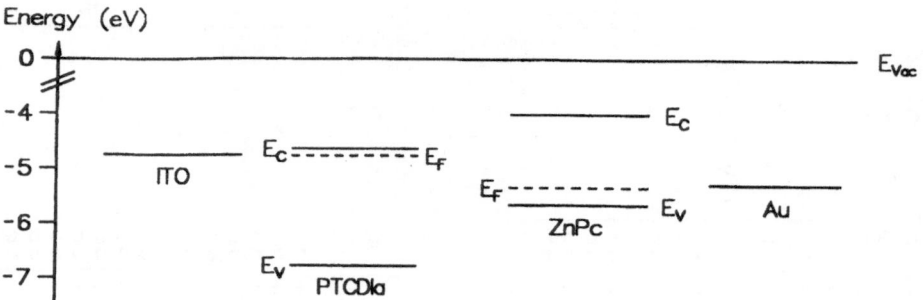

Fig. 5.17. Energy levels of different materials used for an ITO/perylenetetracarboxylic acid diimide/Zn(II)phthalocyanine/Au cell. *Solid lines* represent localized states with weak interactions

Table 5.3. Photovoltaic properties of some organic solar cells at RT under air. V_{OC} open circuit voltage; I_{SC} short circuit current; FF Fill factor; η conversion efficiency

No	Cell configuration	$V_{OC}(V)$	$I_{SC}(\mu A/cm^2)$	FF	$\eta(\%)$	Remark	Reference
1.	ITO*bis*methylperylene-*bis*carboxamide (25 nm)/ZnPc (50 nm)/Au	0.41	3530	0.30	0.43	Irradiation with 100 mW/cm² through ITO	124
2.	Like no. 1	0.26	140	0.42	1.51	Irradiation with 1 mW/cm² through ITO	124
3.	ITO/plasma-polymerized CuPc (100 mm)/Al	1.15	0.37	0.21	$1\,10^{-3}$	Irradiation with 100 mW/cm² through Al	130
4.	ITO/H$_2$Pc, polyvinyliden fluoride (2 μm)/Al	0.50	0.64	0.24	0.78	Irradiation with 10 μW/cm² at 617 nm through Al	128
5.	Like no. 4, MPc in polystyrene	0.41	0.003	0.33	$4 \cdot 10^{-2}$	Irradiation with 10 μW/cm² 617 nm through Al	128
6.	ITO/H$_2$Pc, polycarbonate (1.8 μm)/In	0.45	122	~0.33	0.35	Irradiation with ~5 mW/cm² through In-transmitted irradiance	117, 129
7.	Like no. 6, organic layer with 15% trinitrofluorenone	0.42	420	~0.33	1.33	Irradiation with ~5 mW/cm² through In-transmitted irradiance	117, 129

tained by low-temperature plasma polymerization exhibit in Schottky devices (rectifying contact to Al, Table 5.3, no. 3) excellent open circuit voltage V_{OC}, but low short circuit currents I_{SC} [130]. The reason is not known. It may be that these plasma films contain many defects (traps) produced by this film process. Also, for Schottky cells containing a low molecular weight phthalocyanine incorporated into a polymer by coating (Table 5.3, nos 4–7) V_{OC} is high and partly good I_{SC} are measured (at low light intensity) [128, 129]. It is interesting to note that polar polymers, such as poly(vinylidenfluride) (high dipole moment), and a high amount of acceptors (trinitrofluorenone) improve the photocurrent, which means enhanced charge separation and reduced charge recombination. For better I_{SC} (which will lead also to higher Fill Factor (FF)) a combination of high-ordered metal complexes in a polar environment or doped state have to be found.

5.3.4 Nonlinear Optical Properties

When highly coherent and very intensive light of a laser interacts with certain anisotropic inorganic or organic materials a number of nonlinear optical (NLO) effects can be measured [131, 132]. Linear effects, such as, absorption, are described by $\chi^{(1)}$ and are realized in normal spectroscopy. $\chi^{(2)}$ is responsible, for example, for the generation of new frequencies (such as generation of the second harmonic (SH); the laser light is doubled after passing the medium). The value $\chi^{(3)}$, which does not require a non-centro-symmetrical medium, can also generate new frequencies (such as generation of the third harmonical (TH). The NLO materials may serve as optical devices of various types and in a variety of processes.

Metal-organic NLO materials have been reviewed [133]:

1. Second-order nonlinear optics by ferrocene derivatives, metal-carbonyl, metal pyridine, or bipyridine complexes, metal-organic complexes of thiourea and other inclusion complexes.
2. Third-order nonlinear optics by poly(metal–ethynyl) compounds (Chap. 2.3.4). polysilanes, polygermylenes (Chap. 2.3.3), and polymetallocenes (Chap. 2.3.6). $\chi^{(3)}$ of phthalocyanines are reviewed in [79, 133]. For $\chi^{(3)}$ of these compounds theroretical and experimental conditions are presented [134].

It can be summarized that some promising results are described. Future work should concentrate on asymmetrical substituted planar macrocyclic metal complexes orientated anisotropically.

The third-order hyperpolarizability coefficients $\chi^{(3)}$ of [Al(F)Pc]$_n$ and [Ga(Cl)Pc]$_n$ are compared [135], and the higher $\chi^{(3)}$ value of the Al compound is explained by cofacial stacking (Chap. 2.3.5.2). Cofacially stacked macrocycles with coordinative/coordinative bond between the metal ions (Chap. 2.3.5.3) seem to fulfill good prerequisites for $\chi^{(3)}$ values if the macrocycles are bearing bulky substituents [79]. In [Ru(p-diisocyanobenzene)Pt(t-but.)$_4$]$_n$ the rigid spacer and the large tert-butyl substituents probably ensure decoupling of the phthalocyanine units, thus reducing dipole–dipole interactions. Spin-cast films of an oligomer of this compound were investigated by degenerate four-wave mixing with laser wavelengths varied of the complete Q-band absorption region and by third harmonic generation at 1064 nm [136]. It was found that the optical properties are determined by the individual Pc ring. The magnitude of $\chi^{(3)}$ is comparable to other dye systems and smaller compared with conjugated polymers. Poly-ynes of the structure $[M(P(C_4H_9)_3)_2\text{-}C{\equiv}C\text{-}C{\equiv}C]_n$ (M = Ni, Pt) (Chap. 2.3.4) exhibit $\chi^{(3)}$ values considerably higher than polydiacetylenes [137].

5.3.5 Hole Burning

Persistent spectral hole burning has been used in low-temperature spectroscopy and was suggested as the basis for high-density frequency domain optical storage [138, 139]. Practical application of hole-burning phenomena had been limited, due to liquid helium temperature for a long time. Examples of compounds are metal-free porphyrin or phthalocyanine in amorphous polymers or on the surface of inorganic carriers, such as SiO_2 or alumina, for processes of photochemical hole burning (PHB) [138, 139]. In some cases persistent spectral holes were burned at 80 K. The PHB is assigned to intramolecular tautomerization reaction, of, e.g., H^+ of metal-free porphyrins. Nonphotochemical hole-burning processes (NPHB) are caused by the position of the whole molecule in a matrix.

The NPHB experiments have been carried out with low loadings $(5 \cdot 10^{-7} \text{mol/g})$ Zn(II) chelates of positively charged tetrakis(N-methylpyridyloxy)phthalocyanines and tetrakis(N-ethyl)-2,3-pyridinotetraazaporphyrins in the molecular sieve AlPO (Chap. 2.4.2) [139]. Spectral holes can be burned at high temperatures (Fig. 5.18). Saturation effect side-band enhancement (marked by arrows in Fig. 5.18) are absent and allow reasonable detection of the hole (width 19 cm^{-1}). The NPHB may be due to chromophore coupling of extrinsic two-level systems with high barriers enabling the formation of stable

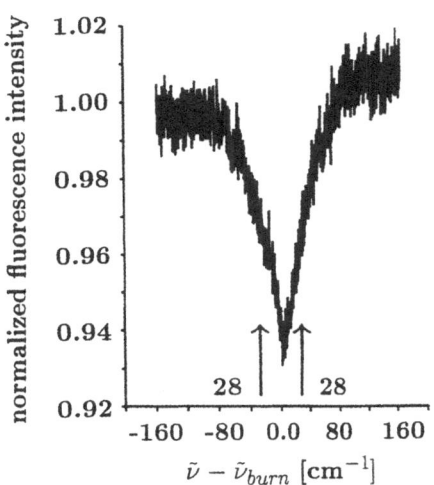

Fig. 5.18. Shallow spectral hole of ZnTaP in hydrated AlPO burned at 80 K and 655 nm with 13 mW/cm^2 für 100 s

products. The high temperature used confirms the dominating influence of matrix stiffness and interaction of the porphyrin with this matrix on the frequency of the coupled phonon. In addition, it is mentioned that for Zn(II) porphyrins in a polymer matrix at 20 K, a two-color irradiation (dye laser and Ar⁺ laser) forms holes 500 times more efficiently than conventional one-color irradiation [140]. Considering results of [139] and [140] progress in hole burning may be possible in the future.

References

1. Kaneko M, Wöhrle D (1987) Adv Polym Sci 84:141. Lin RJ, Kaneko M (1992) Molecular electronics and molecular devices 1:207–241. Yagi M, Nagai K, Onokuba T, Kaneko M (1995) J Electrochem Soc (in press). Ramaraij R, Kaneko M (1995) Adv Polym Sci 123:216
2. Kaneko M, Yamada A (1984) Adv Polym Sci 55:1
3. Siddarth P, Marcus RA (1993) J Phys Chem 97:2400
4. Beratan BN, Onuchic JN, Betts JN, Bowler BE, Gray HB (1990) J Am Chem Soc 112:7915
5. Vassilian A, Wishart JF, van Hemelryck B, Schwarz H, Isied SS (1990) J Am Chem Soc 112:7278
6. Fox MA (1990) Photochem Photobiol 52:617
7. Drain CM, Mauzerall B (1992) Biophys J 63:1556
8. Stern O, Volmer M (1919) Physk Z 20:183
9. Perrin J (1927) CR Acad Sci 184:1097
10. Kalyanasundarum K, Grätzel M (ed) (1993) Photosensitization and photocatalysis using inorganic and organometallic compounds. Kluwer, London
11. Pelizzetti E, Schiavello (ed) (1991) Photochemical conversion and storage of solar energy. Kluwer, London
12. Kaneko M, Yamada A, Tsuchida E, Kurimura Y (1982) J Polym Sci Polym Lett Ed 20:593
13. Hou X-H, Kaneko M, Yamada A (1986) J Polym Sci Polym Lett Ed 24:2749
14. Kaneko M, Yamada A, Tsuchida E, Kurimura Y (1984) J Phys Chem 88:1062
15. Kaneko M, Nakamura H (1987) Macromolecules 20:2265
16. Kaneko M, Hou X-H, Yamada A (1987) Bull Chem Soc Jpn 60:2523
17. Kaneko M (1991) In: Tsuchida E (ed) Macromolecular complexes. VCH Publishers, New York
18. Wöhrle D, Gitzel J, Krawczyk G, Tsuchida E, Ohno H, Okura I, Nishiaka T (1988) J Macromol Sci Chem A25:1227
19. Wöhrle D, Krawczyk G, Paliuras M (1988) Makromol Chem 189:1001; 189:1013
20. Wöhrle D, Krawczyk G (1986) Makromol Chem 187:2535. Wöhrle D, Paliuras M, Okura I (1991) Makromol Chem 192:819
21. Darwent JR, Douglas P, Harriman A, Porter G, Richoux M-C (1982) Coord Chem Rev 44:83
22. Okura I, Hosono H (1992) J Phys Chem 96:4466
23. Näther D, Gilchrist JR, Gensch T, Roeder B (1993) Photochem Photobiol 57:1056
24. Schneider G, Wöhrle D, Spiller W, Stark J, Schulz-Ekloff G (1994) Photochem Photobiol 60:333
25. Wöhrle D, Shopova M, Müller S, Milev AD, Mantareva VN, Krastev KK (1993) J Photochem Photobiol B21:155; (1994) J Photochem Photobiol B23:35
26. Dougherty TJ (1993) Photochem Photobiol 58:895
27. Wöhrle D, Ardeschirpur A, Heuermann A, Müller S, Graschew G, Rinneberg H, Kohl M, Neukammer J (1992) Makromol Chem Macromol Symp 59:17. Sinn HJ, Schrenk HH, Maier-Borst, Friedrich E, Graschew G, Wöhrle D, Klemmer T (1991) European Patent PCT/EP91/00992

28. Kaneko M, Takabayashi N, Yamada A (1982) Chem Lett 2832
29. Kaneko M, Takabayashi N, Yamauchi Y, Yamada A (1984) Bull Chem Soc Jpn 57:156
30. Kaneko M, Hou X-H, Yamada A (1986) J Chem Soc Faraday Trans 1, 82:1637
31. Yuasa M, Nishide H, Tsuchida E, Yamagashi A (1988) J Phys Chem 92:298
32. Tsuchida E, Kaneko M, Nishide H, Hoshino M (1986) J Phys Chem 90:2283
33. Kaneko M, Imai Y, Tsuchida E (1991) J Chem Soc Faraday Trans 1, 87:83
34. Kaneko M, Ochiai M, Yamada A (1982) Macromol Chem Rapid Commun 3:299
35. Kaneko M, Yamada A, Oyama N, Yamaguchi S (1982) Macromol Chem Rapid Commun 3:769
36. Kaneko M, Moriya S, Yamada A, Yamamoto H, Oyama N (1984) Electrochim Acta 29:115
37. Kaneko M, Hara S, Yamada A (1985) J Electroanal Chem 194:164
38. Kaneko M, Yamada A (1986) Electrochim Acta 31:273
39. Kaneko M (1987) J Macromol Sci Chem A24:357
40. Ueno Y, Yamada K, Yokata T, Ikeda K, Takamiya N, Kaneko M (1993) Electrochim Acta 38:129
41. O'Regan B, Grätzel M (1991) Nature 353:737. Nazeeruddin K, Liska P, Moser J, Vlachopoulos N, Grätzel M (1990) Helv Chim Acta 73:1788. O'Regan, Moser J, Anderson M, Grätzel M (1990) J Phys Chem 94:8720. Wagfeldt A, Grätzel M (1985) Chem Rev 95:49
42. Amadelli R, Argazzi R, Bignozzi CA, Scandola F (1990) J Am Chem Soc 112:7099
43. Mallouk TE (1991) Nature 353:698
44. Kaneko M, Wöhrle D, Schlettwein D, Schmidt V (1988) Makromol Chem 189:2419
45. Schlettwein D, Kaneko M, Yamada A, Wöhrle D (1991) J Phys Chem 95:1748
46. Schlettwein D, Jaeger NI, Wöhrle D (1992) Makromol Chem Macromol Symp 59:267
47. Schlettwein D, Jaeger NI (1993) J Phys Chem 97:3333
48. Yamada K, Ueno Y, Ikeda K, Takamiya N, Kaneko M (1989) Denki Kagaku 57:1129
49. Wöhrle D, Schmidt V, Schumann B, Yamada A, Shigehara K (1987) Ber Bunsenges Phys chem 91:975
50. Yanagi H, Wada M, Ueda Y, Ashida M, Wöhrle D (1992) Makromol Chem 193:1903
51. Schlettwein D, Jaeger NI, Wöhrle D (1991) Ber Bunsenges Phys Chem 95:1526. Schlettwein D, Wöhrle D, Karmann E, Melville U (1994) Chem Mater 6:3
52. Lin R-J, Onikubo T, Kaneko M (1993) J Electroanal Chem 348:189
53. Ramaraj R, Kira A, Kaneko M (1993) J Electroanal Chem 348:367
54. Ramaraj R, Kaneko M (1993) J Chem Soc Chem Commun 579
55. Yoshida T, Tsutsumida K, Teratani S, Yasufuku K, Kaneko M (1993) J Chem Soc, Chem Commun 631
56. Yoshida T, Ida T, Shiraagi T, Lin R-J, Kaneko M (1993) J Electroanal Chem 344:355
57. Kaneko M, Yao G-J, Kira A (1989) J Chem Soc, Chem Commun 1338
58. Mizoguchi K (1994) Macromol Symp 80:359
59. Schmid G (1992) Chem Rev 92:1709
60. Schulz-Ekloff G (1995) In: Lehn JM (ed) Comprehensive supramolecular chemistry, vol 11. Pergamon Press, Oxford (in press)
61. Marquardt P, Börngen L, Nimtz G, Gleiter H, Sonnberger R, Zhu J (1986) Phys Lett 114A:39. Nimtz G, Marquardt P, Gleiter H (1988) J Crystal Growth 86:66
62. Schmitt-Rink S, Müller DAB, Chemla DS (1987) Phys Rev B35:8113. Stuky GD, Mac-Dougall JE (1990) Science 247:669
63. Gutierez M, Henglein A J (1993) Phys Chem 97:11368. Theo BK, Keating K, Hao Y-H (1987) J Am Chem Soc 109:3494. Selby K et al. (1989) Z Phys D 12:477
64. Ellis JR (1986) In: Skotheim TA (ed) Handbook of conducting polymers, vol 1. Marcel Dekker, New York, p. 489
65. MacDiarmid (1986) In: Skotheim TA (ed) Handbook of conducting polymers vol 1. Marcel Dekker, New York, p. 689
66. Snow AW, Barger WR (1989) In: Leznoff CC, Lever ABP (eds) Phthalocyanines: properties and applications. VCH Publishers, New York, p. 341
67. Wöhrle D, Meissner D (1991) Adv Mater 3:129

68. Simon J, Andre JJ (1984) Molecular semiconductors. Springer, Berlin Heidelberg New York
69. Natan MJ, Wrighton MS (1989) Prog Inorg Chem 37:391
70. Oyama N, Kitagawa M, Inabe H, Kawase K (1994) Macromol Symp 80:337
71. Nicholson MM (1993) In: Leznoff CC, Lever ABP (eds) Phthalocyanines: properties and applications. VCH Publishers, New York, p. 71
72. Sinclair M, McBrauch D, Moses A, Heeger AJ (1980) Synt Met 27–39:D645. Wu W, Kivelson S Synt Met 27–39:D 575
73. Hmyene M, Yassar A, Escorne M, Garnier F (1994) Adv Mater 6:564
74. Wöhrle D (1983) Adv Polym Sci 50:45
75. Wöhrle D (1989) In: Leznoff CC, Lever ABP (eds) Phthalocyanines: properties and applications. VCH Publishers, New York, p. 55
76. Liao MS, Kuo KT (1993) Polym J 25:947
77. Wöhrle D, Schmidt V, Schumann B, Yamada A, Shigehara K (1987) Ber Bunsenges Phys Chem 91:975
78. Marks TJ (1990) Angew Chem 102:886
79. Schultz H, Lehmann H, Rein M, Hanack M (1991) Structure and bonding. 74:41
80. Hanack M (1994) Macromol Symp 80:83. Hanack M, Lang M (1994) Adv Mater 6:819
81. Diehl BN, Inabe T, Lyding JW, Schoch KF, Kannewurf CR, Marks TJ (1983) J Am Chem Soc 105:1551
82. Inabe T, Gaudiello JG, Moguel MK, Lyding JW, Burton RL, McCarthy WJ, Kannewurf CR, Marks TJ (1986) J Am Chem Soc 108:7595 (1986), Synt Met 13:219
83. Wynne KJ (1985) Inorg Chem 24:1339
84. Wynne KJ, Nohr RS (1983) Mol Cryst Liq Cryst 81:243
85. Djurado D, Hamwi A, Cousseins JC, Bidar H, Fabre C, Berthet G (1985) Synth Met 11:109
86. Orti E, Bredas JL, Piqueras MC (1990) Chem Mater 2:110 (1991), Synt Met 41–43:2647
87. Hanack M et al. (1988) Chem Ber 121:1601, (1989) Synt Met 29:F1, (1983) Angew Chem 95:741
88. Dirk CW, Mintz EA, Schoch KF, Marks TJ (1981) J Macromol Sci Chem A 16:275. Inabe T, Moguel MK, Kannewurf CR, Marks TJ (1983) Mol Cryst Liq Cryst 93:355
89. Inabe T, Moguel MK, Marks TJ, Burton R, Lyding JW, Kannewurf CR (1985) Mol Cryst Liq Cryst 118:349
90. Wynne KJ, Nohr RS (1983) Mol Cryst Liq Cryst 81:243. Nohr RS, Kuznesof PM, Wynne KJ, Kenney ME, Siebenmann KG (1981) J Am Chem Soc 103:4371
91. Diehl BN, Inabe T, Jaggi NK, Lyding JW, Schneider O, Hanack M, Kannewurf CR, Marks TJ, Schwartz LH (1984) J Am Chem Soc 106:3207
92. Kobel W, Hanack M (1986) Inorg Chem 25:103
93. Hanack M, Keppeler U, Schulz HJ (1987) Synt Met 20:347
94. Keppeler U, Deger S, Lange A, Hanack M (1987) Angew Chem 99:349
95. Schwiegk S, Fischer H, Xu Y, Kremer F, Wegner G (1991) Makromol Chem Macromol Symp 46:211
96. van der Linden JH, Schoonman J, Nolte RJM, Drenth W (1984) Recl Trav Chim Pays-Bas 103:260
97. Sielcken OE, van Lindert HCA, Drenth W, Schoonman J, Schram J, Nolte RJM (1989) Ber Bunsenges Phys Chem 93:702
98. Nolte RJM, Drenth W (1992) In: Laine RM (ed) Inorganic and organometallic polymers with special properties. Kluwer, London, p. 223
99. Sielken OE, Nolte RJM, Schoonman J (1990) Recl Trav Chim Pays-Bas 109:230. Sielken OE, Drenth W, Nolte RJM (1990) Recl Trav Chim Pays-Bas 109:425
100. Sielcken OE, van de Kuil LA, Drenth W, Schoonman J, Nolte RJM (1990) J Am Chem Soc 112:3086
101. Roisin P, Wright JD, Nolte RJM, Sielcken OE, Thorpe SC (1992) J Mater Chem 2:131
102. Reynolds JR, Lillya CP, Chien JCW (1987) Macromolecules 20:1184
103. Schumater RR, Engler EM (1977) J Am Chem Soc 99:5521

104. Rivera NM, Engler EM, Schumater RR (1979) J Chem Soc Chem Commun 184
105. Ribas J, Cassaux PC (1981) R Seances Acad Sci 293:665
106. Dirk CW, Mintz EA, Schoch KF, Marks TJ (1986) In:·Carraher CE et al. (eds) Advances in organometallic and inorganic polymer science. Marcel Dekker, New York, p. 225
107. Teo BK, Wudl F, Hauser JJ, Krüger A (1977) J Am Chem Soc 99:4862
108. Böhm MC (1984) Phys Stat Sol (B) 121:255
109. Nishihara H, Shimura T, Ohkubo M, Matsuda N, Aramaki K (1993) Adv Mater 5:752
110. Kawashima M, Saot Y, Sakaguchi M (1991) Polym J 23:37. Wöhrle D, Kaune H, Schumann B (1986) Makromol Chem 187:2947
111. Tawansi A, Soliman MA, Kinawy N, Badr SIM (1988) Polymer Bull 19:289. Biswas M, Moitra S (1989) J Appl Polym Sci 38:1243. Hirai H et al. (1986) Makromol Chem Rapid Commun 7:351
112. Biswas M, Mukherjee (1994) Adv Polym Sci 115:89
113. Hamann C, Heim J, Burghardt H (1981) Organische Leiter, Halbleiter und Photoleiter, Vieweg, Braunschweig, Wiesbaden
114. Meier H, Albrecht W, Wöhrle D, Jahn A (1986) J Phys Chem 90:6349
115. Law K-Y (1993) Chem Rev 93:449
116. Meier H, Albrecht W, Zimmerhackl E, Hanack M, Fischer K (1985) J Mol Electron 1:47, (1985) Synt Met 11:333. Meier H, Albrecht W, Hanack M (1993) Mol Cryst Liq Cryst 228:69
117. Loutfy RO (1982) J Phys Chem 86:3302
118. Allen NS, Richards AM (1990) Eur Polym J 26:1229
119. Lawrence MF, Huang Z, Longford CH, Ordonez I (1993) J Phys Chem 97:944. Huang Z, Ioannidis A, Lawrence MF (1993) J Phys Chem 97:952
120. Fujimaki Y et al. (1990) SPSE Proceedings, The 5th International Congress on Advances in Non-Impact Printing Technologies, p. 37, (1991) 7th International Congress in Advances in Non-Impact Printing Technologies, p. 269
121. Yokoyama K, Yokoyama M (1990) J Imag Technol 16:219
122. Mylnikov V (1994) Adv Polym Sci 115:1
123. Schlettwein D, Wöhrle D, Karmann E, Melville U (1994) Chem Mater 6:3. Schlettwein D, Jaeger NI, Wöhrle D (1991) Ber Bunsenges Phys Chem 95:1527
124. Wöhrle D, Elbe J, Kreienhoop L, Schnurpfeil G, Tennigkeit B, Schlettwein D, Hiller S (1995) J Mater Chem 5:issue 9 in press
125. Hiramoto M, Fujiwara H, Yokoyama M (1992) J Appl Phys 72:3781
126. Whitlock JB, Panayotatos P, Sharma GD, Cox MD, Sauers RR, Bird GR (1993) Opt Engin 32:1921
127. Yanagi H, Douko S, Ueda Y, Ashida M, Wöhrle D (1992) J Phys Chem 96:1366
128. Minami N, Sasaki K, Tsuda K (1983) J Appl Phys 54:6764
129. Loutfy RO, Sharp HJ, Hsiao CK, Ho R (1981) J Appl Phys 52:5218
130. Osada Y, Mizumoto A, Tsuruta H (1987) J Macromol Sci Chem A24:403
131. Shen YR (1984) The principles of non-linear optics. Wiley, New York
132. Chemla DS, Zyss J (eds) (1989) Non-linear optical properties of organic molecules and crystals vols 1, 2. Academic Press, New York
133. Nalwa HS (1991) Appl Organomet Chem 5:349. Long NJ (1995) Angew Chem 107:37
134. Li DQ, Ratner MA, Marks TJ (1988) J Am Chem Soc 110:1707. Auston DH (1987) Appl Opt 26:211
135. Ho ZZ, Ju CY, Herrington WH (1987) J Appl Phys 62:716
136. Grund A, Kaltbeitzel A, Mathy A, Schwarz R, Bubeck C, Vermehren P, Hanack M (1992) J Phys Chem 96:7450
137. Blau WJ, Byrne HJ, Cardin DJ, Daney AP (1991) J Mater Chem 1:245
138. Friedrich J, Haarer D (1984) Angew Chem 96:96. Horie K (1991) In: Honda K (ed) Photochemical processes in organized molecular systems. Elsevier, Amsterdam, p. 451. Horie K (1994) Macromol Symp 80:353
139. Ehrl M, Deeg FW, Bräuchle C, Franke O, Sobbi A, Schulz-Ekloff G, Wöhrle D (1994) J Phys Chem 98:47
140. Machida S, Horie K, Yamashita T (1992) Appl Phys Lett 60:288

6 Outlook

F. Ciardelli, E. Tsuchida and D. Wöhrle

6.1 Structure of MMC

Chapters 2, 3, 4 and 5 of the present book provide an up-to-date description of the complex nature of MMC, which makes it difficult both to define them in a simple unequivocal way and also to characterize them from a structural viewpoint.

In this book it has been acceptable to consider as MMC practically all systems where metals and macromolecules are together in such a way that the material displays properties different from the simple mixture of polymer with a transition metal complex, atomic metal or cluster. Even the concept of the macromolecule has been accepted in a very broad sense, i. e., macromolecule in the subject MMC is everything having on one side a repeating structure along either in one dimension, linear macromolecule, or in two dimensions, polymer films and surfaces, or in three dimensions, networks and crystalline materials (Sect. 2.1).

In this sense it has been necessary therefore to discuss all presently available approaches to prepare the various types of complexes when starting with different ligands and metal derivatives with the related type of chemical or physical interactions leading to the MMC (Chap. 2). The possible routes are numerous and some are very unexpected. Indeed, in some cases the MMC formation is simply due to the fact that starting with bidentate ligands, and in the presence of multiple coordination sites of the metal derivative (Sect. 2.3), the complexation reaction is actually a polymerization reaction. The multimodality of MMC is further enhanced by the complex, and inhomogeneous nature which was observed in all versions. Indeed, practically all different types of MMC can be considered as composite materials where it is possible to identify either molecular or even macroscopic domains having different metal concentrations and various dimensions. In this connection a monotonic trend from almost homogeneous to highly heterogeneous composite materials can be observed when going from well-defined multidentate ligands and stable metal complexes (Sects. 2.2 and 2.3) to polymer-wrapped metal clusters (Sect. 2.4). Moreover, the metal species can be inhomogeneous structurally, a part of the distribution of metal containing domain in the macromolecular matrix. In this last connection the trend is much less monotonic, because structural homogeneity of the metal species can be envisaged either with a well-defined bidentate ligand giving only one type of very stable complex with the metal, and on the other side, when metal clusters are formed, because in this last case the metal atoms are all more or less in the same environment inside the metal-containing domain.

From this summary it is reasonable to conclude that the synthesis of MMC has reached a highly mature level, and many routes are available for a very broadly varied selection of MMC characterized by different and tailor-made structural features.

However, too often it is necessary to assume average structural information, due to the difficulty of characterizing at molecular level MMC systems particularly due to their limited solubility and heterogeneous nature. Therefore, as far as preparation and structure are concerned, the good position reached at present should be substantially improved particularly with regard to the control of the reaction conditions and characterization techniques, rather than with regard to new preparative methods. These can be of great interest not only for the study and characterization of human-made MMC, but also for a better understanding of the role that metals attached to biological macromolecules play in the living processes. Indeed, Chap. 3 has shown that a good level of knowledge has been reached in biorganic MMC, but also point out that additional work needs to be done. On the other side, what we learn from MMC in nature has to be used to develop new, intelligent materials based on polymeric ligands and metal compounds.

6.2 Properties and Applications of MMC

If on one side preparative approaches have reached a mature level, the properties and applications of MMC, on the other side, seem to be still under development, and their future looks open to provide new materials with unique and unexpected features.

Chapters 4 and 5 indicate that the future of materials based on MMC will be mainly in highly specialized applications particularly in electronics and photonics, whereas structural applications seem to be practically negligible, and the use of chemicals as catalysts seems to been concentrated in specific cases, with the most significant being found in living systems as metal enzymes (see Chap. 3). This last remark probably suggests that for chemical applications the obtainment of well-defined homogeneous species is certainly the most important aspect. The use of macromolecular ligands in catalysis was mainly aimed to physically heterogenize homogeneous metal complexes in order to obtain catalytic systems combining the positive aspects of easy separability from reaction products and the high activity and selectivity of monosite metal catalysts.

Ion-conductive materials based on electrolytes in a polymeric matrix have been reviewed in Sect. 4.1, and it was shown how they can help in converting wet-type devices into dry system due to the nature of solid electrolytes. Thus, they can be used for solid batteries, electrochemical displays and molecular devices.

Also, weak coordination of metal species wrapped into macromolecules can be used for specific and selective separation of small molecules from complex mixtures where they may be present even in small concentration (Chap. 4.2). The design of metal environment and ligands can be certainly developed for broadening the application to a larger number of small molecules, whereas the structural characteristics of the material can vary widely by a proper selection of polymeric matrices.

Great interest both fundamental and applied is offered by assembled porphyrins and their interaction with oxygen (Sect. 4.3). Amphiphilic porphyrins
particularly can give highly ordered aggregates of paramount significance for
developing our knowledge and molecular-design capacity of improving oxygen
binding and transporting capability.

Application of MMC in heterogenizing homogeneous catalysts (Sect. 4.4)
appears to have finally assumed a more limited, but balanced, extension. Indeed,
after the explosion of activity in this area approximately 15 years ago, it now
seems to be concentrated on reactions involving very expensive metals which,
however, still maintain a good activity despite steric effects and diffusion problems related to the polymeric matrix. Insteads the study of these polymer-
attached catalytic metal complexes has been of help to start an approach that is
now fully under development as based on inorganic matrices more or less functionalized at the surface with various ligands. Also, interest and large potentiality can be expected for small metal clusters stabilized by the wrapping of macromolecules, which hinders recombination into bulk metal.

A very recent development that is expected to grow in the future concerns
with molecular conversions based on efficient multielectron transfer facilitated
by polynuclear MMC (Sect. 4.5). This should be a very effective system in this
process when the entropy effect of the cooperative interaction is applicable as
typical macromolecular properties. Also, polynuclear complexes, which are
effective in this type of process, seem to be easier to prepare and develop with
macromolecular ligands.

The MMC can also act as new and modulable photosensitive materials
(Chap. 5). Photoinduced electron transport in synthetic polymer-metal complexes
(MMC) can be of great help for elucidating the microenvironmental effect on the
photoprocess by polymer solution, polymer bulk or molecular aggregates. In these
systems it was shown that the metal complex itself can act as the photoexcitation
centre, which is coupled with another electron donor or acceptor, whereas in other
cases the metal complex behaves as electron acceptor/donor in combination with
another photoexcitation centre. These aspects are particularly interesting when the
metal complex is combined into a polymeric membrane where electronic
properties can be modulated and the design of devices can be effective. Solid MMC
(Sect. 5.3) can have interesting electroconductivity, photoconductivity and nonlinear optical properties, which can be modulated for different applications both
at the level of the metal complex and of the polymer matrix.

Many of the properties described are connected with higher-order functions of
MMCs, which are caused by the conjugation of the dynamic interactions and the
electronic processes. In this last connection some potential applications derived
from research studies on macromolecular complexes are illustrated in Fig. 6.1.
They include advanced functions that are derived from molecular-level features of
the macromolecular complexes, and that should be realized in the near future. Most
of them are now under development and promise a great impact on related scientific
and technological fields. For example, some of the macromolecular complexes can
be used as high-performance molecular devices, such as super conductors, organic
ferromagnets and nonlinear optics. This points out further the necessity to promote
studies of the macromolecular complexes as new materials.

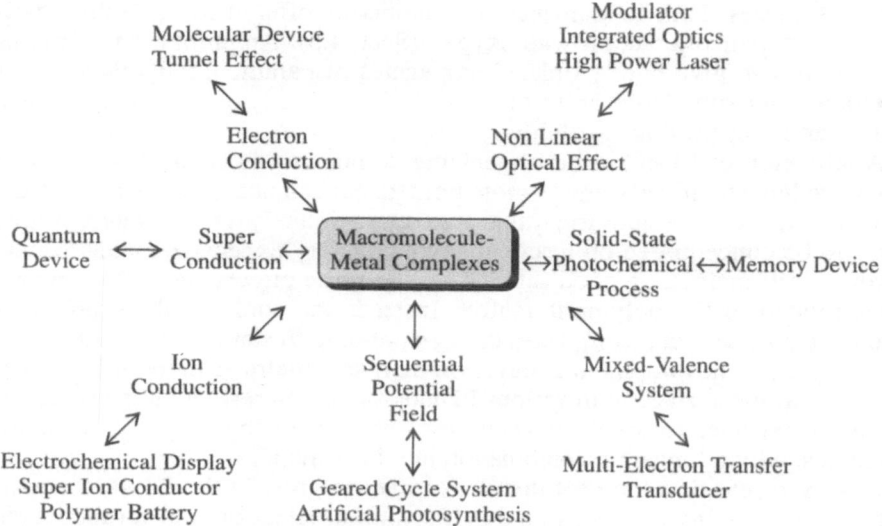

Fig. 6.1. Potential applications of MMC for photoexcitation, electron transfer and molecular conversion

6.3 Conclusion

This book has probably shown that MMC have reached the scientific standard of an interdisciplinary and mature science. Indeed, different expertise have merged in this area from various branches of chemistry, physics, biology and engineering under the effect of the driving force of a common interest consisting of development of materials combining polymer and metal properties. In this way it has been possible to identify a large variety of structurally complex materials with different chemical and physical characteristics, and to indicate synthetic routes to obtain them.

The number of properties and more or less developed applications described in the book is very large and covers different areas. Such extension could not be envisaged with single-component materials and is certainly a good consequence of the nature of MMC, which combine systems that by themselves are effective in well-distinct domains. For these reasons we think that the area of MMC needs to be further cultivated by scientists with the various expertise needed by the interdisciplinary character of the field. Also, they may further open new areas and contribute to developing new processes of fundamental technical importance for human-made materials and biosystems.

Subject Index

L.A. Pilato, M.J. Michno
Advanced Composite Materials
1994. XIII, 208 pp. 50 figs., 49 tabs. Hardcover **DM 138,-**; öS 1076,40; sFr 132,50
ISBN 3-540-57563-4

Advanced composite materials or high performance polymer composites are an unusual class of materials that possess a combination of high strength and modulus and
are substantially superior to structural metals and alloys on an equal weight basis. The
book provides an overview of the key components that are considered in the design of
a composite, of surface chemistry, of analyses/testing, of structure/property relationships with emphasis on compressive strength and damage tolerance. Newly emerging
tests, particularly open hole compression tests are expected to provide greater assu-
rance of composite performance. This publication is an „up-to-date" treatment of
leading edge areas of composite technology with literature reviewed until recently and
includes thermoplastic prepregs/composites and major application areas.

J.W. Buchler (Ed.)
Metal Complexes with Tetrapyrrole Ligands III
With contributions by numerous experts
1995. Approx. 200 pp. 50 figs., 36 tabs. (Structure and Bonding, Vol. 84) Hardcover
DM 198,-; öS 1445,40; sFr 187,- ISBN 3-540-59281-4

Contents: **J.W. Buchler, C. Dreher, F.M. Künzel:** Synthesis and Coordination
Chemistry of Noble Metal Porphyrins.- **S. Licoccia, R. Paolesse:** Metal Complexes of
Corroles and Other Corrinoids.- **J. Sima:** Photochemistry of Tetrapyrrole Complexes.

Prices are subject to change.
In EU countries the local VAT is effective.

Springer-Verlag, Postfach 31 13 40, D-10643 Berlin, Fax 0 30 / 82 07 - 3 01 / 4 48   e-mail: orders@springer.de   tm.BA9508.01

F. Francuskiewicz

Polymer Fractionation

1994. XVII, 215 pp. 32 figs., 42 tabs. (Springer Lab Manual) Hardcover **DM 98,-**;
öS 764,40; sFr 94,50 ISBN 3-540-57539-1

The fractionation of polymers via differences in solubility, especially in a preparative
scale, is an important presupposition for the determination of molecular weight-
dependent polymer properties. In this book, a big variety of fractionation methods,
their theoretical base, applications, equipments, preparatory and fractionation steps
are discussed. The text is focussed on practical aspects of the carrying-out of polymer
fractionations. Each fractionation procedure is completed by practical examples.
Appendices and glossary are a useful supplement. The book will enable all polymer
chemists, physicists and technicians as well as material scientists and students in these
fields to choose the optimal fractionation variant for his problem.

N.A. Peppas, R.S. Langer (Eds.)

Biopolymers II

With contributions by numerous experts
1995. X, 287 pp. 64 figs., 10 tabs. (Advances in Polymer Science, Vol. 122) Hardcover
DM 238,-; öS 1856,40; sFr 224,- ISBN 3-540-58788-8

Contents: **A.B. Scranton, B. Rangarajan, J. Klier,** Biomedical Applications of
Polyelectrolytes.- **D. Putnam, J. Kopecek,** Polymer Conjugates with Anticancer
Activity.- **C.L. Bell, N.A. Peppas,** Biomedical Membranes from Hydrogels and
Interpolymer Complexes.- **K.S. Anseth, S.M. Newman, C.N. Bowman,** Polymeric
Dental Composites: Properties and Reaction Behavior of Multimethacrylate Dental
Restorations.- **I.V. Yannas,** Tissue Regeneration Templates Based on Collagen-
Glycosaminoglycan Copolymers.- **R.C. Thomson, M.C. Wake, M.J. Yaszemski,
A.G. Mikos,** Biodegradable Polymer Scaffolds to Regenerate Organs.

■ ■ ■ ■ ■ ■ ■ ■ ■ ■ ■

Prices are subject to change.
In EU countries the local VAT is effective.

Springer-Verlag, Postfach 31 13 40, D-10643 Berlin, Fax 0 30 / 82 07 - 3 01 / 4 48   e-mail: orders@springer.de   tm.BA95087.1